AF540867

MATHEMATICS FOR MANAGEMENT

MATHEMATICS FOR MANAGEMENT

By

Dr. R.K. Rajput

DISCOVERY PUBLISHING HOUSE PVT. LTD.
NEW DELHI-110 002

First Published - 2008

Reprinted - 2017

ISBN: 978-81-8356-335-2

Mathematics for Management

Published by:

DISCOVERY PUBLISHING HOUSE PVT. LTD.
4383/4B, Ansari Road, Darya Ganj
New Delhi-110 002 (India)
Phone: +91-11-23279245, 43596064-65
Fax: +91-11-23253475
E-mail: discoverypublishinghouse@gmail.com
sales@discoverypublishinggroup.com
web: www.discoverypublishinggroup.com

Printed at:
Infinity Imaging Systems
Delhi

Preface

This book "Mathematics for Management" has been written to provide a comprehensive and updated foundation course in Mathematics and its varied applications in management, commerce and economics.

This book has been written from studies point of view, so that they can easily understand various mathematical concepts, techniques and tool needed for their course. Each new concept and technique is properly supported by suitable solved examples. In solving the questions cave has been taken to explain each step so that students can follow the subject matter themself without even consulting others.

I believe that the book should serve specific need of the students appearing in B.Com. (Hons.), B.A. (Eco. Hons.), M.Com., (M.A.) (Eco.) B.B.A., B.C.A., M.B.A. and other competitive examinations.

Suggestions and comments for the improvement of the book from readers will be thankfully received and duly incorporated in the subsequent editions.

Author

CONTENTS

1

Introductory Set Theory

INTRODUCTION

In relation to the current research activity in algebra, it could be described as "not to abstract" from the point of view of some one schooled in the calculus and who is seeing the present material for the first time, it may very be described as "quite abstruct" The algebra which has evolved as an out growth of all this is not only a subject with an independent life and vigor, it is one of the important current research areas in mathematics, but it also series as the unifying thread which interlaces almost all of mathematics— geometry, number theory, analysis topology and even applied mathematics. An introduction to that part of mathematics that today goes by the name of abstract algebra. The term "abstract" is a highly subjective one; what is abstract to one person is very often concrete and down to earth to another and *vice-versa.* Be that as it may, we shall concern ourselves with the introduction and development of some of the important algebriac system-group, rings, vector space, fields.

STATEMENTS

We communicate our ideas to others through sentences. In our mathematical language, *the sentences which can be judged to be true or false are known as statements.* For Example $2 + 3 = 5$ is a true statement while, $(3 + 4) < 6$ is a false statement. Also, Ram is a handsome boy, is a sentence, not a statement. We use connectives *'and'* and *'or'* to combine two statements to form a compound statement.

We know that for any three numbers a, b, c we always have $a (a + b) = a.b + a.c.$

We say that in numbers, multiplication distributes addition.

But a + (bc) ≠ (a + b), (a + c) *i.e.,* addition does not distribute multiplication is numbers. However, is statements 'and' *distributes* 'or' and 'or' *distributes* 'and'.

Thus for any statements p, q and r we have,

p and (q or r) = (p & q) or (p & r)

and p or (q & r) = (p or q) and (P or r).

We shall be using the symbols '∀' and '∃' for *'for all values of'* and *'there exists'* respectively. Also,

(i) If (p ⇒ q and q ⇒ p), the we write, **'p ⇒ q'** *i.e.,* p *implies and is implied by q.*

(ii) we write, **'p ⇒ q'**, if whenever p is true, the q is true and we say that, *p implies q.*

Tautologies

A statement which is always true is called tautology. Thus, for a statement to be a tautology its truth value should be T for all its entries in the truth value.

EQUIVALENCE CLASSES

Let R be an equivalence relation on a set A. Let a ∈ A. Then, the set of all those elements of A, which are related to a is called an equivalence class determined by a and it is denoted by [a] or $\bar{a}$.

Thus, [a] = {b ∈ A: (a, b, ∈ R} = {b ∈ A: aRb].

Quotient Set

Let S be a non-empty set and R be an equivalance relation defined in S. The set of mutually disjoint equivalance classes in which S is partitioned relative to the equivalance relation R is said to be quotient set of S for the equivalance relation R, and is denoted by S/R or By S.

PROPERTIES OF EQUIVALENCE CLASSES

Example 1:

On the set Z of all integers, consider the relation

R = {(a, b): a ≡ b (mod 3)}.

This has earlier been shown that the above relation is an equivalence relation.

The equivalence classes determined by the integers 1, 2, and 3 are the set consisting of integers related to 1, 2, and 3 respectively.

Now $[1] = \{..., -8, -5, -2, 1, 4, 7, 10,...,\}$

$[2] = \{..., -7, -4, -1, 2, 5, 8, 11,...,\}$

and $[3] = \{..., -6, -3, 0, 3, 6, 9, 12, 15,...,\}$.

It may be noted here that, $Z = [1] \cup [2] \cup [3]$, where [1], [2] and [3] are mutually disjoint equivalence classes.

In the above case, we say that the sets [1], [2] [3] and constitute the partition of Z.

Partitions: Let S be a non exmpty set A set P = {A, B, C...} of non empty subset of S will be called a partition of S if:

(i) $A \cup B \cup C \cup ... = S$ *i.e.* The set S is the union of the sets in P and

(ii) The intersection of every pair of definite subset of $S \in P$ is the null set *i.e.*, if A and $B \in P$ then either $A = B$ or $A \cap B = \phi$.

Theorem 2(a):

An equivalence relation R on set A, partitions it into mutually disjoint equivalence classes in such a way that any two elements from the same class are not related to each other and the elements from different classes are not related to each other.

Conversely, *every partition of a set A determines an equivalence relation on A.*

Proof:

Let R be an equivalence relation on a set A. Let a be an arbitrary element of A. Then, by reflexivity $a R a \ \forall\ a \in A$ and therefore, $a \in [a]$.

Thus, every element of A belongs to some equivalence class i.e, A is the union of some equivalence classes. Moreover, any two equivalence classes are either disjoint or identical.

Thus, A is expressible as the union of some mutually disjoint equivalence classes.

Now, let x and y be any two elements from the same class [a].

Then $x \in [a]$ and $y \in [a]$

$\Rightarrow$ x R a and y R a

$\Rightarrow$ x R a and a R y [by symmetry]

$\Rightarrow$ x R y [by transitivity].

This has that any two elements from the same class are related to each other.

Again, let $[a] \neq [b]$ and let $x \in [a]$ and $y \in [b]$, and we must show that x is not related to y.

Now $x \in [a]$ and $y \in [b]$

$\Rightarrow$ x R a and y R b.

Now, if possible, let x R y. Then

x R y, y R b $\Rightarrow$ x R b.

And x R a, x R b

$\Rightarrow$ a R x, x R b [by symmetry]

$\Rightarrow$ a R b

$\Rightarrow a \in [b]$

$\Rightarrow [a] = [b]$.

Hence it is contradiction.

Thus, it follows that any two elements from two different classes are not related to each other.

As a consequence of the above results, it follows that A is expressible as the union of mutually disjoint equivalence classes satisfying the requisite properties and therefore the theorem follows.

Conversely, let $\{A_1, A_2, A_3,...\}$ constitute the partition of A, so that A'^3 are mutually disjoint and $A = A_1 \cup A_2 \cup A_3 \cup ...$ Consider the relation R on A defined by,

$$a \text{ R } b \Leftrightarrow a, b \in A_j \text{ for a fixed } j.$$

Then, R is

(i) *Reflexive,* for if a is an arbitrary element of A, then a is contained in some A, and since any two elements from the same A, are related to each other, so in particular aRa.

(ii) *Symmetric,* for, if a R b, then

a R b

$\Rightarrow$ a, b $\in A_j$ for a fixed j

$\Rightarrow$ b, a $\in A_j$ for a fixed j

$\Rightarrow$ a, R a.

(iii) *Transitive,* for, if a R b and b R c, then

a R be and b R c

$\Rightarrow$ a, b $\in A_j$ for some fixed j and b, c $\in A_j$ for some fixed j.

$\Rightarrow$ a, c $\in A_j$ for some fixed j [$\because A_j \cap A_j = \phi$ for $i \neq j$]

$\Rightarrow$ a R c.

Thus, R is an equivalence relation on A and A's are precisely the equivalence classes determined by the elements of A.

Theorem 2(b):

Let R be an equivalence relation on a set A and let us denote by [a], the equivalence class determined by the element a, then (i) $a \in [a]$, (ii) $a \in [b] \Leftrightarrow [a] = [b]$,

(iii) *The equivalence classes determined by two elements are either disjoint or identical.*

i.e., either $[a] \cap [b] = \phi$ or $[a] = [b]$.

Proof:

(i) The relation R being reflexive, we have

$a \; R \; a \; \forall \; a \in A$, Hence relative R being reflexive

And therefore, $a \in [a] \; \forall \; a \in A$.

(ii) Let $a \in [b]$, then a R b.

So, $x \in [a] \Leftrightarrow x \; R \; a$

$\Leftrightarrow x \; R \; a$ and $a \; R \; b$ (given)

$\Leftrightarrow x \; R \; b$ (by transivity)

$\Leftrightarrow x \in [b]$.

$\therefore \; = [a] = [b]$.

Thus $a \in [b] \Leftrightarrow [a] = [b]$.

Again, let $[a] = [b]$

Then from (i), $a \in [a]$ and so $a \in [b]$ $(\because [a] = [b])$

Hence $a \in [b] \Leftrightarrow [a] = [b]$.

(iii) For any two classes, [a] and [b], if $[a] \cap [b] = \phi$, then the result follows. So , let us consider the case, when $[a] \cap [b] \neq \phi$.

Let $x \in [a] \cap [b]$.

Then $x \in [a] \cap [b]$.

$\Rightarrow x \in [a]$ and $x \in [b]$

$\Rightarrow x \; R \; a$ and $x \; R \; b$

$\Rightarrow a \; R \; x$ and $x \; R \; b$ [by symmetry]

$\Rightarrow a \; R \; b$ (by transitivity)

$\Rightarrow a \in [b] \Rightarrow a \; [a] = [b]$.

Thus, either $[a] \cap [b] = \phi$ or $[a] = [b]$.

BINARY OPERATION OR BINARY COMPOSITION ON A SET

Let G be a non empty set. Then an operation $*$ on a non-empty set G is said to be binary, if

$$a \in G, b \in G \Rightarrow a * b \in G \ \forall \ b \in G.$$

The above property is called the closure property and if this is satisfied, we say that G is closed under the operation

Example:

Addition operation (+) as well as multiplication operation (×) is a binary operation on the set N of all natural numbers, since

$$a \in N, b \in N \Rightarrow a + b \in N \ \forall \ a, b \in N$$

and $$a \in N, b \in N \Rightarrow a \times b \in N \ \forall \ a, b \in N.$$

Similarly each one of addition and multiplication operations, is a binary operation on each of the sets, the set Z of integers, the set Q of rational numbers, the set R of all real numbers and the set C of all complex numbers.

However, the set P of all irrationals, is not closed for multiplication, since $\sqrt{2} \in P$, $\sqrt{2} \in P$, but $\sqrt{2} \times \sqrt{2} = 2 \notin P$.

Similarly, subtraction operation (–) on N is not a binary composition and so is the case with the division operation (÷) on all the above sets.

Exponential operation $[(a, b) \to a^b]$ is a binary composition on each of the sets N, Q and R, but it is not a binary composition on the set Z of integers, since

$$2 \in Z, -2 \in Z \text{ but } 2^{-2} = \frac{1}{2^2} \notin Z.$$

LAWS OF BINARY COMPOSITION

A binary $*$ composition * on a non-empty set G is said

(i) **Commutative,** let

$$a * b = b * a \ \forall \ a, b \in G;$$

(ii) **Associative,** let

$$(a * b) * c = (b * c) \ \forall \ a, b, c, \in G;$$

(iii) *Also, if e is an identity element, with respect to* $*$ *in G, then an element* $a \in G$ *is said to be* **invertible,** *if there exists an element* $b \in G$ *such that,*

$$a * b = b * a = e$$

and in this case a and b are said to be the inverses of each other, written as, $a^{-1} = b$ *and* $b^{-1} = a$*;*

(iv) **to have an identity element,** *if there exists an element denoted by e in G such that,*

$$a * e = e * a \ \forall \ a, \in G;$$

(v) *to satisfy the* **left cancellation law,** *if*

$$a * b = a * c \Rightarrow b = c \ \forall \ a, b, c \in G;$$

(vi) *to satisfy the* **left cancellation law, if**

$$b * a = c * a \Rightarrow b = c \ \forall \ a, b, c \in G;$$

(vii) to be **right distributive** over another composition o on G. if

$$a * (boc) = (a * b) \text{ o } (a * c) \ \forall \ a, b, c \in G;$$

and (vii) to be **left distributive** over o, if

Algebraic Structures: A non-empty set G along with one or more binary compositions on G is called an algebraic structure or an *algebraic system.*

Thus, each on of the systems (N, +), (Q, ×) (R, +, ×) and [P (A), ∪, ∩)] is an algebraic structure.

SET

Definition: *A set as a well defined collection of objects* by the term *'well* **defined'**, we mean that we must given a rule with the help of which we should able to tell whatever a given object belongs or does not belongs to that given collection.

The object in a set are called its members or elements. We usually denote sets by capital letters and their elements by small letters.

If an object x is a member of a set A, we write x ∈ A, which means that 'x belongs to A' or that 'x is an element of A'. On the other hand, if x does not belong to A, we write x ∉ A.

Example of Sets

(i) The collection of vowels in English alphabet is a set, containing five elements, namely a, e, i, o, u.

(ii) The collection of four fourth roots of unity, constitutes a set with 1, –1, i, –i as its elements,

(iii) The collection of first four counting numbers is a set containing 1, 2, 3, 4.

Note: The terms good, bad, rich, poor, honest etc. are vague and these are not well defined. Thus.

HOW TO DESCRIBE OR SPECIFY A SET

There are two methods for describing a set:

(1) **Tabulation of Roaster Method:** *Under this method we just make a list of the objects of the set and put it within braces { }.* or

In tabulation each element is seperated by commas and inclosed in brackets.

Examples 1:

(i) If B = {3, 5, 7, 9, 11}, then we may write.

B = {x: x = 2n + 1, where n is a counting number, n < 6}.

(ii) If A is the set of first four counting numbers, then A = {x: x is a counting number, x < 5}.

Examples 2:

(i) If B is the set of four fourth roots of unity, then B = {1, –1, i, –i}.

(ii) If A is the set of vowels in English alphabet, then A = {a, e, i, o, u}

(iii) **Description or Set Builder Method:** *This method consists in the listing of the property or properties satisfied by the elements of the set.* We write, {x: x *satisfied Properties P}, i.e.,* the set of all those elements such that each element x satisfies the properties P.

SETS OF NUMBERS

(a) **Natural Numbers:** *Counting numbers are called natural numbers.* Thus, the set N = {1, 2, 3, 4, ...} is the set of all natural numbers. Note that $0 \notin N$

(b) **Integers:** *The set Z consisting of all natural numbers, 0 and negatives of natural numbers, is the set of all integers.*

Thus, Z = {0, –3, –2, –1, 0, 1, 2, ...}.

Remarks:

(a) The integers divisible by 2 are *even integers,* while the integers not divisible by 2 are *odd integers.*

(b) The set of +ve *integers is,* Z^+ = {1, 2, 3, 4, ...}

The set of –*ve integers is,* Z^- = {–1, –2, –3, ...}

Note: 0 is neither +ve nor –ve.

(c) **Whole Numbers:** *The set obtained by adjoining 0 to the set of all natural numbers, is the set W of* ***whole numbers.***

Thus, W = {0, 1, 2, 3, 5}.

(d) **Real Number:** *The totality of retionals and irrationals forms the set R of all* **real numbers.**

(e) **Rational Numbers:** *The set Q of all numbers of the form (p/q) where p and q are integers and q ≠ 0, is the set of* **rationals.**

Thus, *Q = {p/q: p and q are integers q ≠ 0}.*

(f) **Irrational Numbers:** *The set of all numbers which can be expressed in decimal form in non-terminating and non-repeating form only, is the set of irrationals. e.g.,* $\sqrt{2}$, $\sqrt{3}$, $\sqrt{5}$, π, e etc. are irrationals.

(g) **Intervals:** If a and b are real numbers such that a , b, then

(a) the set $\{ x \in R: a \leq x \leq b\}$ is called a *closed interval* denoted by [a, b];

(b) the set $\{x \in R: a < x < b\}$ is called an *open interval* denoted by]a, b[;

(c) the set $\{x \in R: a < x \leq b\}$ is called a *left half open interval*, denoted by]a, b]; and

(d) the set $\{x \in R: a \leq x < b\}$ is called a *right half open interval*, denoted by [a, b[.

(h) **Complex Number:** The set

Z = {a + ib: a and b are real and $i = \sqrt{-1}$ is the set of all *Complex Numbers.*

EMPTY SET

*A set consisting of no element at all is called an **empty*** or a ***void*** or a ***null*** set and it is denoted by ϕ.

In Roaster method, it is denoted by { }.

Examples:

Each one of the sets given below is an empty set.

(i) {x: x is a natural number, 2 < x < 3}

(ii) {x: x is a real number, $x^2 < 0$}

(iii) $\{x: x \neq x\}$.

Singleton Set: *A set consisting of a single point x is called a* **Singleton Set,** *to be denoted by {x}.* Thus {x: x + 5 = 5} = {0} is a singleton set, whose only member is 0.

FINITE AND INFINITE SETS

A set is said to be finite if it consists of a specific number of different element i.e., counting process can come to an end otherwise a set is said to be infinite.

Example 1:

Each one of the sets given below is a finite set.

(i) Set of all persons on earth;

(ii) {x: x is a natural number, x < 5 crores}
(iii) Set of all rivers in India.

Example 2:

Each one of the sets given below is an infinite set.

(i) Set of all points on the arc of a circle.
(ii) {x: x is a rational number, 0 < x < 1};
(iii) {x: x is multiple of 2, x is an integer};
(iv) Set of all concentric circles;

EQUAL SETS

Two sets A and B are said to be equal, written as A = B, if every element of A is in B and every element of B is in A, if $A \subset B$ and $B \subset A \Rightarrow A = B$. What is meant by the equality of two sets? For us this is always mean that they contain the same elements that is every element which is in one is in other, and vice-versa.

Notes:

(i) The repetition of elements in a set is meaningless.
Thus {x: x is a letter in 'follow'} = {x: x is a letter in 'wolf' *i.e.*,
{f, o, *l*, *l*, o, w} = {w, o, *l*, f}.

(ii) ϕ, {0} and 0 are all different, since ϕ is a set containing no element at all, {0} is a set containing one element, namely 0 and 0 is a number, not a set.

(iii) The elements of a set may be listed in any order. Thus {p, a, t} = {a, t, p}{t, p, a} etc.

EQUIVALENT SETS

The number of distinct element contained by a finite set A is called the **cardinal number** *of A, to be denoted by n (A). Two finite sets are said to be* **equivalent,** *if they have the same number of distinct elements.*

Remark:

Equivalent sets are not always equal, but equal sets are always equivalent e.g. {1, 2, 3} and {a, b, c} are equivalent, but not equal.

SUBSETS

If A and B be two sets given in such a way that every element of A is in B, then we say that A is a **subset** *of B and write,* $A \subseteq B$. *Also, if* $A \subseteq B$, *then we say that B is a* **superset** *of a and write,* $B \supseteq A$. *If* $A \subseteq B$ *but* $A \neq B$, *then A is called a* **proper subset** *of B, written as,* $A \subset B$.

Two sets A and B are said to be comparable if A is the subset of B or B is the subset of A, that is if $A \subseteq B$ of $B \supseteq A$.

Notes:

(i) Every set is a subset of itself.

(ii) If $\exists$ even a single element in A which is not in B, then A is not a subset of B and we write, $A \not\subseteq B$.

(iii) If A and B are two sets given in such a way that no element is common to both A and B, then A and b are said to be *disjoint* sets.

(iv) ϕ has no proper subset.

Examples:

(i) $N \subset W \subset Z \subset Q \subset R \subset C$ but $W \not\subseteq Z^+$, $\{0\} \not\subseteq Z^+$,

(ii) $\{x \in z\text{: x is a multiple of } 6\} \subset \{x \in Z\text{: x is a multiple of } 3\}$

(iii) $\{1, 2\} \not\subseteq \{2, 4, 6, 8\}$

(iv) $\{1\} \subset \{1, 2, 3\}$.

Axiom of Specification: If A is a set and P(x) is a statement then there exist a subset B of A whose elements are exactly those elements x of A for which the statement P(x) is true.

We say that the subset B is specified by the statement P(x) notice that this subset B is unique statement P(x), then by the axim of identity B = B'.

We write $B = \{x\text{: } P(x) . x \in A\}$,

or simply $B = \{x\text{: } P(x)\}$.

Example 1:

(i) Let A = {x: x is even}, *i.e.*, A = {2, 4, 6, . . .) and let B = {x: x is a positive poweer of 2}, *i.e.*, B = {2, 4, 8, 16}. Then $A \subset B$, *i.e.*, A is contained in B.

(ii) If P be the set of all parallelograms and S is the set of all squares in a plane, *i.e.*,

P = {all parallelograms in the plane}

and S = {all squares in the plane}

Then S is a subset of P, *i.e.*,

$S \subset P$.

Remark 1: *If we have to prove $A \subset B$, then we should prove that $x \in A \Rightarrow x \in B$.*

Symbolically, $A \subset B$ iff $(x \in A \Rightarrow x \in B)$.

Remark 2: *If we have to prove that $A \not\subset B$ then we should show that there exists atleast one element x such that $x \in A$ but $x \notin B$.*

Symbolically, $A \not\subset B$ iff $\{\exists\ x \in A$ s.t. $x \notin B)$.

The symbol $\exists$ stands for 'there exists' and s.t. for 'such that'.

Example 2:

Write the subset of the following set:

$$A = \{1, 2, 3\}.$$

Solution:

The required subsets are:

$$\phi, \{1\}, \{2\}, \{3\}, \{1, 2\}, \{1, 3\}, \{2, 3\}, \{1, 2, 3\}.$$

Equality of Sets: Set A is said to be equal to set B if both of them have the same members, *i.e.*, every element which belongs to A also belongs to B ($A \subset B$) and every element which belongs to B also belongs to A ($B \subset A$). We denote the equality of sets A and B by

$$\Rightarrow A = B.$$

Proper Subsets: Let A and B be two sets. If $B \subset A$ and $B \neq A$, then B is said to be a proper subset of A.

For example, if A {1, 2, 3} and B = { 2, 3} then B is proper subset of A because all the elements of B are in A but one element 1 of A is not in B.

Theorem 1:

Null set ϕ is a subset of every set.

Proof:

Let A be any given set. We shall prove that $\phi \subset A$. Let us suppose, on contrary, that $\phi \not\subset A$.

$\phi \not\subset A \Rightarrow$ there is atleast one element x such that $x \in \phi$

and $\quad x \notin A$ ...(1)

But since ϕ is a null set,

therefore $\quad x \notin \phi$ ...(2)

Therefore, (1) and (2) $\Rightarrow x \in \phi$ and $x \notin \phi$, which is absurd and therefore, our assumption is wrong. Consequently $\phi \subset A$.

Since A is an arbitrary set ϕ is a subset of every set.

Number of Subsets of a Finite Set: Consider a set {a}. It has two possible subsets {a} and ϕ. Again for the set {a, b}, the possible subsets are ϕ, {a}, {b}, {a, b} which are 4 subsets in number. A set {a, b, c} has as many as the following subsets.

ϕ, {a}, {b}, {c}, {a, b}, {b, c}, {a, c}, {a, b, c}.

which are eight in numbers.

Thus we can generalize the result as follows:

A set with 1 element has 2^1 subsets.

A set with 2 element has 2^2 subsets.

A set with 3 element has 2^3 subsets.

Theorem 2:

Every set is a subset of itself.

Proof:

Let A be any set. Then each element of A is clearly in A itself. Hence $A \subset A$.

Theorem 3:

If $A \subset B$ *and* $B \subset C$ *then* $A \subset C$.

Proof:

We must show that every element in A is also an element is C. Let x be an element of A, *i.e.*,

$$x \in A.$$

But it is given that $A \subset B$.

$$\therefore \quad x \in A \Rightarrow x \in B \qquad ...(1)$$

Also $\quad B \subset C$

$$\therefore \quad x \in B \Rightarrow x \in C \qquad ...(2)$$

Hence from (1) and (2)

$$x \in A \Rightarrow x \in C$$

$$\therefore \quad A \subset C.$$

Theorem 4:

The total number of subsets of a finite set containing n elements is 2^n.

Let A be a finite set containing n elements. Let $O \le r \le n$. Consider those subsets of A that have r elements each. We know that the number of ways in which r elements can be chosen out of n elements is ${}^nC_r$. Therefore, the number of subsets of A having r elements each is ${}^nC_r$.

Hence, the total number of subsets of A.

$$= {}^nC_0 + {}^nC_1 + {}^nC_2 + \ldots + {}^nC_r + \ldots + {}^nC_n$$

$$= (1 + 1)^n = 2^n.$$

SOME RESULTS ON SUBSETS

Theorem 1:

The total number of subsets of a given set containing n elements is 2^n.

Proof:

Let A be an arbitrary set containing n elements. Then, one of its subsets is the empty set. Apart from this,

the number of singleton subsets of A = n = ${}^nc_1$,

the number of subsets of A, each containing 2 elements = ${}^nc_2$,

the number of subsets of A, each containing 3 elements = ${}^nc_3$,

....

the number of subsets of A, each containing (n – 1) elements = ${}^nc_{n-1}$,

the number of subsets of A, each containing n elements = ${}^nc_n$,

total number of subsets of A

$$= 1 + {}^nc_1 + {}^nc_2{}^nc_3 + \ldots + {}^nc_{n-1} + {}^nc_n,$$

$$= (1 + 1)^2 = 2^n$$ [Using Binomial Theorem] **Hence Proved.**

Theorem 2:

The empty set is a subset of every set.

Proof:

Let A be an arbitrary set. Then, in order to show that $\phi \subset A$, we must show that there is no element of ϕ which is not contained in A and since ϕ contains no element at all, no such element can therefore be found. Hence $\phi \subset A$. **Hence Proved.**

RELATIONS

Let A and B be two non-empty sets. Then, a relation from A to B is a subset of A × B. Thus.

R is a relation from A to B $\Leftrightarrow R \subseteq (A \times B)$.

It A = B we say that R is the relation on A we write a Rb iff $(a, b) \in R$ and say that a is related to b or that b is relative of a.

We also write a (~R)b when a is not R-relative to b.

It a consist of m elements and B consist of n elements then A × B consist of m × n elements and so (P(A × B) will have 2^{mn} elements hence the total number of different relation from A to B is 2^{mn}.

Notes:

(i) Let $R \subseteq A \times B$ be a relation from A to b. Now, if $(a, b) \in R$, then we say that 'a is R – related to b', and we write, a R b.

On the other hand, if (a, b) ∉ R, then 'a is not R-related to b', and we write, a K b.

(ii) The sets A × B and ϕ being subsets of A × B, it follows that each one is a relation from A to B. More-over, A × B is known as a universal relation, while ϕ is called an empty relation.

Domain and Range of a Relation: If R is a relation from a set A to another set B. Then, we domain of R is the set of all first element of R and the range of R is the set of all second coordinate of the member of R.

Domain (R) = {a: (a, b) ∈ R} and **range (R)** = {b: (a, b) ∈ R}.

Example:

Let A = {1, 2, 3,} and B = {2, 4, 5}, then

A × B = {(1, 2), (1, 4), (1, 5), (2, 2), (2, 4), (2, 5), (3, 2), (3, 4), (3, 5)}

Let R = {1, 2), (1, 5), (3, 4)}.

Then, R being a subset of A × B, it is a relation from A to B.

Here (1 R 2), (1 R 5), (3 R 4).

Apart from these, no element of A is related to an element of B.

Clearly, 2 R 1, 3 R 1, 4 R 2 etc.

Here, **Dom (R)** = {1, 3} and **range (R)** = {2, 4, 5}.

The Indentity Relation: Let A be any set, then the identity relation IA or (diagonal relation) on A is defined by setting (a, b) IA iff a = b the domain and range of IA are both A.

Example:

Let A = {1, 2, 3, 4} then the identity relation IA = {(1, 1), (2, 2), (3, 3), (4, 4)}

INVERSE RELATION

Let R ⊆ A × B be a relation from A to B. Then, the inverse of R, denoted by R^{-1} is a relation from B to A, defined as

R^{-1} = {b, a): ∈ R}.

Clearly, the domain and range of R^{-1} are respectively the range and domain of R.

Example:

Let A {0, 1, 2} and B {1, 3, 4}.

Now, If R = {(0, 1), (1,1), (1, 3), (2, 3)},

Then $R^{-1} = \{(1, 0), (1, 1) (3, 1), (3, 2)\}$.

Clearly, dom $(R^{-1}) = \{1, 3\}$ = range (R)

and range $(R^{-1}) = \{0, 1, 2\}$ = dom (R).

COMPOSITE RELATION

Let R and S be the relations from the sets A to B and B to C respectively. Then, the composite of R and S is a relation from A to C, denoted by SoR and defined by

$SoR = \{(a, c): \exists\, b \in B$ such that $(a, b) \in R$ and $(b, c) \in S\}$.

Thus, $(a, b) \in R, (b, c) \in S \Rightarrow (a, c) \in SoR$.

Example:

If two relations R and S be given as

$R = \{(1, 3), (2, 1), (3, 4), (4, 2)\}$

and $S = \{1, 2), (2, 3), (3, 4), (4, 1)\}$

find SoR

Solution:

Clearly, dom (SoR) = dom R.

Now $(1, 3) \in R, (3, 4) \in S \Rightarrow (1, 4) \in SoR$;

$(2, 1) \in R, (1, 2) \in S \Rightarrow (2, 2) \in SoR$;

$(3, 4) \in R, (4, 1) \in S \Rightarrow (3, 1) \in SoR$;

$(4, 2) \in R, (2, 3) \in S \Rightarrow (4, 3) \in SoR$;

$\therefore$ $SoR = \{(1, 4), (2, 2), (3, 1), (4, 3)\}$.

Theorem:

Let A, B and C be non empty sets and let $R \subseteq A \times B$ and $S \subseteq B \times C$, then $(SoR)^{-1} = (R^{-1}oS^{-1})$.

Proof:

By definition we have

$R \subseteq A \times B, S \subseteq B \times C \Rightarrow SoR \subseteq A \times C \Rightarrow (SoR)^{-1} \subseteq C \times A$.

Let (c, a) be an arbitrary element of $(SoR)^{-1}$. Then

$(c, a) \in (SoR)^{-1}$

$\Rightarrow$ $(a, c) \in (SoR)$

$\Rightarrow$ $\exists\, b \in B$ such that $(a, b) \in R$ and $(b, c) \in S$

$\Rightarrow$ $(b, a) \in R^{-1}$ and $(c, b) \in S^{-1}$

$\Rightarrow$ $(c, a) \in R^{-1}oS^{-1}$.

Thus, $(SoR)^{-1} \subseteq R^{-1}oS^{-1}$.

Similarly, $R^{-1}oS^{-1} \subseteq (SoR)^{-1}$.

Hence $(SoR)^{-1} = R^{-1}oS^{-1}$.

Hence $(SoR)^{-1} = R^{-1}S^{-1}$.

Universal Relation: Let A be any set and R be the set A = A then R is called the universal relation on A.

Void Reation in a Set: Every subset of A × A is a relation on A. Since ϕ is also a subset of A × A. Therefore the null set ϕ is also a relation on A the relation is called the empty relation or void relation on A.

BINARY RELATION

Let A be a non-empty set. Then, a subset of A × A is called a binary relation on A.

The set $I_A = \{(a, a): a \in A\}$ is know as a **identity relation** on A.

UNIVERSAL SET

While making a study of sets, there happens to be a set, which is a superset of each one of the sets under consideration. Such a set is known as the **Universal Set.**

Examples:

(i) If A = {1, 2, 3, 4}, B = {2, 4, 6, 8}, C = {1, 3, 5, 7} then
X = {1, 2, 3, 4, 5, 6, 7, 8} is clearly a universal set.

Power Set: *The family or collection of all subsets of a given set A, is called the* **power set** *of A and it is denoted by* P(A).

(ii) If B = {1, {2}} then, we may write B = {1, A} where A = {2}

$\therefore$ P (B) = {ϕ, {1}, {A}, {1, A}}
= {ϕ, {1}, {{2}, {1, {2}}}

(iii) If C = {{1}, {2, 3}} then we may write,
C = {A, B}, where A = {1}, B = {2, 3}

$\therefore$ P(C) = {ϕ, {A}, {B}, {A, B}}
= {ϕ, {{1}}, {{2, 3}}, {{1}, {2, 3}}}.

OPERATION ON SETS

1. **Union of Sets.** *The union of two sets A and B denoted by A ∪ B is the set constituting of all those elements, each one of which is contained either in A or in B or in both A and B.*

Thus $A \cup B = \{x: x \in A \text{ or } x \in B\}$

Consequently, $x \in A \cup B \Rightarrow x \in A \text{ or } x \in B.$

And $x \notin A \cup B \Rightarrow x \notin A \text{ and } x \notin B.$

Note: A wood about use of "or" in ordinary English when we say that something is one of the other we imply that it is not both. The mathematical "or" is quite different, at least when we are speaking in set theory. For when we say that x is in A or X is in B we mean x is in atleast one of A or B, and may be both.

Example:

(i) If $A = \{x: x \text{ is a negative integer}\}$

and $B = \{x: x \text{ is a positive integer}\}$

then $A \cup B = \{x: x \text{ is an integer}, x \neq 0\}.$

2. **Intersection of Set:** *The intersection of two sets A and B denoted by $A \cap B$ is the set consisting of all those elements which are common to both A and B.*

Thus $A \cap B = \{x: x \in \text{ and } x \in B\}$

Consequently, $x \in A \cap B \Rightarrow x \in A \text{ and } x \in B.$

And $x \notin A \cap B \Rightarrow x \in A \text{ or } x \notin B.$

When the sets A and B are **disjoint**, no element is common to both A and B and thus in this case, $A \cap B = \phi$.

When $A \cap B \neq \phi$, the sets A and B are said to be intersecting.

Example:

(i) Let $A = \{1, 2, 3\}$

and $B = \{2, 4, 6\}$,

then $A \cup B = (1, 2, 3, 4, 6\}.$

(ii) If $A = \{x: x = 2n + 1, n \in Z\}$

and $B = \{x: x = 2n, n \in Z\}.$

then $A \cup B = \{x: x \text{ is an odd integer}\}$

$\cup\{x: x \text{ is an even integer}\}$

$= \{x: x \text{ is an integer}\} = Z.$

Note 1: Each element in a set is listed once only because the repetition of elements is meaningless in a set. If A is a set of cricket players, B is a set of tennis players and some players are common to both the teams, then

$A \cup B$ = Set of all players in the two teams.

Note 2: From the definition it is clear that A and B are the subsets of $A \cup B$.

Symbolically, $A \subset (A \cup B)$ and $B \subset (A \cup B)$.

Note 3: The union of a finite number of sets $A_1, A_2, \ldots, A_n$ is denoted by $A_1 \cup A_2 \cup A_3, \ldots \cup A_n$

or by $\bigcup_{i=1}^{n} A_1$.

Example:

(i) If $A = \{x: x \text{ is a +ve integer, } x \text{ is a multiple of } 4\}$

and $B = \{x: x \text{ is a +ve integer, } x \text{ is a multiple of } 6\}$

then $A \cap B = \{x: x \text{ is a +ve integer, } x \text{ is a multiple of } 12\}$.

Examples:

(i) If $A = \{a, b, i, z\}$

$B = \{b, c, i, u, v\}$

Then $A \cap B = \{b, i\}$.

(ii) If $A = \{x: x = 2n, n \in Z\}$ and

$B = \{x: x = 3n, n \in Z\}$,

Then $A \cap B = \{x: x = 2n, n \in Z\} \cap \{x: x = 3n, n \in Z\}$

$= \{\ldots, -4, -2, 0, 2, 4, 6, \ldots\} \cap \{\ldots, -9, -6, -3, 0, 3, 6, 9, \ldots\}$

$= \{\ldots, -6, 0, 6, 12, \ldots\} = \{x: x = 6n, n \in Z\}$.

Note 1: If A and B are any two sets, then $A \cap B \subset A$ and $A \cap B \subset B$.

Note 2: The intersection of finite number of sets $A_1, A_2, \ldots, A_n$ is denoted by

$$A_1 \cap A_2 \cap \ldots \cap A_n$$

or by $\bigcap_{i=1}^{n} A_1$

3. **Difference of Sets:** *If A and B are two sets, then their difference denoted by A – B is defined by*

$A - B = \{x: x \in A \text{ and } x \notin B\}$

Thus $x \in A - B \Rightarrow x \in A$ and $x \notin B$.

Also the *symmetric difference* $A \Delta B$ is defined as

$$A \Delta B = (A - B) \cup (B - A).$$

Example:

If $A = \{a, b, c, d\}$ and $B = \{b, d, e, f\}$

then $A - B = \{a, c\}$ and $B - A = \{e, f\}$

and $A \Delta B = \{a, c, e, f\}$

Note that in general, $(A - B) \neq (B - A)$.

4. **Disjoint Sets:** *Two sets A and B are said to be disjoint sets if they have no element in common, i.e., if their intersection is a null set, i.e.,*

$$A \cap B = \phi.$$

Thus, $A \cap B = \phi \Rightarrow$ A and B are disjoint. For example, if

$$A = \{a, b, c\} \text{ and } B = \{p, q, r\}$$

then A and B are disjoint sets as there is no element common to both of them.

Difference of Two Sets: Let A and B be two sets. The difference of A and B, written as A–B, is the set of all those elements of A which do not belong to B.

Thus, $A - B = \{ x: x \in A \text{ and } x \notin B\}$

or $A - B = \{ x \in A: \text{ and } x \notin B\}$.

In Fig. (1.1) the shaded part represents A–B.

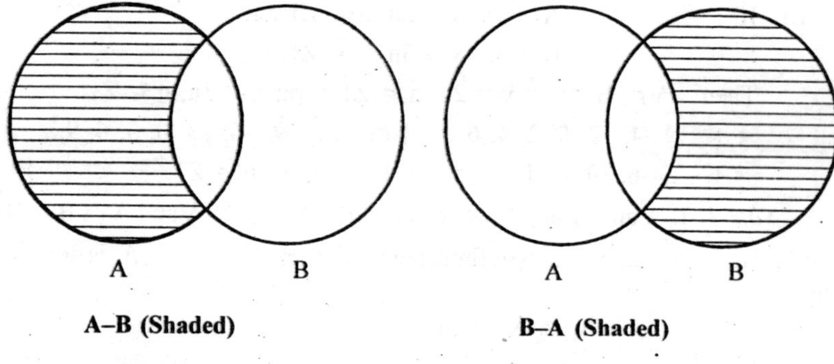

A–B (Shaded) **B–A (Shaded)**

Fig. 1.1 **Fig. 1.2**

Similarly the difference B–A is the set of all those elements of B that do not belong to A, *i.e.*,

$$B - A = \{x \in B: x \notin A\}$$

In Fig (7.2) the shaded part represents B–A.

Example: If $A = \{2, 3, 4, 5, 6, 7\}$

and $B = \{ 3, 5, 7, 9, 11, 13\}$,

then $A - B = \{2, 4, 6\}$

and $B - A = \{9, 11, 13\}$.

Symmetric Difference of Two Sets: Let A and B be two sets. The symmetric difference of sets A and B is the set $(A - B) \cup (B - A)$ and is denoted by $A \Delta B$.

Thus, $A \Delta B = (A - B) \cup (B - A)$

$= \{x: x A \cap B\}$

In following figure shaded part represents $A \Delta B$.

For example, If $A = \{1, 2, 3\}$

$B = \{3, 4, 5)$

then $A - B = \{1, 2\}$

$B - A = \{4, 5\}$

$\therefore A \Delta B = \{1, 2, 4, 5\}$.

COMPLEMENT OF A SET

The complement of a set B relation to antother set A is the set of all elements which belongs to A but which do not belong to B and is denoted by A~B or by A – B, it sometime called the difference of A and B and read as "A difference B". Or,

Let A be a subset of a universal set X. Then the complement of A in X, written as X ~ A or N is defined as $A' = \{x \in X: x \notin A\}$

Thus $x \in A' \Leftrightarrow x \notin A$

Example 1:

(i) If $X = \{x: x \text{ is a letter in the English alphabet}\}$

and $A = \{a, e, i, o, u\}$

then $A' = \{x: x \text{ is a consonant in the English alphabet}\}$

(ii) If $X = \{a, b, c, d, e\}$ and $A = \{b, d\}$

then $A' = \{a, c, e\}$.

Example 2: Le N, the set of all natural numbers be taken as the universal set and let

$A = \{x: x \text{ is an even number and } x \in N\}$,

then $A' = \{x: x \text{ is an odd number, } x \in N\}$.

Note 1: The set A and its complement A' are disjoint sets.

Note 2: Let A and B be two sets. The set $X = \{x: x \in A, x \notin B)$ is called the complement of the set B with respect to A (*i.e.*, A – B).

SOME RESULTS ON COMPLEMENTATION

(i) $\phi' = \{x \in X: x \notin \phi\} = X$

(ii) $X' = \{x \in X: x \notin X\} = \phi$

(iii) $(A')' = \{x \in X: x \notin A'\} = \{x \in X: x \in A\} = A$

(iv) $A \cap A' = \{x \in X: x \in A\} \cap \{x \in X: x \notin A\} = \phi$

(v) $A \cup A' = \{x \in X: x \in A\} \cup \{x \in X: x \notin A\} = X$.

LAW OF OPERATIONS

If A, B, C are any subsets of a set X, then

(a) De-Morgan's laws

(i) $(A \cup B)' = A' \cap B'$

(ii) $(A \cap B)' = A' \cup B'$.

(b) Distributive laws

(i) $A \cup (B \cap C) = (A \cup B) \cap (A \cup C)$

(ii) $A \cap (B \cup C) = (A \cap B) \cup (A \cap C)$

(c) Associative laws

(i) $(A \cup B) \cup C = A \cup (B \cup C)$

(ii) $(A \cap B) \cap C = A \cap (B \cap C)$

(d) Commutative laws

(i) $A \cup B = B \cap A$

(ii) $A \cap B = B \cap A$.

(e) Identity laws

(i) $A \cup \phi = A$

(ii) $A \cap X = A$

(f) idempotent laws

(i) $A \cup A = A$

(ii) $A \cap A = A$.

Venn-Euler Diagrams

A clerke known as Venn-Euler diagram is very often used to assist thinking on the relation which may exist between some subset of the universal set.

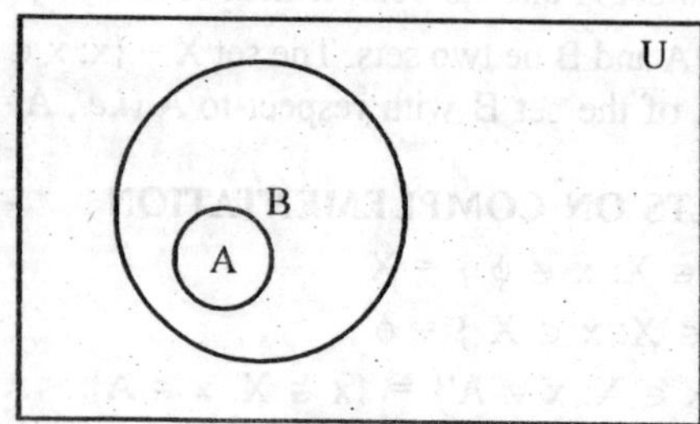

Fig. 1.3

A Venn diagram is a schematic representation of set by set of points. The universal set is represetation by points either a rectangle and a subset A of the universal set U, is represented by the interior of a circle. Disjoint set are depicted by non-developing resions and set in class on is depicted by taking one region lying entirely within the others.

The idea of $A \subset B$ and $A \neq B$ can be represented by the following diagram.

Example 1:

To prove: $(A \cap B)' = A' \cup B.$

Proof:

Let x be an arbitrary element of $(A \cap B)'$.

Then we have

$$
\begin{aligned}
x \in (A \cap B)' &\Rightarrow x \notin (A \cap B) \\
&\Rightarrow x \notin A \text{ or } x \notin B \\
&\Rightarrow x \in A' \text{ or } x \in B' \\
&\Rightarrow x \in (A' \cup B')
\end{aligned}
$$

Thus $(A \cap B)' \subseteq (A' \cup B')$

Similarly, $(A' \cup B') \subseteq (A \cap B)'$

Hence $(A \cap B)' = (A' \cup B')$ **Hence Proved.**

Example 2(a):

To prove: $A \cup B = B \cup A.$

Proof:

Let x be an arbitrary element of $A \cup B$.

$$
\begin{aligned}
\text{Then } x \in A \cup B &\Rightarrow x \in A \text{ or } x \in B \\
&\Rightarrow x \in B \text{ or } x \in A \\
&\Rightarrow x \in B \cup A
\end{aligned}
$$

$\therefore \quad A \cup B \subseteq B \cup A.$

Similarly, $B \cup A \subseteq A \cup B.$

Hence $A \cup B = B \cup A.$ **Hence Proved.**

Example 2(b):

To prove: $A \cup (B \cap C) = (A \cup B) \cap (A \cup C).$

Proof:

Let x be an arbitrary element of $A \cup (B \cap C)$.

Then $x \in A \cup (B \cap C)$

$\Rightarrow x \in A$ or $x \in (B \cap C)$

$\Rightarrow x \in A$ or ($x \in B$ and $x \in C$)

$\Rightarrow$ ($x \in A$ or $x \in B$)

and ($x \in A$ or $x \in C$) [$\because$ 'or' distributes 'and']

$\Rightarrow x \in (A \cup B)$ and $x \in (A \cup C)$

$\Rightarrow x \in (A \cup B) \cap (A \cup C)$

$\therefore \quad A \cup (B \cap C) \subseteq (A \cup B) \cap (A \cup C)$

Similarly, $(A \cup B) \cap (A \cup C) \subseteq A \cup (B \cap C)$.

Hence $A \cup (B \cap C) = (A \cup B) \cap (A \cup C)$. **Hence Proved.**

Example 3:

To prove: $(A \cap B) \cap C = A \cap (B \cap C)$.

Proof:

Let x be an arbitrary element of $(A \cap B) \cap C$.

Then $x \in (A \cap B) \cap C \Rightarrow x \in (A \cap B)$ and $x \in C$

$\Rightarrow$ ($x \in A$ and $x \in B$) and $x \in C$

$\Rightarrow x \in A$ and ($x \in B$ and $x \in C$)

$\Rightarrow x \in A$ and $x \in (B \cap C)$

$\Rightarrow x \in A \cap (B \cap C)$

$\therefore \quad (A \cap B) \cap C \subseteq A \cap (B \cap C)$

Similarly, $A \cap (B \cap C) \subseteq (A \cap B) \cap C$.

Hence $(A \cap B) \cap C = A \cap (B \cap C)$. **Hence Proved.**

SOME OTHER RESULTS ON OPERATIONS ON SETS

Theorem 1:

For any sets A and B, prove that:

(i) $A - B = A \cap B'$

(ii) $(A - B) \cup B = A \cup B$.

(iii) $(A - B) \cap B = \phi$.

Proof:

(i) Let x be an arbitrary element of $(A - B)$.

Then, $x \in (A - B) \Rightarrow x \in A$ and $x \notin B$

$\Rightarrow x \in A$ and $x \in B'$

$\Rightarrow x \in (A \cap B')$.

$\therefore \quad (A - B) \subseteq (A \cap B')$

Similarly, $(A \cap B') \subseteq (A - B)$

Hence $(A - B) = (A \cap B')$.

(ii) Let x be an arbitrary element of $(A - B) \cup B$

Then $x \in (A - B) \cup B$

$\Rightarrow x \in (A - B) \Rightarrow x \in B$

$\Rightarrow (x \in A \;\&\; x \notin B) \Rightarrow (x \in B)$

$\Rightarrow (x \in A \Rightarrow x \in B)$ and $(x \notin B \Rightarrow x \in B)$ [$\because$ 'or' distributes '&']

$\Rightarrow x \in (A \cup B)$

$\therefore \quad (A - B) \cup B \subseteq (A \cup B)$ ...(i)

Again, let y be an arbitrarily element of $(A \cup B)$. Then

$y \in (A \cup B)$

$\Rightarrow (y \in A \Rightarrow y \in B)$

$\Rightarrow (y \in A \Rightarrow y \in B)$ and $(y \notin B \Rightarrow y \in B)$ **[Note]**

$\Rightarrow (y \in A \;\&\; y \notin B) \Rightarrow (y \in B)$ ['or' distributes '&']

$\Rightarrow y \in (A - B) \Rightarrow (y \in B)$

$\Rightarrow y \in (A - B) \cup B$ ...(ii)

Hence from (i) and (ii), we have $(A - B) \cup B = (A \cup B)$.

(iii) If possible, let $(A - B) \cap B \neq \phi$ and let $x \in (A - B) \cap B$.

Then, $x \in (A - B) \cap B$

$\Rightarrow x \in (A - B)$ and $x \in B$

$\Rightarrow (x \in A \;\&\; x \notin B)$ and $(x \in B)$

$\Rightarrow x \in A$ and $(x \notin A \;\&\; x \in B)$

But, $x \notin B$ and $x \in B$ both can never hold simultaneously.

Thus, we arrive at a contradiction.

Since the contradiction arises by assuming that $(A - B) \cap B \neq \phi$, hence $(A - B)\square \cap B \neq \phi$. **Hence Proved.**

Theorem 2:

Using various laws on operations on sets, prove the following:

(i) $A \cap (B \Delta C) = (A \cap B) \Delta (A \cap C)$

(ii) $A - (A - B) = A \cap B$

(iii) $(A \cup B) - (A \cap B) = (A - B) \cap (B - A)$.

Proof:

(i) We have $A \cap (B \Delta C)$

$= A \cap [(B - C) \cup (C - B)]$

$= [A \cap (B - C)] \cup [A \cap (C - B)]$ (distributive law)

$= [(A \cap B) - (A \cap C)] \cap (A \cap C) - (A \cap B)]$

$= (A \cap B) \Delta (A \cap C).$

(ii) We have

$A - (A - B) = A - (A \cap B')$ $[\because A - B = A \cap B']$

$= A \cap (A \cap B')'$

$= A \cap (A' \cup B')$ [by De-Morgan's law]

$= (A \cap A') \cup (A \cap B)$

$= \phi \cup (A \cap B) = A \cap B.$

(iii) $(A \cup B) - (A \cap B)$

$= (A \cup B) \cap (A \cap B)'$

$= (A \cap B) (A' \cup B')$ [De-Morgan's law]

$= [A \cap (A' \cup B')' \cup [B \cap (A' \cup B')]$ (distributive law)

$= [(A \cap A) \cup (A \cap B')] \cup [(B \cap A') \cup (B \cap B')]$

(distributive law)

$= [\phi \cup (A \cap B')] \cup [(B \cup A') \cup \phi]$

$= (A \cap B') \cup (B \cap A')$

$= (A - B) \cup (B - A).$ **Hence Proved.**

Example:

Let A = N and for each $\lambda \in A$.

Let $A_\lambda = \{1, 1/2, 1/3\}$

Let $U = \{x \in R, 0 \leq x \leq 1\}$

Find $U\{A_\lambda: \lambda \in A$.

Solution:

Here $\lambda = 1, 2, 3, 4,...$, so that we have

$A_1 = \{1\}, A_2 = \{1, 1/2\}. A_3 = \{1, 1/2, 1/3\}$ etc.

it is clear that

$U\{A_\lambda: \lambda \in A\} = \{1, 1/2, 1/3,...1/\lambda, 1/\lambda+1\}$

ORDERED PAIR

By the definition of equality of sets, for any two objects a and b we have, {a, b} = {b, a}. However, if we keep in mind the order in which the elements are being listed, then *a set of two elements whose elements have been listed in a specific order is called an* **ordered pair** to be denoted by (a, b). Thus for different elements a and b we have, $(a, b) \neq (b, a)$. In general, $(a_1, b_1) = (a_2, b_2) \Leftrightarrow a_1 = a_2$ and $b_1 = b_2$.

CARTESIAN PRODUCT OR DIRECT PRODUCT OF SETS

If A and B are two non-empty sets, then the set of all ordered pairs (a, b) such that $a \in A$ and $b \in B$ is called the Cartesian product of A and B, to be denoted by $A \times B$.

Thus $A \times B = \{(a, b): a \in A \text{ and } b \in B\}$

In general, $A \times B \neq B \times A$.

Example:

If $A = \{0, 1\}$ and $B = \{1, 2, 3\}$, then we have

$A \times B = \{(0, 1)\ (0, 2)\ (0, 3)\ (1, 1)\ (1, 2)\ (1, 3)\}$

$B \times A = \{(1, 0)\ (1, 1)\ (2, 0)\ (2, 1)\ (3, 0)\ (3, 1)\}$

Clearly, $A \times B \neq B \times A$.

Notes:

It may be noted that:

(i) If either A or B is an infinite set and the another one is a non-empty set, the $A \times B$ is an infinite set.

(ii) If $A = \phi$ or $B\ \phi$, then $A \times B = \phi$.

(iii) If A has m elements and B has n elements, then $A \times B$ has mn elements.

Like ordered pairs we may define ordered triples as a set of three elements listed in a specific order and thus we define, $A \times B \times C = \{(a, b, c) : a \in A, b \in B \text{ and } c \in C\}$.

In a similar fashion, we define an n-tuple as a set of n elements listed in a specific order as $(a_1, a_2,..., a_n)$. Thus the product of n sets may be defined as,

$$A_1 \times A_2 \times ...A_n = \{(a_1, a_2,..., a_n): a_1 \in A_i \text{ for } i = 1, 2,..., n\}.$$

SOME RESULTS ON CARTESIAN PRODUCT OF SETS

Theorem 1:

For any three sets A, B and C we have

(i) $A \times (B \cup C) = (A \times B) \cup (A \times C)$.

(ii) $A \times (B \cap C) = (A \times B) \cap (A \times C)$.

(iii) $A \subseteq B \Rightarrow A \times C \subseteq B \times C$.

(iv) $A \times (B - C) = (A \times B)\ (A \times C)$. **Hence Proved.**

Proof:

(i), (ii), (iii) Do yourself.

(iv) Let (a, b) be an arbitrary element of $A \times (B - C)$. Then

$(a, b) \in A \times (B - C)$

$\Rightarrow$ $a \in A$ and $b \in (B - C)$

$\Rightarrow$ $a \in A$ and ($b \in B$ and $b \notin C$)

$\Rightarrow$ ($a \in A$ and $b \in B$) and ($a \in A$ and $b \notin C$)

$\Rightarrow$ $(a, b) \in (A \times B)$ and $(a, b) \notin (A \times C)$

$\Rightarrow$ $(a, b) \in (A \times B) - (A \times C)$

$\therefore$ $A \times (B - C) \subseteq (A \times B) - (A \times C)$

Similarly, $(A \times B) - (A \times C) \subseteq A \times (B - C)$

Hence $A \times (B - C) = (A \times B) - (A \times C)$. **Hence Proved.**

Theorem 2:

For any sets A, B, C and D we have

(i) $(A \times B) \cap (C \times D) - (A \cap C) \times (B \cap D)$.

(ii) $A \subseteq B$ and $C \subseteq D \Rightarrow (A \times C) = (B \times D)$

Proof:

Do yourself.

Theorem 3:

If A and B are two non-empty sets having n elements in common, then $A \times B$ and $B \times A$ have n^2 elements in common.

Proof:

Let $C = A \cap B$. Then, we claim that

$C \times C = \{A \times B) \cap (B \times A)$, Since

$(a, b) \in C \times C$

$\Leftrightarrow$ $a \in C$ and $b \in C$

$\Leftrightarrow$ $a \in A \cap B \in A \cap B$ [$\because C = A \cap B$]

$\Leftrightarrow$ ($a \in A$ & $b \in B$) and ($a \in B$ & $b \in A$)

$\Leftrightarrow$ $(a, b) \in (A \times B)$ and $(a, b) \in (B \times A)$

$\Leftrightarrow$ $(a, b) \in (A \times B) \cap (B \times A)$.

Thus $C \times C$ $(A \times B) \cap (B \times A)$.

And, since $C \times C$ has n^2 elements, so $(A \times B) \cap (B \times A)$ has n^2 elements *i.e.*, $A \times B$ and $B \times A$ have n^2 elements in common. **Hence Proved.**

VARIOUS TYPES OF RELATIONS ON A SET

Let R be a binary relation on a set A. Then R is said to be

(i) **Reflexive,** if a R a $\forall$ a $\in$ A, *i.e.*, (a, a) $\in$ R $\forall$ a $\in$ A;

(ii) **Symmetric,** if a R b $\Rightarrow$ bRa *i.e.*, (a, b) $\in$ R $\Rightarrow$ (b, a) $\in$ R;

(iii) **Transitive,** if a R b, b R c $\Rightarrow$ a R c
i.e., (a, b) $\in$ R, (b, c) $\in$ R $\Rightarrow$ (a, c) $\in$ R;

(iv) **Antisymmetric,** if a R b, b R a $\Rightarrow$ a = b.
i.e., (a, b) $\in$ R, (b, a) $\in$ R $\Rightarrow$ a = b.

Equivalence Relation: *A relation which is simultaneously reflexive, symmetric and transitive is known as an equivalence relation.*

Partial Order Relation: *A relation which is reflexive, antisymmetric and transitive is known as a partial order relation.*

Order Relation: *A relation r on A is said to be an order relation, if*

(i) a, b $\in$ A and a $\neq$ b $\Rightarrow$ a R b or b R a,

(ii) a R b $\Rightarrow$ a $\neq$ b,

(iii) a R b and b R c $\Rightarrow$ a R c.

Example:

Let m be an arbitrary but a fixed positive integer. For any two integers a and b, define a relation,

'Congruence modulo m', by

a $\equiv$ b (mod m) $\Leftrightarrow$ (a – b) is divisible by m.

Then, the relation R = {(a, b): a, b $\in$ Z and a $\equiv$ b (mod m)} is

(i) Reflexive, since a – a = 0 is divisible by m
$\Rightarrow$ a $\equiv$ a (mod m) $\forall$ a $\in$ Z.

(ii) Symmetric, since a – b is divisible by m
$\Rightarrow$ (b – a) is divisible by m.

(iii) Transitive, since (a – b) is divisible by m and (b – c) is divisible by m
$\Rightarrow$ (a – b + b – c) = (a – c) is divisible by m.

So, r is an equivalence relation on Z.

Theorem:

A relation R on a set A is

(i) **Reflexive** $\Leftrightarrow$ $I_A \subseteq R$, where I_A = {(a, a): a $\in$ A};

(ii) **Symmetric** $\Leftrightarrow$ $R^{-1} = R$;

(iii) **Transitive** $\Leftrightarrow$ RoR $\subseteq$ R.

Proof:

(i) by definition of reflexivity, we have

R is reflexive $\Leftrightarrow$ a R a $\forall$ a $\in$ A

$\Leftrightarrow$ (a, a) $\in$ R a $\forall$ a $\in$ A

$\Leftrightarrow I_A \subseteq R$.

(ii) Let R be symmetric. Then,

(a, b) $\in$ R $\Leftrightarrow$ (b, a) $\in$ R [$\because$ R is symmetric]

$\Leftrightarrow$ (a, b) $\in R^{-1}$

Thus $R = R^{-1}$,

So, when R is symmetric, then $R = R^{-1}$.

Again, let $R^{-1} = R$. Then

(a, b) $\in$ R $\Leftrightarrow$ (b, a) $\in R^{-1}$

$\Leftrightarrow$ (b, a) $\in$ R [$\because R^{-1} = R$]

This show that whenever $R^{-1} = R$, then R is symmetric.

Hence R is symmetric $\Leftrightarrow R^{-1} = R$.

(iii) Let R be transitive and let (a, c) be an arbitrary element of RoR. Then,

(a, c) $\in$ RoR

$\Rightarrow$ $\exists$ b $\in$ A such that (a, b) $\in$ R and (b, c) $\in$ R

$\Rightarrow$ (a, c) $\in$ R [by transitivity of R]

Thus, RoR $\subseteq$ R.

So, whenever R is transitive, then RoR □$\subseteq$ R.

Again, let RoR $\subseteq$ R and let (a, b) $\in$ R, (b, c) $\in$ R, then

(a, b) $\in$ R, (b, c) $\in$ R $\Rightarrow$ (a, c) $\in$ RoR

$\Rightarrow$ (a, c) $\in$ R [$\because$ RoR $\subseteq$ R]

This shows that R is transitive.

Thus, whenever RoR $\subseteq$ R, then R is transitive.

Hence, R is transitive $\Leftrightarrow$ RoR = R.

SOLVED EXAMPLES

Example 1:

Two finite sets have m and n elements. The total number of subsets of the first set is 56 more than the total number of subsets of the second set. Find the values of m and n.

Solution:

Let A and B be two sets having m and n elements respectively. Then,

Number of subsets of A = 2^m,
Number of subsets of B = 2^n,
It is given that $2^m - 2^n = 56$

$\Rightarrow \quad 2^n(2^{m-n} - 1) = 2^3(2^3 - 1)$

$\Rightarrow \quad n = 3$ and $m - n = 3$

$\Rightarrow \quad n = 3$ and $m = 6$.

Example 2:

By mathematical induction, prove that $7^{2n} + (2^{3n-3})3^{n-1}$ is divisible by 25, $n \in N$.

Solution:

Let the statement P(n) be defined as

$$P(n) = \text{“}7^{2n} + (2^{3n-3})\, 3^{n-1} \text{ is divisible by 25”}$$

When n = 1, we get

$$P(1) = 7^2 + 1\ (1) = 50$$

Which is divisible by 25,

Let the result P(r) be true.

That is $7^{2r} + (2^{3r-3})\, 3^{r-1}$ is divisible by 25

Let $\quad 7^{2r} + (2^{3r-3})\, 3^{r-1} = 25\, k,\ k \in N \qquad \text{....(1)}$

Now $\quad P(r+1) = 7^{2r+2} + (2^{3r})\, 3^r$

$$= 7^{2r}\,(49) + (2^{3r-3}\,.\,2^3)\, 3^{r-1}\,.\,3$$
$$= 49\,(7^{2r}) + 24\,(2^{3r-3})\, 3^{r-1}$$
$$= (50 - 1)\,(7^{2r}) + (25 - 1)\,(2^{3r-3})3^{r-1}$$
$$= 50\,(7^{2r}) + 25\,(2^{3r-3})\, 3^{r-1} - [7^{2r} + (2^{3r-3})\, 3^{r-1}]$$
$$= 25\,[2\,(7^{2r}) + (2^{3r-3})\, 3^{r-1}] - 25\,k \qquad \text{(from (1))}$$
$$= 25\,[2\,(7^{2r}) + (2^{3r-3})\, 3^{r-4} - k]$$
$$= \text{divisible by 25.}$$

Therefore, P (r + 1) is also true. By mathematical induction, P(n) is true for all $n \in N$.

Example 3:

Using mathematical induction prove that for every integer $n \geq 1$, $\left(3^{2^n} - 1\right)$ is divisible by 2^{n+2} but not by 2^{n+3}.

Solution:

Let P(n) be the statement P(n): $3^{2^n} - 1$ is divisible by 2^{n+2} but not by 2^{n+3}. For n = 1, we have $P(1) = 3^2 - 1 = 8 = 2^3$, which is divisible by 2^3 but not by 2^4.

Hence P(1) is true. Let the statement P(r) be true. That is

$$P(r): 3^{2^r} - 1 \text{ is divisible by } 2^{r+2} \text{ but not by } 2^{r+3}$$

Therefore, we can write

$$P(r): 3^{2^r} - 1 = k \cdot 2^{r+2}$$

where k is odd (if k is even, then $k.2^{r+2}$ is divisible by 2^{r+3}).

Hence $3^{2^r} = 1 + k\,2^{r+2}$, k odd.

Now, $P(r+1): 3^{2^{r+1}} - 1 = 3^{(2^r, 2)} - 1$

$$= \left(3^{2^r}\right)^2 - 1 = (1 + k \cdot 2^{r+2})^2 - 1$$
$$= k^2 \cdot 2^{2(r+2)} + 2k \cdot 2\,2^{r+2} + 1 - 1$$
$$= k^2 \cdot 2^{2r+4} + k \cdot 2^{r+3} = k^2 \cdot 2^{r+3} \cdot 2^{r+1} + k \cdot 2^{r+3}$$
$$= k \cdot 2^{2r+3} [1 + k \cdot 2^{r+1}]$$

Which is divisible by 2^{r+3}, but not by 2^{r+4} as k is odd. Hence P(r + 1) is true. By mathematical induction, P(n) is true fro all $n \in N$..

Examples 4:

The necessary and sufficient condition for a set Y to be a subset of X is that $X \cup Y = X$.

Solution:

Let $Y \subset X$, then

$$x \in Y \Rightarrow x \in X \quad \text{...(i)}$$

Now, $x \in X \cup Y \Rightarrow x \in X \text{ or } x \in Y$

$$\Rightarrow x \in X$$

$$\therefore \quad X \cup Y \subset X \quad \text{...(ii)}$$

Also, we know that

$$X \subset X \cup Y \quad \text{...(iii)}$$

(ii) and (iii) $\Rightarrow X \cup Y = X$.

Conversely,

if $X \cup Y = X$, we have to prove that $Y \subset X$.

Now, $X \cup Y = X \Rightarrow X \cup Y \subset X$

and $X \subset X \cup Y$

$\Rightarrow X \cup Y \subset X$

$\Rightarrow X \subset X$ and $Y \subset X$

$\Rightarrow Y \subset X$.

Examples 5:

Prove that $(A')' = A$.

Solution:

We have to prove $(A')' \subset A$ and $A \subset (A')'$.

Let $x \in (A')'$ then

$$x \in (A')' \Leftrightarrow x \notin A'$$
$$\Leftrightarrow x \in A$$

Thus $x \in (A')' \Leftrightarrow x \in A$

$$\Rightarrow (A')' \subset A \text{ and } (A')' \supset A$$
$$\Rightarrow (A')' = A.$$

Examples 6:

Let A, B, C be three sets then

$$(A - B) \cap (A - C) = A - (B \cup C)$$

Solution:

Let $x \in (A - B) \cap (A - C)$ then

$$x \in (A - B) \cap (A - C) \Leftrightarrow x \in (A - B) \text{ and } x \in (A - C)$$
$$\Leftrightarrow x \in A, x \notin B \text{ and } x \in A, x \notin C$$
$$\Leftrightarrow x \in A, x \notin B \text{ and } x \notin C$$
$$\Leftrightarrow x \in A, x \notin B \cup C$$
$$\Leftrightarrow x \in A - (B \cup C)$$

Which implies that $(A - B) \cap (A - C) = A - (B \cup C)$

Example 7:

Prove, using mathematical induction that $10^{2n-1} + 1$ is divisible by 11 for all $n \in N$.

Solution:

Let P(n) be the statement.

P(n): "$10^{2n-1} + 1$ is divisible by 11."

When n = 1, we get $10^{2n-1} + 1 = 10 + 1 = 11$,

Which is divisible by 11. Hence, P(1) is true.

Let the result be true for P(r), *i.e.*, $10^{2r-1} + 1$ is divisible by 11. Therefore,

$$10^{2r-1} + 1 = 11k, k \in N.$$

or $$10^{2r-1} = 11k - 1.$$

We are now to prove that P(r + 1) is true.

Now, $= P(r + 1)\text{: } 10^{(2r + 1) - 1} - 1 = 10^{2r + 1} + 1$

$= 10^{2r - 1}.10^2 + 1+(11k-1)\ 100 + 1=11(100k - 9)$

= multiple of 11.

Hence, 10^{2r+1} is divisible by 11. That is, P(r + 1) is true. By the principle of mathematical induction P(n) is true for all $n \in N$.

Example 8:

Using mathematical induction prove that $(1 + x)^n > 1 + nx$, *for* $n \geq 2$ *and* $x > -1$, $(\neq 0)$.

Solution:

Let the given result be denoted by P(n).

When n = 1, we get

$$(1 + x) > 1 + x$$

Which is not true (this is given in the problem).

When n = 2, we get

$$(1 + x)^2 > 1 + 2x + x^2 > 1 + 2x$$

Which is true. Therefore, P(2) is true.

Let the result P(r) be true, *i.e.*,

$$(1 + x)^r > 1 + rx \quad ...(1)$$

We are to prove that P (r + 1) is true,

i.e., $(1 + x)^{r+1} > 1 + (r + 1) x$

Now $(1 + x)^{r+1} = (1 + x)^r (1 + x)$

$$> (1 + rx)(1 + x) \quad \text{(using (1))}$$
$$= 1 + rx + x + rx^2$$
$$= 1 + (r + 1) x + rx^2$$
$$> 1 + (r + 1) x$$

Therefore P (r + 1) is also true. Hence, P (n) is true for all $n \geq 2$, $n \in N$.

EXERCISES

1. Define a set and give examples to illustrate the difference between a collection and a set. What are different ways to specify a set? Give examples.
2. If A and B are any two sets, then prove that A – B, A ∩ B and B – A are pairwise disjoint.
3. For any two sets A and B, show that A × B and B × A have one element in common iff A and B have one element in common.
4. If A and B are any subsets, then prove that A ∩ (B – A) = ϕ.
5. For any two sets A and B, prove that A' – B' = B – A.
6. Find out which of the following sets are null or singleton.

   (i) A = {x: x is a letter before a in alphabets}

(ii) A = {x: x + 16 = 16}

(iii) A = {x: x^2 = 4, x is an odd integer}

7. N be the set of all natural numbers. Prove that each one of the relations R_1 and R_2 on N × N, defined by

   (a, b) R_1 (c, d) ⇔ a + d = b + c;

   and (a, b) R_2 (c, d) ⇔ ad = bc

   is (a, b) equivalence relation.

8. If A = {0, 1}, B = {1, 2, 3,}, show that A × B = B × A.

9. Prove that (A ∪ B) ∩ B' = A, if and only if A ∩ B = ϕ.

10. If A = {x: x is a factor of 20}

    B = {x: x is a multiple of 5 and x < 20}

    Find A – B and B – A.

2

Mapping or Functions

INTRODUCTION

Let A = {a, b, c, d} and B = {{x, y, z}. Suppose, by some rule or other, we assign to each element of A a unique element of B. Suppose a is associated to x, b is associated to y, c is associated to x and d is associated to x. The set

{(a, x), (b, y), (c, x), (d, x)}

of such assignments is called a function or a mapping from A to B. If we denote this set by f, then we write

$f : A \rightarrow B$

Which is read as "f is a function of A to B" or "f is a mapping from A to B".

It must be noted that there may be some elements of the set B which are not associated to any element of the set A. Each element of the set A must be associated to one and only one element of the set B. Two or more elements of the set A may be associated to the same element of the set B.

IDENTITY MAPPING

Let a be any set. Let the function $f : A \rightarrow A$ be defined by the formula $f(x) = x \ \forall x \in A$, that is, let each element of A be mapped on itself. Then f is called the identity function, or the identity transformation on A. We denote this function by I_A. Thus, if I_A denotes the identity mapping of a set A, we have $I_A(x) = x \ \forall \ x \in A$.

Example:

Let A = {1, 2, 3, 4, 5}. Then f = {(1, 1), (2, 2), (3, 3), (4, 4), (5, 5)} is an identity mapping of the set A.

Identity mapping is always one-one and onto.

CONSTANT FUNCTION

A function $f : A \rightarrow B$ is called a constant function if the some element $b \in B$ is assigned to every element in A. In other words, $f : A \rightarrow B$ is a constant function if the range of f consists of only one element.

Example:

Let $f : R \rightarrow R$ be defined by the formula $f(x) = 5$. Then f is a constant function since 5 is assigned to every element.

ONLY ONE-ONE AND ONTO MAPPINGS POSSESS INVERSE MAPPINGS

If the mapping $f : A \rightarrow B$ is not onto, the correspondence f^{-1} will not be a mapping from B to A. The reason is that in this case there will be some elements in B which will have no f^{-1} image in A. Again if $f : A \rightarrow B$ is not one-one, even then the correspondence f^{-1} will not be a mapping from B to A. The reason is that in this case some elements in B will be associated to more than one element in A. Thus, in this case some elements in B will not have a unique f^{-1} image in A.

Theorem 1:

If $f : A \rightarrow B$ be one-one and onto, then the inverse mapping of f is unique.

Proof:

Let $f : A \rightarrow B$ be one-one and onto. Let $g : B \rightarrow A$ and $h : B \rightarrow A$ be two inverse mappings of A. To prove that $g = h$.

Let b be any arbitrary element $\in$ B. Let $g(b) = a_1$ and $h(b) = a_2$. Since g is an inverse mapping of f, therefore, $g(b) = a_1 \Rightarrow f(a_1) = b$. Also since h is an inverse mapping of f, therefore $h(b) = a_2 \Rightarrow f(a_2) = b$. But f is a one-one mapping.

$\therefore f(a_1) = b$ and $f(a_2) = b \Rightarrow a_1 = a_2 \Rightarrow g(b) = h(b)$. Hence $g = h$.

Theorem 2:

Let A and B be two sets. If $f : A \rightarrow B$ is one-one onto, then $f^{1} : B \rightarrow A$ is also one-one onto.

Proof:

First, let us prove that the mapping f^{-1} is one-one. Let y_1 and y_2 be any two element of B.

Suppose $f^{-1}(y_1) = x_1$, and $f^{-1}(y_2) = x_2$ where $x_1, x_2 \in A$. Then by the definition of the mapping f^{-1}, $f(x_1) = y_1$, and $f(x_2) = y_2$.

Now $f^{-1}(y_1) = f^{-1}(y_2) \Rightarrow x_1 = x_2$

$\Rightarrow f(x_1) = f(x_2)$ [$\because$ f is a mapping from A to B]

$\Rightarrow y_1 = y_2$.

$\therefore$ The mapping f^{-1} is one-one.

Now to prove that the mapping f^{-1} is onto A. Let x be any arbitrary element of A. Since f is a mapping from A to B, therefore there exists an element $y \in B$ such that $y = f(x)$ or $x = f^{-1}(y)$. Thus x is the f^{-1} image of the element $y \in B$. Hence the mapping f^{-1} is also onto.

SOME SETS OF NUMBERS

We shall here given a list of important sets of numbers along with the symbols we shall often use to denote them.

1. The set N = {1, 2, 3, 4, 5,...} of all natural numbers. The natural numbers are also called counting numbers. The number 0 is not an element of the set of natural numbers.
2. The set I = {..., –3, –2, –1, 0, 1, 2, 3,...} of all integers. The integers are also called whole numbers. Then set of integers is also denoted by Z.
3. The set I_+ = {1, 2, 3, 4,...} of all positive integers. The number 0 is neither positive nor negative.
4. The set $Q = \{x : x = p/q$, where p and q are integers and $q \neq 0\}$ of all rational numbers.
5. The set Q_0 of all non-zero rational numbers.
6. The set R of all real numbers *i.e.,*

   $$R = \{x : x = ab_1b_2b_2...b_n...\},$$

   where $a \in I$ and b's are any of 0, 1, 2, 3,..., 9; n can be as large as we please and can go upto infinity. Here every real number has been expressed in the form of decimals with ten as base.
7. The set c of all complex numbers, *i.e.,* $C = \{x : x = a + ib$, where a and b are real numbers and $i = \sqrt{(-1)}\}$.

INTERVALS DEFINED AS SETS OF REAL NUMBERS

Consider the following sets of real numbers:

$$A = \{x : x \in R \text{ and } 2 \leq x \leq 7\}$$

$$B = \{x : x \in R \text{ and } 2 < x < 7\}$$

$$C = \{x : x \in R \text{ and } 2 < x \leq 7\}$$

$$D = \{x : x \in R \text{ and } 2 \leq x < 7\}.$$

Such sets are called intervals.

The set A is a closed interval and is denoted by A = [2, 7]. The set B is an open interval and is denoted by B = (2, 7) or by B = [2, 7]. The set C is an open closed interval and is denoted by C = (2, 7) or by C = [2, 7]. The set D is a closed open interval and is denoted by D = [2, 7] or by D = [2, 7].

FUNCTION

Definition: *Let A and B be two given sets. Suppose there exists a correspondence denoted by f, which associates to each member of A a unique member of b. Then f is called a function or a mapping of A to B. The mapping f of A to B is denoted by*

$$f : A \rightarrow B \Rightarrow \text{by } A \xrightarrow{f} B$$

The set A is called the domain of the function f, and B is called the *co-domain* of f. Further if a ∈ A, then the element in B which is assigned to a is called the f image of a or the value of the function f for a and is denoted by f (a). The element a may be referred to as the **pre-image** of f (a).

RANGE OF A FUNCTION

Let f be a mapping of A to B *i.e.,* let f : A → B. Each element of A has unique image and each element in B need not appear as the image of an element in A. There can be more than one elements of A which have the same image in B. We define the *range* of f to consist of those elements in B which appear as the image of at least one element in A. Hence, we often speak of the range of a function as the *image of its domain.*

We denote the range of f : A → B by f (A).

Thus f (A) = {f (x) : x ∈ A}. Obviously f (A) ⊆ B.

Also we can say that

range of f = f (A) = {y : y ∈ B and y = f (x) for some x ∈ A}.

IMAGE OF A SUBSET

Definition:

Let be a function of A to B. Let x be any subset of A. Then the set Y = {y : ∈ B and y = f (x) for some x ∈ X} is called the image of X under f and we write f (x) = Y.

TRANSFORMATIONS OR OPERATORS

If the domain and co-domain of a function f are both the same set, say f : A → A, then f is often called an operator or a transformation on A.

FUNCTIONS AS SETS OF ORDERED PAIRS

If A and Bare any two non-empty sets, then a mapping f of A to B is a subset f of A × B satisfying the following conditions:

(i) for each $a \in A$, $(a, b) \in f$ for some $b \in B$;

(ii) if $(a, b) \in f$ and $(a, b') \in f$, then $b = b'$.

The first condition ensures that we have a rule that assigns to each element $a \in A$ some element $b \in B$. Thus, each element in A will have image. The second condition guarantees that the image is unique. Accordingly f is a function from A to B.

Note: If $f : A \to B$, it is important to distinguish between a function f and the value f (x) of f for any element x. While f is a subset of A × B, f (x) is an element of the set B.

Example 1:

Let A = {a, b, c, d} and B = {a, b, c}. Then f = {(a, b), (b, c) (c, c), (d, b)} is a function from A to B, since to each element ∈ A we have assigned a unique element of B. Here f (a) = b = c, f (c) = c and f (d) = b. Then domain of f is the set A and the range of f i.e., f (A) = {b, c}.

Example 2:

Let A = {a = {a, b, c} and B = {x, y, z}. Then

$$f = \{(a, x), (b, y), (a, z), (c, z)\}$$

is not a function from A to B. The reason is that two elements, x and z∈B are assigned to the same element a∈A.

Example 3:

Let A = {1, 2, 3} and B ={1, 2, 3, 4, 5, 6}. Let f assign to each member in A its square. Then f is not a function from A to B since no member of B is assigned to the element 3 ∈ A.

EQUALITY OF TWO FUNCTIONS

Definition: Two functions f and g of $A \to B$ are said to be equal iff $f(x) = g(x)\ \forall\ x \in A$ and we write f = g. For two unequal mappings from A to B, here must exist at least on element $x \in A$ such that $f(x) \neq g(x)$.

DIAGRAMMATIC REPRESENTATION OF A FUNCTION

Sometimes a function may be represented by a diagram as will be obvious from the following example:

Let A {a, b, c, d} and B = {t, x, y, z}. Let $f : A \to B$ be defined by the correspondence f (a) = y, f (b) = x, f (c) = z and f (d) = y.

We represent the two sets A and B by the interiors of two circles. The unction $f : A \to B$ is represented by means of a collection of arrows joining the points which represent the elements of A to points representing the corresponding elements of B. By the definition of function, it is obvious that

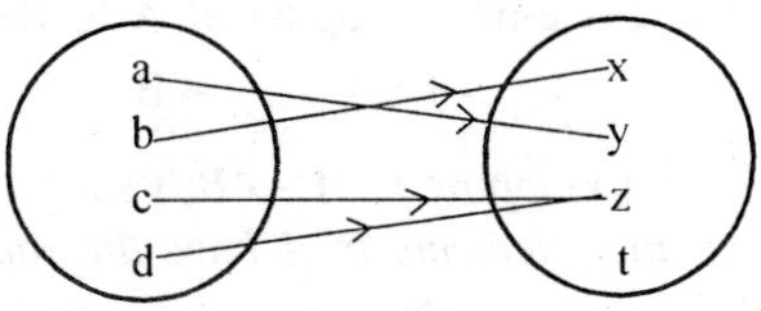

Fig. 2.1

(i) Every point in A is joined to some point in B by an arrow;

(ii) A point in a cannot be joined to two or more distinct points in B;

(iii) Two or more points in A may be joined to the same point in B;

(iv) There may be some points in B which are not joined to any point in A.

TYPES OF FUNCTIONS

(a) 'Into' and 'Onto' Mapping

If the mapping $f : A \to B$ is such that there is at least one element in B which is not the f-image of any element in A, then we say that f is a mapping of A 'into' B. In this case the range of f is a proper subset of the co-domain of f *i.e.,* $f(A) \subseteq B$. Thus in 'into' mappings at least one element of the co-domain B is left uncovered by the f-images of the domain A.

If the mapping $f : A \to B$ is such that each element in B is the f-image of at least one element in A, then we say that f is a mapping of A 'onto' B. In this case the range of f is equal to the co-domain of f i.e., $f(A) = B$. Thus in 'onto' mappings the co-domain B is completely covered by the f-images of the domain A. If f is a mapping of A onto B, we write

$$f : A \xrightarrow{\text{onto}} B.$$

An onto mapping is also called as surjection.

Note: It should be noted that in common use whenever we say that f is a mapping of A into B, we usually include in it the possibility that the mapping f may be onto B also.

(b) 'One-one' and 'Many-one' Mappings

A mapping $f : A \to B$ is said to be one-one or one-to-one (abbreviated 1-1) if different elements in A have different f-images in B i.e., if

$$f(x) = f(x') \Rightarrow x = x' \ (x \text{ and } x' \in A).$$

One-to-one mapping is also sometimes known as injection.

In one-one mappings an element in B has only one pre-image in A. If f is a one-one mapping of A to B, we write

$$f : A \xrightarrow[1-1]{} B.$$

A mapping $f : A \to B$ is said to be many-one if two (or more than two) distinct elements in A have the same f-image in B i.e.,

$$f(x) = f(x'),\ x \neq x'.$$

In many-one mappings some elements in B have more than one pre-image in A.

One-one Onto Mappings: *If $f : A \to B$ is 1-1 and onto B, then f is called a one-to-one correspondence between A and B.*

One-one onto mapping is sometimes also known as *bijection.* If f : A → B 1-1 and onto B, we write

$$f : A \xrightarrow[1-1]{\text{onto}} B.$$

COUNTING PRINCIPLE

If A, B and C are finite sets, and U be the finite universal set, then

(i) $n(A \cup B) = n(A) + n(B) - n(A \cap B)$

(ii) $n(A \cup B) = n(A) + n(B) \Leftrightarrow$ A, B are disjoint non-void sets

(iii) $n(A - B) = n(A) - n(A \cap B)$, *i.e.,*

$n(A - B) + n(A \cap B) = n(A)$.

(iv) $n(A \Delta B)$ = No. of elements which belong to exactly one of A or B

$= n((A - B) È (B - A)$

$= n(A - B) + n(B - A)$ ((A – B) and (B – A) and disjoint)

$= n(A) - n(A \cap B) + n(B) - n(A \cap B)$

$= n(A) + n(B) - 2n(A \cap B)$

(v) $n(A \cup B \cup C) = n(A) + n(B) + n(C) - n(A \cap B) - n(B \cap C) - n(A \cap C) + n(A \cap B \cap C)$

(vi) No. of elements in exactly two of the sets A, B, C

$= n(A \cap B) + n(B \cap C) + n(C \cap A) - 3n(A \cap B \cap C)$

(vii) No. of elements in exactly one of the sets A, B, C.

$= n(A) + n(B) + n(C) - 2n(A \cap B) - 2n(B \cap C) - 2n(A \cap C) + 3n(A \cap B \cap C)$

(viii) $n(A' \cup B') = n((A \cap B)')$

$= n(U) - n(A \cap B)$

(ix) $n(A' \cap B') = n((A \cup B)')$

$= n(U) - n(A \cup B)$.

DUALITY PRINCIPLE

It may be noticed that if any law of the algebra of sets, universal set U is replaced by ϕ, ϕ by U, $\cup$ by $\cap$ and $\cap$ by $\cup$ wherever these occur, the new statement thus, obtained is also a law of the algebra of sets. This fact is known as the duality principle, and any law obtained as a result of its application is called the dual of the original law. It is sometimes called primal law. For example, the dual of the law.

$$A \cup (B \cap C) = (A \cup B) \cap (A \cup C)$$

is $$A \cap (B \cup C) = (A \cap B) \cup (A \cap C)$$

It will be observed that this principle is always true.

DIAGRAMMATIC REPRESENTATION OF DIFFERENT TYPES OF MAPPINGS

(a) The mapping g : P → Q in the adjoining diagram is one-one into. Different points in P are joined to different points in Q and there are some points in Q and there are some points in Q which are not joined to any point in P.

(b) The mapping f : A → B in the adjoining diagram is many-one into. Two or more points in A are joined to the same point in B and there are some points in B which are not joined to any point in A.

(c) The mapping i : X→Y in the adjoining diagram is one-one onto. Different points in X are joined to different points in Y and no point in Y is left vacant.

(d) The mapping h : S→T in the adjoining diagram is many-one onto. Each point in T is joined to at least one point in S and two or more points in S are joined to the same point in T.

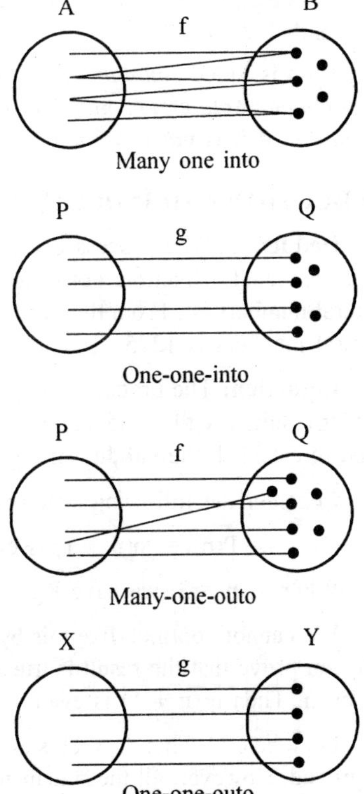

Fig. 2.2

INVERSE IMAGE OF AN ELEMENT

Let f be a function of A to B and let b ∈ B. Then the inverse image of the element be under f denoted by f^{-1} (b) consists of those elements in A which have b as their f-image. Symbolically, if f : A → B, then f^{-1} (b) = {x : x ∈ A and f (x) = b}.

f^{-1} is read as "f inverse". f^{-1} (b) is always a subset of A.

CARDINALLY EQUIVALENT SETS

Two sets A and B are said to have the same number of elements iff a one-to-one mapping of A onto B exists. Such sets are said to be cardinally equivalent and we write A~B. Cardinally equivalent sets are to have the same cardinal number or the same cardinality.

DENUMERABLE SETS

We are familiar with the set of natural numbers, N = {1, 2, 3, 4,...}. If a set A is equivalent to N. *i.e.,* if ∃ a one-one mapping from A onto N, then A is called *denumerable.*

A set is called *countable* if it is finite or denumerable and a set is called *non-denumerable* or *uncountable* if it is infinite and if it is not equivalent to N, *i.e.,* if it is not *countable.*

DEDUCTION AND INDUCTION

Deduction: The process of deducing particular results from a general result is called deduction. For example, we know that the sum of the first n natural numbers is . From this result, we deduce that the sum of the first 50 natural numbers is 1275.

Induction: The process of establishing a valid general result from particular results is called induction. The principle of mathematical indiction is used to establish the validity of a general result involving natural numbers.

Consider the following statement :

P(n) = "n(n + 1) is even".

When n = 2, we have P(2) = 2(3) = 6 is even.

We cannot continue like this by testing for all natural numbers n ∈ N. We can prove that the result is true by using the following argument. Let n be even. Then n (n + 1) is even.

Let n be odd. Then (n + 1) is even. Therefore n(n +1) is even. The result is proved. However, all the statements as given above may not be as simple to prove.

PRINCIPLE OF MATHEMATICAL INDUCTION

Let P(n), $n \in N$ be a statement such that

(i) P(1) is true, and (ii) truth of P(r) implie the truth of P(r + 1). Then, by the principle of mathematical induction, the statement P(n) is true for all $n \in N$. Obviously, the principle of mathematical induction involves the following steps :

(i) First, we prove that the result is true for n = 1,

(ii) Then, we assume that the result is true for n = r.

(iii) Finally, we prove that the result is true for n = r + 1.

Then, we conclude by the principle of mathematical induction that the statement is true for all $n \in N$.

Example 1:

Let R be the set of real numbers and let $f : R \to R$ be defined by the formula $f(x) = x^2$. Then $f^{-1}(9) = \{3, -3\}$, since 9 is the f-image of both 3 and –3 and there is no other real number whose square is 9. Also we observe that $f^{-1}(-5) = \varnothing$, since there is no real number whose square is –5.

Example 2:

Let A = {a, b, c, d, e}, and B = {x, y, z}.

Let $f : A \to B$ be defined by

$f = \{(a, x), (b, x), (c, y), (d, x), (e, y)\}$.

Then $f^{-1}(z)$ is the null set $\varnothing$, since no element of A has z as its f-image.

$f^{-1}(y) = \{c, e\}$ since both c and e have y as their f-image.

Also $f^{-1}(x) = \{a, b, d\}$.

INVERSE IMAGE OF A SUBSET

Let $f : A \to B$ and let C be a subset of B *i.e.*, $C \subseteq B$. Then the inverse image of C under f, denoted by $f^{-1}(C)$, consists of those elements in A which are mapped into some element in C. Symbolically,

$$f^{-1}(C) = \{x : x \in A \text{ and } f(x) \in C\}.$$

$f^{-1}(C)$ is always a subset of A. In particular $f^{-1}[f(A)] = A$, where f (A) is the range of the function $f : A \to B$. Also $f^{-1}(B) = A$.

INVERSE FUNCTION

Let $f : A \to B$ be a one-one onto mapping. Let b be any arbitrary element of B. Since the mapping f is onto B, therefore there will be at least one element in A, say a, such that b = f (a), b ∈ B, a ∈ A.

Since the mapping f is also one-one, therefore there will be only one element a in A such that b = f (a). Let us denote a by f^{-1} (b). Thus we see that if f : A → B is one-one onto we can define a new correspondence, denoted by f^{-1}, which associates to each element in B a unique element in A. Accordingly f^{-1} is a function of B to A.

Hence, if f : A → B is one-one onto, the f^{-1} B → A. The mapping f^{-r} is called the inverse mapping of the mapping f.

Definition: Let f : A → B be a one-one onto mapping. Then the mapping, f^{-1} : B → A, which associates to each element b ∈ B the element a ∈ A, such that f (a) = b, is called the inverse mapping of the mapping f : A → B.

HOW TO PROVE THAT A GIVEN FUNCTION F IS ONE-ONE

If we are given a function f and we are to prove that it is one-one, we can do so by showing that if $f(x_1) = f(x_2)$ then $x_1 = x_2$, where x_1 and x_2 are arbitrary elements of the domain of f. We can also prove it by showing that if $x_1 \neq x_2$, then $f(x_1) \neq f(x_2)$.

HOW TO PROVE THAT A GIVEN FUNCTION F : A → B IS ONTO

To prove that f is onto, we show that if y ∈ B, ∃ x ∈ A such that y = f (x). Then y ∈ B ⇒ y ∈ f (A). Having chosen y arbitrarily, every element of B is an element of f (A) and hence B ⊆ f (A). But f (A) ⊆ B. Therefore B = f (A) and the function f is onto B.

SOLVED EXAMPLES

Example 1:

Show that the mapping f : R → R defined by

$$f(x) = \cos x \ \forall x \in R$$

is neither one-one nor onto. Modify the domain and co-domain of this mapping so that it may be both one-to-one and onto.

Solution:

We have f : R → R such that

$$f(x) = \cos x \ \forall \ x \in R.$$

f is not one-one. We have 0, 2π ∈ R

and $f(0) = \cos 0 = 1,\ f(2\pi) = \cos 2\pi = 1.$

Thus, 0 and 2π are two distinct elements in the domain R of f and they have the same f-image 1 in the co-domain R of f. Therefore f is may-one and not one-one.

f is not onto: We know that if $x \in R$, then $-1 \leq \cos x \leq 1$. Therefore if $y \in$ the co-domain R of f and $y > 1$ or $y < -1$, then there exists no $x \in$ the domain R of f such that $f(x) = \cos x = y$. Therefore the mapping f is into and not onto.

Now let $X = \{x : x \in R \text{ and } 0 \leq x \leq \pi\}$

and $\quad Y = \{y : y \in R \text{ and } -1 \leq y \leq 1\}$.

If we take X as the domain of f and Y as the co-domain of f, then the mapping $f : X \to Y$ such that

$$f(x) = \cos x \ \forall \ x \in X$$

is one-one and onto.

Example 2:

If $f : X \to Y$ and $A, B \subseteq X$. then prove that

$$f(A \cup B) = f(A) \cup f(B).$$

Solution:

Let y be any arbitrary element of $f(A \cup B)$

Then $\quad y \in f(A \cup B)$

$\Rightarrow \exists \ x \in A \cup B$ such that $f(x) = y$

$\Rightarrow \exists \ x \in A$ or $x \in B$ such that $f(x) = y$

$\Rightarrow y \in f(A)$ or $y \in f(B)$

$\Rightarrow y \in f(A) \cup f(B)$.

$\therefore \ f(A \cup B) \subseteq f(A) \cup f(B)$. ...(1)

Again let y be any arbitrary element of $f(A) \cup f(B)$.

Then $\quad y \in f(A) \cup f(B)$

$\Rightarrow y \in f(A)$ or $y \in f(B)$

$\Rightarrow \exists \ x \in A$ or $x \in B$ such that $f(x) = y$

$\Rightarrow \exists \ x \in A \cup B$ such that $f(x) = y$

$\Rightarrow y \in f(A \cup B)$

$\therefore \ f(A) \cup f(B) \subseteq f(A \cup B)$. ...(2)

From (1) and (2), we conclude that

$$f(A \cup B) = f(A) \cup f(B).$$

Example 3:

Let A = (–2, –1, 0, 1, 2}. Let the function f : A → R be defined by the promuls f (x) = x^2 + 1. Find the range of f.

Solution:

The range of f consists of those elements of R which appear as f-image of different elements of A. So we calculate the f-image of each element of A.

$$f(-2) = (-2)^2 + 1 = 5,\ f(-1) = (-1)^2 + 1 = 2,$$
$$f(0) = (0)^2 + 1 = 1,\ f(1) = (1)^2 + 1 + 1 = 2.$$
$$f(2) = (2)^2 + 1 = 4 + 1 = 5.$$

Thus, the range of f is the set {5, 2, 1, 2, 5}, *i.e.,* the set {5, 2, 1}.

Example 4(a):

Let N be the set of natural numbers and A be the set of even natural numbers.

Let f : N → A be defined by the formula f (x) = 2x, x ∈ N. Show that the mapping f is one-one onto. Find the formula that defines the inverse function f^{-1}.

Solution:

First to prove that f is a one-one mapping.

Suppose m and n are any two different elements ∈ N.

Then $m \neq n \Rightarrow 2m \neq 2n \Rightarrow f(m) \neq f(n)$.

Thus different elements belonging to N have different f-images in A. Hence f is one-one.

Now to prove the f is an onto mapping. Let y be any arbitrary element in A *i.e.,* let y be any even natural number. Then f (y/2) = y and y/2 is a natural number *i.e.,* y/2 ∈ N. Thus each element in A is the f-image of some element in N. Hence f is onto.

Since f : N → A is one-one onto, therefore f has an inverse function f^{-1} : A → N.

Let y be the image of x under the function f. Then

$$y = f(x) = 2x.$$

Consequently, x will be the image of y under the inverse function f^{-1} *i.e.,*

$$x = f^{-1}(y).$$

Solving for x in terms of y in the equation y = 2x, we get

$$x = y/2.$$

Then $$f^{-1}(y) = y/2.$$

Hence $f^{-1}(y) = y/2$, $(y \in A)$ is a formula defining the inverse function $f^{-1} : A \to N$.

Example 4(b):

Let S be the set of all triangles and R_+ be the set of positive real numbers. Prove that the map $f : S \to R_+$ given by

$f(\Delta)$ = area of the Δ, $(\Delta \in S)$, is many-one and onto.

Solution:

Let y be any arbitrary element in R_+ *i.e.*, let y be any positive real number. Then $\exists$ a triangle whose area is equal to y. Thus, every element in R_+ is the f-image of some element in S. Therefore f is onto.

Again there can be two or more triangles which have the same area. Therefore f is many-one.

Hence $f : S \to R_+$ is many-one onto.

Example 5:

Let Q be the set of rational numbers. Let $f : Q \to Q$ be defined by

$$f(x) = 2x + 3, \ (x \in Q).$$

Show that f is one-one and onto. Also find a formula that defines the inverse function f^{-1}.

Solution:

Let m and n be any two different elements in Q.

Then $m \neq n \Rightarrow 2m \neq 2n$

$$\Rightarrow 2m + 3 \neq 2n + 3 \Rightarrow f(m) \neq f(n).$$

Thus different elements in Q have different f-images in Q. Hence f is one-one.

Let y be any arbitrary element in Q. If $y = 2x + 3$, we have

$$x = (y - 3)/2 \text{ which is also a rational number.}$$

Thus $f\left(\dfrac{y-3}{2}\right) = y$ *i.e.*, any arbitrary element y in Q is the f-image of the element $(y - 2)/2 \in Q$. Hence f is onto

Since $f : Q \to Q$ is one-one onto, therefore f has an inverse function $f^{-1} : Q \to Q$.

Let y be the image of x under the function f. Then

$$y = f(x) = 2x + 3.$$

Consequently, x will be the image of y under the inverse function f^{-1} *i.e.,* $x = f^{-1}(y)$.

Solving for x in terms of y in the equation $y = 2x + 3$, we get

$$x = (y - 3)/2.$$

Thus, $f^{-1}(y) = (y - 3)/2$, $(y \in Q)$ is the formula defining the inverse function $f^{-1} : Q \to Q$.

Note: In order to prove that the mapping f is one-one, we can also argue like this:

Let m and n be any two elements in Q. Then

$$f(m) = f(n) \Rightarrow 2m + 3 = 2n + 3 \Rightarrow 2m = 2n \Rightarrow m = n.$$

Therefore f is one-one.

Example 6(a):

Let R_0 be the set of all non-zero real numbers. Let $f : R_0 \to R_0$ be defined by the formula $f(x) = 1/x$, $x \in R_0$. Show that f is one-one and onto mapping.

Solution:

Let m and n be any two different elements in R_0.

Then $m \neq n \Rightarrow 1/m \neq 1/n \Rightarrow f(m) \neq f(n)$.

Thus different elements in R_0 have different f-images in R_0. Hence f is one-one.

Now let y be an arbitrary element in R_0. If y 1/x, we have $x = 1/y$ which is a non-zero real number if y is a non-zero real number. Thus $f(1/y) = y$ *i.e.,* any arbitrary element y in R_0 is the f-image of the element 1/y in R_0. Hence f is onto.

Thus $f : R_0 \to R_0$ is one-one onto.

It should be noted that $f(x) = 1/x$ cannot be a mapping from the set of real numbers to the set of real numbers. Then reason is that the image of 0 will not exist because $f(0) = 1/0$ and 1/0 is not a real number.

Example 6(b):

Let C be the set of complex numbers. Prove that the map $f : C \to R$ given by $f(z) = |z|$, $z \in C$ is neither one-one nor onto.

Solution:

Let $z = x + iy$ be any complex number, where $x, y \in R$ and i $\sqrt{(-1)}$.

Then | z | = √($x^2 + y^2$) Also | z | is always a non-negative real number *i.e.*, | z | ≥ 0.

If m be any negative real number ∈ R, then there exists no complex number z ∈ C such that | z | = m. Thus m is not the f-image of any complex number ∈ C. Hence f is not onto but is into.

Also $z_1 = 2 + i3 \in C$ and $z_2 = 2 - i3 \in C$.

Then | 2 + i3 | = √13 and | 2 – i3 | = √13.

Thus $z_1 \neq z_2$, although $| z_1 | = | z_2 |$ *i.e.*, $f(z_1) = f(z_2)$.

Thus we see that two different complex numbers z_1 and $z_2 \in C$ have the same f-image in R. Hence f is not one-one.

Example 6(c):

Let $X = \{x : x \in R$ *and* $- \pi/2 \leq x \leq \pi/2$ *i.e., let*

$X = [- \pi/2, \pi/2]$ *and*

$Y = (y : y \in R$ *and* $- 1 \leq y \leq 1\}$ *i.e., let* $Y = [-1, 1]$.

Show that the funtion $f : X \to Y$ *defined by*

$f(x) = \sin x$, $(x \in X'$, *is one-one onto.*

Also give the inverse mapp $f^{1} : Y \to X$.

Solution:

Let m and n be any two different real numbers lying in the closed interval [– π/2, π/2]. We know that any two different real numbers lying in the closed interval [– π/2, π/2] have not the same sine.

∴ m ≠ n ⇒ sin m ≠ sin n ⇒ f (m) ≠ f (n). Hence f is one-one.

Again if y is any arbitrary real number lying in the closed interval [–1, 1], ∃ a real number x lying in the closed interval [– π/2, π/2] such that isn x = y.

Thus, every element y in Y is the f-image of some element x in X. Hence f is onto.

Thus f : X → Y is one-one onto, therefore f has an inverse function $f^{-1} : X \to Y$.

Let y be the image of x under the function f. Then

$$y = f(x) = \sin$$

Consequently, x will be the image of y under the inverse function f^{-1} *i.e.*,

$$x = f^{-1}(y).$$

Solving for x in terms of y in the equation y = sin x, we get

$x = \sin^{-1} y$.

Thus $f^{-1}(y) = \sin^{-1} y$, $(y \in Y)$ is the formula defining the inverse function $f^{-1} : Y \to X$.

Example 7:

Decide whether or not the following are functions from A to B where A = {1, 2, 3, 4, 5} and B = {a, b, c, d, e}.

If they are functions, give the range of each. If they are not, tell, why?

1. f = ({1, a), (2, b), (3, b), (5, e)}.
2. g = {(1, e), (5, d), (3, a), (2, b), (1, d), (4, a)}.
3. h = {(5, a), (1, e), (4, b), (3, c), (2, d)}.

Solution:

1. Since the element $4 \in A$ is not associated to any element $\in B$, therefore f is not a function from A to B.
2. The element $1 \in A$ is asociated to two different elements e and $d \in B$. Therefore g is not a function from A to B.
3. Each element of A is associated to a unique element of B. Therefore h is a function from $A \to B$. The range of h is the set of the h-images of all elements of A. So range of h = h (A) = {a, e, b, c, d} = B.

Example 8:

Show that the mapping $f : I_+ \to I_-$ defined by $f(x) = x^2$, $x \in I_+$ where I_+ is the set of positive integers, is one-one into.

Solution:

It is given that f is a mapping from I_+ to I_+ defined by

$$f(x) = x^2 \ \forall \ x \in I_+.$$

We have $I_+ = \{1, 2, 3, 4, ...\}$.

f is one-one. Let $x_1\ x_2 \in I_+$. We have

$$f(x_1) = f(x_2) \Rightarrow x_1^2 = x_2^2 \text{ [by def. of f]}$$

$$\Rightarrow x_1 = x_2 \Rightarrow \text{f is one-one.}$$

f is into. Now we shall prove that f is into and not onto. We have $2 \in I_+$. Now there exists no $x \in I_+$ such that $x^2 = 2$. Thus $2 \in I_+$ is not the f-image of any $x \in I_+$. Therefore the mapping f is into and not onto.

Example 9:

Each of the following formulas defines a function from R to R. Find the range of each function.

1. $f(x) = x^3$;
2. $g(x) = \sin x$;
3. $h(x) = x^2 + 1$.

Solution:

1. We know that every real number a has a real cube root $\left(\sqrt[3]{a}\right)$. Therefore if a be any arbitrary element $\in$ R, then $f\left(\sqrt[3]{a}\right) = \left(\sqrt[3]{a}\right)^3 = a$ *i.e.,* $\sqrt[3]{a} \in R$ is the pre-image of $a \in R$. Since a is any arbitrary element $\in$ R, therefore the range of f is R. This function will be an onto function.
2. The sine of any real number lies in the closed interval $[-1, 1]$. Also all the numbers in this interval will be the sine of some real numbers. Hence the range of g is the closed interval $[-1, 1]$. This function will be an into function.
3. If we add 1 to the square of each real number, we get the set of numbers which are greater then or equal to 1. Let y be any real number greater than or equal to 1 and let $y = h(x) = x^2 + 1$. Then $x = \pm \sqrt{(y-1)}$ are also real numbers.

Thus, every real number which is greater than or equal to 1 is the h-image of some or other real number. Hence the range of h is the infinite interval $(1, \infty)$.

Example 10(a):

Let R be the set of all real numbers. Using the fact that every cubic equation with real coefficients has a real root, show that $x \to x^2 - x$ defines a mapping of R onto R. Is this a one-one mapping ?

Solution:

If $x \in R$, then $x^3 - x \in R$ and is unique. Therefore $x \to x^2 - x$ defines a mapping of R to R.

Let y be any arbitrary element $\in$ R *i.e.,* let y be any real number. Then $x^3 - x = y$ is a cubic equation with real coefficients. It will have at least one real root. Thus for any $y \in R$ there exists $x \in R$ such that $x^3 - x = y$. Therefore $x \to x^2 - x$ defines a mapping of R onto R.

Again $\quad 1 \to 1^3 - 1$ *i.e.,* $1 \to 0$

and $\quad -1 \to (-1)^3 - (-1)$ *i.e.,* $-1 \to 0$.

Thus, the two elements 1 and $-1 \in R$ map onto the same element $0 \in R$. Hence the mapping is not one-one.

Example 10(b):

Classify the following as mappings of $R \to R$:

(i) $f(x) = \log x$,

(ii) $f(x) = +\sqrt{x}$.

(iii) $f(x) = \tan x$, $(x \in R)$.

Solution:

(i) Let $x \in R$ and let x be a negative real number. Then log x is not a real number, *i.e.,* log x is not an element of R. Thus the negative real numbers in R will have no f-image in R. Also $\log 0 = -\infty$ which is not an element of R. Thus, the real number 0 in R has no f-image in R. Hence f is not a function from R to R. However f is a function from R_+ to R.

(ii) Let x be a negative real number. Then $\sqrt{x}$ is not a real number *i.e.,* $\sqrt{x} \notin R$. Thus the negative real numbers in R will have no f-images in R. Hence f is not a function from R to R. If S denotes the set of all non-negative real numbers, then f defined by this formula is a function from S to R.

(iii) Let $x = \pi/2$. Then $\tan \pi/2 =$ which is not an element of R. Thus $\pi/2 \in R$ has no f-image in R. In general if $x \in R$ is such that $x = n.\frac{\pi}{2}$ Where n is an odd integer, then x has no f-image in R. Hence f is not a function from R to R.

However, if $X = \{x : x \in R \text{ and } -\pi/2 < x < \pi/2\}$, then f is a function from X to R, Obviously this function will be one-one and onto.

Example 10(c):

If R is the set of real numbers discuss the mapping

$f : R \to R$ where $f(x) = x^2$, $x \in R$.

What is the domain and the range of the mapping f ?

Solution:

If x is any real number, the x^2 is also a real number and it will be unique. Thus each element $x \in R$ has a unique f-image in R. Therefore $f : R \to R$.

Since there is no real number whose square is negative, therefore any negative number in R is not the f-image of any element in R. Consequently f is a mapping of R into R and not from R onto R.

Also we see that $(2)^2 = 4$ and $(-2)^2 = 4$. Thus the elements 2 and –2 in R have the same f-image 4 in R. Hence f is a many-one into mapping of R to R.

Domain of f: Since $f : R \to R$, therefore domain of f is R.

Range of f: If $x \in R$, then $f(x) = x^2$ is a non-negative real number. Further if y is any non-negative real number, then $\exists$ a real number $\sqrt{y}$ such that $f(\sqrt{y}) = (\sqrt{y})^2 = y$. Thus, every non-negative real number is the f-image of some real number. Therefore the range of f is the set of non-negative real numbers.

3

Some Basic Concept of Matrix

INTRODUCTION

Let us consider the system of equations

$$x + y + z = 1$$

$$x + 2y 3+ z = 4$$

$$x + 3y + 5z = 7.$$

Here x, y, z are all unknown and their co-efficient are numbers. Arranging the coefficients in the order in which they occur in the equations and enclosing them in square brackets. We get a rectangular array of the form,

$$\begin{vmatrix} 1 & 1 & 1 \\ 1 & 2 & 3 \\ 1 & 3 & 5 \end{vmatrix}$$

The rectangular array is an example of a matrix. The horizontal lines (→) are called rows or row vectors, and the vertice lines are called column. There are three rows and three column in this matrix. Hence, it is a matrix of order 3 × 3.

We shall use the capital letters to denote matrix,

Thus $$A = \begin{bmatrix} 5 & 1 & 2 \\ 2 & 1 & 3 \end{bmatrix}_{2\times 3}$$

and $$B = \begin{bmatrix} 1 & 2 & 3 \\ 3 & 2 & 1 \\ 2 & 1 & 3 \end{bmatrix}_{3\times 3}$$

are both matrix. They are of the type 2 × 3 and 3 × 3 respectively.

UNIT MATRIX OR IDENTITY MATRIX

Definition: *If in a scalar martix the diagonal element a = 1, then the matrix is called the unit matrix or identity matrix and is denoted by* $\boldsymbol{I_n}$ *in the case of n × n matrix. Thus a square matrix* $A = [a_{ij}]$ *n × n is called an identity or unit matrix iff* $a_{ij}\begin{cases} 1 & \text{when } i=j \\ 0 & \text{when } i\neq j \end{cases}$

For example $I_4 = \begin{bmatrix} 1 & 0 & 0 & 0 \\ 0 & 1 & 0 & 0 \\ 0 & 0 & I & 0 \\ 0 & 0 & 0 & 1 \end{bmatrix}$

Example 1:

If A be any n × n matrix and I_n *is the identity martix of order n × n, then prove that* $AI_n = I_nA = A$.

Solution:

Let us suppose that

$$A = \begin{bmatrix} a_{11} & a_{12} & \dots & a_{1n} \\ a_{21} & a_{22} & \dots & a_{2n} \\ \dots & \dots & \dots & \dots \\ a_{n1} & a_{n2} & \dots & a_{nn} \end{bmatrix} \text{ and } I_n = \begin{bmatrix} 1 & 0 & \dots & 0 \\ 0 & 1 & \dots & 0 \\ \dots & \dots & \dots & 0 \\ 0 & 0 & \dots & 1 \end{bmatrix}$$

$$\therefore A \cdot \mathbf{I}_n = \begin{bmatrix} a_{11} & a_{12} & \dots & a_{1n} \\ a_{21} & a_{22} & \dots & a_{2n} \\ \dots & \dots & \dots & \dots \\ a_{n1} & a_{n2} & \dots & a_{nn} \end{bmatrix} \times \begin{bmatrix} 1 & 0 & \dots & 0 \\ 0 & 1 & \dots & 0 \\ \dots & \dots & \dots & 0 \\ 0 & 0 & \dots & 1 \end{bmatrix}$$

$$= \begin{bmatrix} a_{11}.1 + a_{12}.0 +\dots+ a_{1n}.0 & a_{11}.0 + a_{12}.1 +\dots+ a_{1n}.0 & \dots & \dots & a_{11}.0 + a_{12}.0 +\dots+ a_{1n}.1 \\ a_{21}.1 + a_{22}.0 +\dots+ a_{2n}.0 & a_{21}.0 + a_{22}.1 +\dots+ a_{2n}.0 & \dots & \dots & a_{21}.0 + a_{22}.0 +\dots+ a_{1n}.1 \\ \dots & \dots & \dots & \dots & \dots \quad \dots \\ a_{n1}.1 + a_{n2}.0 +\dots+ a_{nn}.0 & a_{n1}.0 + a_{n2}.1 +\dots+ a_{nn}.0 & \dots & \dots & a_{n1}.0 + a_{n2}.0 +\dots+ a_{nn}.1 \end{bmatrix}$$

$$= \begin{bmatrix} a_{11} & a_{12} & \dots & a_{1n} \\ a_{21} & a_{22} & \dots & a_{2n} \\ \dots & \dots & \dots & \dots \\ a_{n1} & a_{n2} & \dots & a_{nn} \end{bmatrix} = A$$

Similarly we can show that $\mathbf{I}_n \cdot A = A$.

Hence we have $A \cdot \mathbf{I}_n = \mathbf{I}_n \cdot A = A$.

Example 1:

Prove that $I^m = I^{m-1} = ... = I^2 = I$, where m is any positive integer and I_n is the unit matrix of order $n \times n$.

Solution:

Let A be any $n \times n$ matrix and **I** be the nuit martrix of order $n \times n$ *i.e.*, $\mathbf{I} = \mathbf{I}_n$.

Now we know that $A\mathbf{I}_n = \mathbf{I}_n A = A$

But $\mathbf{I}_n = \mathbf{I}$. ...(1)

$\therefore A\mathbf{I} = \mathbf{I}A = A$

Taking $A = \mathbf{I}$, we have $\mathbf{I} \cdot \mathbf{I} = \mathbf{I} \Rightarrow \mathbf{I}^2 = \mathbf{I}$ ...(2)

Again form (1), taking $A = \mathbf{I}^2$, where $\mathbf{I}^2 = \mathbf{I}$ (proved), we get

$\mathbf{I}^2 \cdot \mathbf{I} = \mathbf{I}^3 \quad \Rightarrow \quad \mathbf{I}^3 = \mathbf{I}^2 = \mathbf{I}$, from (2).

Proceeding in this way, we can prove that

$\mathbf{I}^m = \mathbf{I}^{m-1} = ... = \mathbf{I}^3 = \mathbf{I}^3 = \mathbf{I}$, where m is any positive integer.

PERIODIC MATRIX

Definition: *A square matrix A is called periodic, if $A^{k+1} = A$, where k is a positive integer.*

If k is the least positive integer for which $A^{k+1} = A$, then A is said to be of **period k.**

Idempotent Matrix

Definition: A square matrix A is called idempotent provided it satisfies the relation $A^2 = A$.

Symmetric Idempotent Matrix

Definition: *A square matrix A is called symmetric idempotent if $A = A'$ and $A^2 = A$, where A' is the transposed matrix of A.*

Example 1:

Show that if A and B are matrices of order $n \times n$ and such that $AB = A$ and $BA = B$, then A and B are idempotent martices.

Solution:

We have $\mathbf{ABA = (AB)\ A = (A)\ A}$, $\because$ $\mathbf{AB = A}$ (given)

$\Rightarrow \quad \mathbf{ABA = A^2}$ ...(1)

Also $\quad \mathbf{ABA = A\ (BA) = A\ (B)}$ $\quad \because$ $\mathbf{BA = B}$ (given)

$\mathbf{= AB = A}$, $\quad \because$ $\mathbf{AB = A}$ (given)

$\Rightarrow \quad \mathbf{ABA = A}$ ...(2)

From (1) and (2), we have $A^2 = A$ *i.e.,* A is idempotent.

In a similar manner, we can prove that

$\mathbf{BAB = B\ (AB) = B\ (A)}$, $\quad \because$ $\mathbf{AB = A}$ (given)

$\mathbf{= BA = B}$, $\because$ $\mathbf{BA = B}$ (given)

$\Rightarrow \quad \mathbf{BAB = B}$ ...(3)

Also $\quad \mathbf{BAB = (BA)\ B = (B)\ B}$, $\because$ $\mathbf{BA = B}$ (given)

$\Rightarrow \quad \mathbf{BAB = B^2}$ ...(4)

$\therefore$ From (3) and (4), we have $B^2 = B$ *i.e.,* B is idempotent.

Hence proved.

Example 2:

If a is an idempotent matrix, then the matrix $B = I - A$ is idempotent and $AB = O = BA$.

Solution:

We know $\mathbf{IA = AI = A}$. ...(1)

Also **A** being an idempotent matrix, we have $\mathbf{A^2 = A}$. ...(2)

Since **I** and A are square matrices, so $\mathbf{I - A}$ is also a square matrix and therefore we have

$\mathbf{(I - A)^2 = (I - A)\ (I - A)}$

$\mathbf{= (I - A)\ I - (I - A)\ A}$, by distributive law

$\mathbf{= I^2 - AI - IA + A^2}$

$\mathbf{= I - A - A + A}$, from (1), (2) and $\mathbf{I^2 = I}$

$\Rightarrow \quad \mathbf{(I - A)^2 = I - A}$, *i.e.,* $\mathbf{I - A}$ or **B** is an idempotent matrix by definition.

Again $\mathbf{AB = A\ (I - A) = AI - A^2}$, by distributive law

$= \mathbf{A} - \mathbf{A}$, from (1) and (2)

i.e., $\mathbf{AB} = \mathbf{O}$.

And $\mathbf{BA} = (\mathbf{I} - \mathbf{A})\,\mathbf{A} = \mathbf{IA} - \mathbf{A}^2$, by distributive law

$= \mathbf{A} - \mathbf{A}$, from (1) and (2)

$\Rightarrow$ $\mathbf{BA} = \mathbf{O}$. **Hence proved.**

Example 3(a):

If A and B are idempotent martices, then show that AB is idempotent if A and B commute.

Solution:

If **A** is idempotent, then $\mathbf{A}^2 = \mathbf{A}$ and if **B** is idempotent, then $\mathbf{B}^2 = \mathbf{B}$. ...(1)

And if **A** and **B** commute, then $\mathbf{AB} = \mathbf{BA}$ ...(2)

Now $(\mathbf{AB})^2 = (\mathbf{AB}).(\mathbf{AB})$

$= \mathbf{A}\,(\mathbf{BA})\,\mathbf{B}$, by associative law

$= \mathbf{A}\,(\mathbf{AB})\,\mathbf{B}$, from (2)

$= (\mathbf{AA})\,(\mathbf{BB})$, by associative law

$= \mathbf{A}^2\mathbf{B}^2$

$= \mathbf{AB}$, by (1)

Hence **AB** is idempotent.

Example 3(b):

Show that the matrix $A = \begin{bmatrix} 2 & -2 & -4 \\ -1 & 3 & 4 \\ 1 & -2 & -3 \end{bmatrix}$ *is idempotent.*

Solution:

$$A^2 = A \cdot A = \begin{bmatrix} 2 & -2 & -4 \\ -1 & 3 & 4 \\ 1 & -2 & -3 \end{bmatrix} \times \begin{bmatrix} 2 & -2 & -4 \\ -1 & 3 & 4 \\ 1 & -2 & -3 \end{bmatrix}$$

$$= \begin{bmatrix} 2.2 - 2\,(-1) - 4.1 & 2\,(-2) - 2.3 - 4\,(-2) \\ -1.2 + 3\,(-1) + 4.1 & -1\,(-2) + 3.3 + 4\,(-2) \\ 1.2 - 2\,(-1) - 3.1 & 1\,(-2) - 2.3 - 3\,(-2) \end{bmatrix}$$

$$\begin{matrix} 2(-4) - 2.4 - 4(-3) \\ -1(-4) + 3.4 + 4(-3) \\ 1(-4) - 2.4 - 3(-3) \end{matrix}\Bigg]$$

$$= \begin{bmatrix} 2 & -2 & -4 \\ -1 & 3 & 4 \\ 1 & -2 & -3 \end{bmatrix} = A$$

Hence the matrix A is idempotent.

PROPERTIES OF MATRIX ADDITION

Property I: Addition of Matrices is Commutative

$(A + B) = (B + A)$

i.e., $[a_{ij}] + [b_{ij}] = [b_{ij}] + [a_{ij}]$,

where $[a_{ij}]$ and $[b_{ij}]$ are any two $m \times n$ matrices *i.e.,* matrices of the same order.

Proof:

$[a_{ij}] + [b_{ij}] = [a_{ij} + b_{ij}]$, by definition of addition

$= [b_{ij} + a_{ij}]$, ∵ addition of numbers (elements) is commutative

$= [b_{ij} + a_{ij}]$

i.e., $[a_{ij}] + [b_{ij}] = [b_{ij}] + [a_{ij}]$. Hence the theorem.

Property II: Addition of Matrices is Associative

$[(A + B) + C] = A + [B + C]$

i.e., $\{[a_{ij}] + [b_{ij}]\} + [c_{ij}] = [a_{ij}] + [b_{ij}] + [c_{ij}]\}$,

where $[a_{ij}]$, $[b_{ij}]$ and $[c_{ij}]$ are any three matrices of the same order $m \times n$, say.

Proof:

$\{[a_{ij}] + [b_{ij}]\} + [c_{ij}]$

$= [a_{ij} + b_{ij}] + [c_{ij}]$, by law of addition for matrices

$= [(a_{ij} + b_{ij}) + c_{ij}]$, by law of addition for matrices

$= [a_{ij} + (b_{ij} + c_{ij})]$, ∵ addition of numbers is associative

$= [a_{ij}] + [b_{ij} + c_{ij}]$

$= [a_{ij}] + \{[b_{ij}] + [c_{ij}]\}$. Hence the theorem.

Property III: Addition of Matrices Obey the Distributive Law

$$k [A + B] = kA + kB$$

i.e., $k ([a_{ij}] + [b_{ij}]) = k [a_{ij}] + 0 [b_{ij}]$

where $[a_{ij}]$ and $[b_{ij}]$ are any two matrices of the same order $m \times n$, say.

Proof:

$k ([a_{ij}] + [b_{ij}] = k [a_{ij} + b_{ij}]$, by law of addition

$= [k (a_{ij} + b_{ij})]$, by law of scalar multiplication

$= (ka_{ij} + kb_{ij})]$, by distributive law for numbers

$= [ka_{ij}] + [kb_{ij}]$

$= k [a_{ij}] + k [b_{ij}]$. Hence the theorem.

Property IV: Existence of Additive Identity:

If $A = [a_{ij}]$ be any $m \times n$ matrix and O be the $m \times n$ null matrix, then

$$A + O = A = O + A.$$

Proof:

Is obvious, since $a_{ij} + 0 = a_{ij} = 0 + a_{ij}$

Property V: Existence of Additive Inverse

If $A [a_{ij}]$ be any $m \times n$ matrix, then there exists another $m \times n$ matrix B such that

$$A + B + O = B = A,$$

where O is the $m \times n$ null matrix.

Here the matrix B is called the additive inverse of the matrix A or the negative of A.

Also the (i, j) the element of b is $-a_{ij}$ if $A = [a_{ij}]$

Example 1:

If $A = \begin{bmatrix} 1 & 5 & 6 \\ -6 & 7 & 0 \end{bmatrix}$ and $B = \begin{bmatrix} 1 & -5 & 7 \\ 8 & -7 & 7 \end{bmatrix}$ *then find $A + B$ and $A - B$.*

Solution:

$$A + B = \begin{bmatrix} 1 & 5 & 6 \\ -6 & 7 & 0 \end{bmatrix} + \begin{bmatrix} 1 & -5 & 7 \\ 8 & -7 & 7 \end{bmatrix}$$

$$= \begin{bmatrix} 1+1 & 5-5 & 6+7 \\ -6+8 & 7-7 & 0+7 \end{bmatrix} = \begin{bmatrix} 2 & 0 & 13 \\ 2 & 0 & 7 \end{bmatrix}$$ **Ans.**

and $A - B = \begin{bmatrix} 1 & 5 & 6 \\ -6 & 7 & 0 \end{bmatrix} - \begin{bmatrix} 1 & -5 & 7 \\ 8 & -7 & 7 \end{bmatrix}$

$$= \begin{bmatrix} 1-1 & 5-(-5) & 6-7 \\ -6-7 & 8-7-(-7) & 0-7 \end{bmatrix} = \begin{bmatrix} 0 & 10 & -1 \\ -14 & 14 & -7 \end{bmatrix}$$ **Ans.**

Example 2(a):

Solve the following equations for A and B:

$2A - B = \begin{bmatrix} 3 & -3 & 0 \\ 3 & 3 & 2 \end{bmatrix}$, $2B + A = \begin{bmatrix} 4 & 1 & 5 \\ -1 & 4 & -4 \end{bmatrix}$

Solution:

Given $2A - B = \begin{bmatrix} 3 & -3 & 0 \\ 3 & 3 & 2 \end{bmatrix}$

Multiplying both sides by 2, we get

$$4A - 2B = 2\begin{bmatrix} 3 & -3 & 0 \\ 3 & 3 & 2 \end{bmatrix} = \begin{bmatrix} 6 & -6 & 0 \\ 6 & 6 & 4 \end{bmatrix} \quad ...(1)$$

Also given that $2B + A = \begin{bmatrix} 4 & 1 & 5 \\ -1 & 4 & -4 \end{bmatrix} \quad ...(2)$

Adding (1) and (2) we get

$$5A = \begin{bmatrix} 6 & -6 & 0 \\ 6 & 6 & 4 \end{bmatrix} + \begin{bmatrix} 4 & 1 & 5 \\ -1 & 4 & -4 \end{bmatrix}$$

$$= \begin{bmatrix} 6+4 & -6+1 & 0+5 \\ 6-1 & 6+4 & 4+4 \end{bmatrix} = \begin{bmatrix} 10 & -5 & 5 \\ 5 & 10 & 0 \end{bmatrix}$$

$$\Rightarrow \quad A = \frac{1}{2}\begin{bmatrix} 10 & -5 & 5 \\ 5 & 10 & 0 \end{bmatrix} = \begin{bmatrix} 2 & -1 & 1 \\ 1 & 2 & 0 \end{bmatrix}$$

Again from (2) we get

$$2B = \begin{bmatrix} 4 & 1 & 5 \\ -1 & 4 & -4 \end{bmatrix} - A$$

$$\Rightarrow \quad 2B = \begin{bmatrix} 4 & 1 & 5 \\ -1 & 4 & -4 \end{bmatrix} - \begin{bmatrix} 2 & -1 & 1 \\ 1 & 2 & 0 \end{bmatrix}$$

$$= \begin{bmatrix} 4-2 & 1+1 & 5-1 \\ -1-1 & 4-2 & -4-0 \end{bmatrix} = \begin{bmatrix} 2 & 2 & 4 \\ -2 & 2 & -4 \end{bmatrix}$$

$$\Rightarrow \quad B = \frac{1}{2}\begin{bmatrix} 2 & 2 & 4 \\ -2 & 2 & -4 \end{bmatrix} = \begin{bmatrix} 1 & 1 & 2 \\ -1 & 1 & -2 \end{bmatrix}.$$ **Ans.**

Example 2(b):

Given $A = \begin{bmatrix} 1 & 2 & -3 \\ 5 & 0 & 2 \\ 1 & -1 & 1 \end{bmatrix}$ and $B = \begin{bmatrix} 3 & -1 & 2 \\ 4 & 2 & 5 \\ 2 & 0 & 3 \end{bmatrix}$,

find the matrix C such that A + 2C = B.

Solution:

Given that A + 2C = B or 2= B – A

$$\Rightarrow \quad 2C = \begin{bmatrix} 3 & -1 & 2 \\ 4 & 2 & 5 \\ 2 & 0 & 3 \end{bmatrix} - \begin{bmatrix} 1 & 2 & -3 \\ 5 & 0 & 2 \\ 1 & -1 & 1 \end{bmatrix}$$

$$= \begin{bmatrix} 3-1 & -1-2 & 2-(-3) \\ 4-5 & 2-0 & 5-2 \\ 2-1 & 0-(-1) & 3-1 \end{bmatrix} = \begin{bmatrix} 2 & -3 & 5 \\ -1 & 2 & 3 \\ 1 & 1 & 2 \end{bmatrix}$$

$$\Rightarrow \quad C = \frac{1}{2}\begin{bmatrix} 2 & -3 & 0 \\ -1 & 2 & 3 \\ 1 & 1 & 2 \end{bmatrix} = \begin{bmatrix} 1 & -\frac{3}{2} & \frac{5}{2} \\ -\frac{1}{2} & 1 & \frac{3}{2} \\ \frac{1}{2} & \frac{1}{2} & 1 \end{bmatrix}$$

Example 3:

If $A = \begin{bmatrix} 2 & 3 & 1 \\ 0 & -1 & 5 \end{bmatrix}$ and $B = \begin{bmatrix} 1 & 2 & -6 \\ 0 & -1 & 3 \end{bmatrix}$ *evaluate 3A – 4B.*

Solution:

$$2A - 4B = 3\begin{bmatrix} 2 & 3 & 1 \\ 0 & -1 & 5 \end{bmatrix} - 4\begin{bmatrix} 1 & 2 & -6 \\ 0 & -1 & 3 \end{bmatrix}$$

$$= \begin{bmatrix} 6 & 9 & 3 \\ 0 & -3 & 15 \end{bmatrix} - \begin{bmatrix} 4 & 8 & -24 \\ 0 & -4 & 12 \end{bmatrix}$$

$$= \begin{bmatrix} 6-4 & 9-8 & 3-(-24) \\ 0-0 & -3-(-4) & 15-12 \end{bmatrix}$$

$$= \begin{bmatrix} 2 & 1 & 27 \\ 0 & 1 & 3 \end{bmatrix}.$$

MULTIPLICATION OF MATRICES

Definition: *Let* $A = [a_{ij}]$ $m \times n$ *and* $B = B = [b_{ik}]$ $n \times p$ *be two matrices such that the number of column in A is equal to the number of row in B.*

Then the matrix $[c_{ik}]$, $m \times p$ *such that* $c_{ik} = \sum_{j=1}^{n} a_{ij} b_{ik}$ *is called the product of the matrices A ande B in that order and we write = AB.*

As an example, consider the matrices

$$A = \begin{bmatrix} 1 & 2 & 3 \\ 4 & 5 & 6 \end{bmatrix} \text{ and B } \begin{bmatrix} 7 & 8 \\ 9 & 10 \\ 11 & 12 \end{bmatrix}$$

Here the number of columns in A = 3 = then number of rows in B and thus we can evaluate AB.

Let Ab = $[c_{ij}]$, where $[c_{ij}]$ is 2×2 matrix.

Now to write c_{11}, we take the elements of the first row of A *viz.*, 1, 2, 3 in this order and the elements of the first column of B *viz.*, 7, 9, 11 in this order and form the products 1.7, 2.9, 3.11 and finally add them.

i.e., $c_{11} = 1.7 + 2.9 + 3.11 = 58$

Similarly $c_{12} = 1.8 + 2.10 + 3.12 = 64;$

$c_{21} = 4.7 + 5.9 + 6.11 = 139$

and $c_{22} = 4.8 + 5.10 + 6.12 = 154$

Hence $AB = [c_{ij}] = \begin{bmatrix} c_{11} & c_{12} \\ c_{21} & c_{12} \end{bmatrix} = \begin{bmatrix} 58 & 64 \\ 139 & 154 \end{bmatrix}.$

Note: The product AB can be calculated only if the number of columns in A be equal to the number of rows in b. The two matrices A and B satisfying this condition are called conformable to multiplication.

Post-Multiplication and Pre-Multiplication of Matrices

The matrix AB is the matrix A post-multiplied by B whereas the matrix BA is the matrix A pre-multiplied by B.

In the product AB, the matrix A is know as the pre-factor and the matrix B is know as the post-factor.

The product in both the above the above cases viz. AB and BA may or may not exist and may be equal or different.

i.e., we say $AB \neq BA$ in general.

The same is discussed on the next page:

***Case* 1:** If the matrix A is m × n and the matrix B is n × k, then the product AB exists whereas BA does not exist, since we know that AB can be calculated only if the numbers of columns in A is equal to the number of rows in B.

Case 2: If the matrix A is m × n and the matrix B is n × m, then both AB and BA exist, but the matrix AB is m × m while the matrix BA is n × n.

Hence AB ≠ BA thought AB and BA exist.

Case 3: If both A and B are square matrices of the same order, then AB as well as BA exist but are not necessarily equal *i.e.,* if

$$A = \begin{bmatrix} 1 & 2 \\ 3 & 4 \end{bmatrix} \text{ and } B = \begin{bmatrix} 3 & 1 \\ 4 & 7 \end{bmatrix}$$

then $$AB = \begin{bmatrix} 1 & 2 \\ 3 & 4 \end{bmatrix} \times \begin{bmatrix} 3 & 1 \\ 4 & 7 \end{bmatrix} = \begin{bmatrix} 1.3 + 2.4 & 1.1 + 2.7 \\ 3.3 + 4.4 & 3.1 + 4.7 \end{bmatrix}$$

$$= \begin{bmatrix} 11 & 25 \\ 25 & 31 \end{bmatrix}$$

and $$BA = \begin{bmatrix} 3 & 1 \\ 4 & 7 \end{bmatrix} \times \begin{bmatrix} 1 & 2 \\ 3 & 4 \end{bmatrix} = \begin{bmatrix} 3.1 + 1.3 & 3.2 + 1.4 \\ 4.1 + 7.3 & 4.2 + 7.4 \end{bmatrix}$$

$$= \begin{bmatrix} 6 & 10 \\ 25 & 36 \end{bmatrix}$$

∴ AB ≠ BA.

But if $$A = \begin{bmatrix} 1 & 0 \\ 0 & -2 \end{bmatrix} \text{ and } B = \begin{bmatrix} 1 & 0 \\ 0 & 4 \end{bmatrix}$$

then $$AB = \begin{bmatrix} 1 & 0 \\ 0 & -2 \end{bmatrix} \times \begin{bmatrix} 1 & 0 \\ 0 & 4 \end{bmatrix} = \begin{bmatrix} 1.1 + 0.0 & 1.0 + 0.4 \\ 0.1 - 2.0 & 0.0 - 2.4 \end{bmatrix}$$

$$= \begin{bmatrix} 1 & 0 \\ 0 & -8 \end{bmatrix}$$

and $$BA = \begin{bmatrix} 1 & 0 \\ 0 & 4 \end{bmatrix} \times \begin{bmatrix} 1 & 0 \\ 0 & -2 \end{bmatrix} = \begin{bmatrix} 1.1 + 0.0 & 1.0 + 0\,(-2) \\ 0.1 + 4.0 & 0.0 + 4\,(-2) \end{bmatrix}$$

$$= \begin{bmatrix} 1 & 0 \\ 0 & -8 \end{bmatrix}$$

∴ AB = BA.

Hence in general AB ≠ BA.

Note 1: If AB = BA, then matrices A and B are said to commit. If AB = – BA, the matrices A and B are said to anticommute.

Note 2: If $A = \begin{bmatrix} 1 & 1 \\ 1 & 1 \end{bmatrix}$ and $B = \begin{bmatrix} 1 & 0 \\ -1 & 0 \end{bmatrix}$,

then $$AB = \begin{bmatrix} 1 & 1 \\ 1 & 1 \end{bmatrix} \times \begin{bmatrix} 1 & 0 \\ -1 & 0 \end{bmatrix}$$

$$= \begin{bmatrix} 1.1 + 1.(-1) & 1.0 + 1.0 \\ 1.1 + 1.(-1) & 1.0 + 1.0 \end{bmatrix} = \begin{bmatrix} 0 & 0 \\ 0 & 0 \end{bmatrix}$$

i.e., AB is zero matrix (or null matrix) whereas neither A nor B is a zero matrix.

∴ AB = O does not imply that either A = 0 or B = O.

Here $$BA = \begin{bmatrix} 1 & 0 \\ -1 & 0 \end{bmatrix} \times \begin{bmatrix} 1 & 1 \\ 1 & 1 \end{bmatrix}$$

$$= \begin{bmatrix} 1.1 + 0.1 & 1.1 + 0.1 \\ -1.1 + 0.1 & -1.1 + 0.1 \end{bmatrix} = \begin{bmatrix} 1 & 1 \\ -1 & -1 \end{bmatrix}$$

i.e., BA ≠ O

If $A = \begin{bmatrix} 4 & 4 \\ 3 & 3 \end{bmatrix}$ and $B = \begin{bmatrix} -1 & 1 \\ 1 & -1 \end{bmatrix}$, then

$$AB = \begin{bmatrix} 4 & 4 \\ 3 & 3 \end{bmatrix} \times \begin{bmatrix} -1 & 1 \\ 1 & -1 \end{bmatrix}$$

$$= \begin{bmatrix} 4(-1) + 4(1) & 4(1) + 4(-1) \\ 3(-1) + 3(1) & 3(1) + 3(-1) \end{bmatrix} = \begin{bmatrix} 0 & 0 \\ 0 & 0 \end{bmatrix} = 0$$

i.e., the product of two non-zero square matrices can be a zero matrix.

and $$BA = \begin{bmatrix} -1 & 1 \\ 1 & -1 \end{bmatrix} \times \begin{bmatrix} 4 & 4 \\ 3 & 3 \end{bmatrix}$$

$$= \begin{bmatrix} (-1).4 + 1.3 & (-1).4 + 1.3 \\ 1.4 + (-1).3 & 1.4 + (-1).3 \end{bmatrix} = \begin{bmatrix} -1 & -1 \\ 1 & 1 \end{bmatrix} \neq 0$$

DEFINITION OF A MATRIX

A system of any mn numbers arranged in a rectangular array of m rows and n columns is called a matrix of order m × n or an m × n matrix (which is read as m by n matrix).

Or

A set of mn elements of a set S arranged in a rectangular array of m rows and n columns is called an m × n matrix over S.

A m × n matrix is usually written as

$$\begin{bmatrix} a_{11} & a_{12} & a_{13} & \cdots & a_{1n} \\ a_{21} & a_{22} & a_{22} & \cdots & a_{2n} \\ \cdots & \cdots & \cdots & \cdots & \cdots \\ a_{m1} & a_{m2} & a_{m3} & \cdots & a_{mn} \end{bmatrix} \text{ is an } m \times n \text{ matrix.}$$

where the symbols a_{ij} represent any numbers (a_{ij} lies in the ith row and jth column).

Note 1: A compact form of the above matrix may be represented by the symbols $[a_{ij}]$, (a_{ij}) $\|a_{ij}\|$ or by a single capital letter A, say.

The element a_{ij} belong to ith row and th column and sometime called the (;)th element of the matrix.

Note 2: Each of the mn numbers constituting an m × n matrix is known as an *element of the matrix.*

The elements of matrix may be scalar or vector quantities.

Note 3: The plural of 'matrix' is 'matrices'.

Example 1:

The results of a music competition are given in the following matrix:

$$\begin{bmatrix} 3 & 2 & 1 & 0 \\ 0 & 3 & 2 & 4 \\ 5 & 0 & 3 & 0 \\ 2 & 1 & 4 & 3 \end{bmatrix}$$

Here the rows represent the teams A, B, C, D in heat order and the columns represent the number of wins, first place, second plane third place and fourth place scored by the teams.

From the above matrix find (a) How many events did the team A win? (b) How may first places did the team B win? (c) How many third places did the team C win? (d) what does 0 represent in second row?

Solution:

(a) ∵ The first row represents the team A. So the required number = sum of the elements of first row = 3 + 2 + 1 = 6. **Ans.**

(b) As first element of second row (which represents the team B) is zero, os the team B did not win any first place.

(c) The third row represents the team C and third column represents the third place scored by the teams, so the number of third places won by the team C is 3.

(d) The second row represents the team B and the first column represents the first place scored by teams. So 0 in the second row represents that the team B did not score any first place.

Example 2:

Write down the orders of the matrices:

(a) $\begin{bmatrix} 2 & 3 & 5 \\ 1 & 0 & 3 \end{bmatrix}$; *(b)* $\begin{bmatrix} 2 \\ 3 \end{bmatrix}$,

(c) [3, 4, 5]; *(d) [1].*

Solution:

The order of the given matrix are:

(a) 2×3; (b) 2×1; (c) 1×3; (d) 1×1. **Ans.**

TYPES OF MATRICES

(a) *Horizontal Matrix:* If in a matrix the number of columns is more than the number of rows then it is called a horizontal matrix.

For example $\begin{bmatrix} 1 & 3 & 2 & 3 \\ 2 & 5 & 7 & 9 \end{bmatrix}$ is a horizontal matrix.

(b) *Vertical Matrix:* If in a matrix the number of rows is more than the number of columns it is called a vertical matrix.

For example $\begin{bmatrix} 2 & 3 \\ 3 & 5 \\ 4 & 6 \\ 5 & 7 \end{bmatrix}$ is a vertical matrix.

(c) *Column Matrix:* If there if only one column in a matrix, it is called a column matrix.

For example $\begin{bmatrix} 2 \\ 3 \\ 4 \end{bmatrix}$ This is also called column vector.

(d) *Square Matrix:* If m = n *i.e.,* the number of rows and columns of a matrix are equal, then the matrix is of order n ′ n and is called a square matrix of order n.

For example $\begin{bmatrix} 2 & 3 & 1 \\ 1 & 5 & 2 \\ 7 & 6 & 9 \end{bmatrix}$ is a square matrix and $\begin{bmatrix} 1 & 3 & 2 & 3 \\ 2 & 5 & 7 & 9 \end{bmatrix}$

is a rectangular matrix.

(e) *Rectangular Matrices:* When the number of rows and columns of the array are not equal, then the matrix is known as a rectangular matrix.

(f) *Row Matrix:* If in a matrix, there is only one row it is called a row matrix. For example [1, 2, 3]. This is also called a row vector.

(g) *Null (or zero) Matrix:* If all the elements of an m × n matrix are zero, then it is called a null or zero matrix and denoted by $O_{m \times n}$ or simply O.

For example $\begin{bmatrix} 0 & 0 & 0 \\ 0 & 0 & 0 \end{bmatrix}$ is the 2 × 3 mull matrix.

(h) *Unit Matrix:* A square matrix having unity for its elements in the leading diagonal and all the other elements as zero is called an unit matrix.

For example $\begin{bmatrix} 1 & 0 & 0 & 0 \\ 0 & 0 & 0 & 0 \\ 0 & 0 & 1 & 0 \\ 0 & 0 & 0 & 1 \end{bmatrix}$ is unit matrix of order 4 × 4 denote it byI_4.

∴ an a-rowed square matrix $[a_{ij}]$ is called a unit matrix provided

$$a_{ij} = 1, \text{ whenever } i = j$$
$$= 0, \text{ whenever } i \neq j.$$

(i) *Equal Matrix:* Two matrices A $[(a_{ij})]$ are said to be equal if (a) they are of the same type *i.e.,* if they have same number of rows and columns and (b) the elements in the corresponding positions of the two matrices are equal.

From the definition given above it is evident that

1. If A = B, then B = A (Symmetry)
2. A = A, where A is any matrix. (Reflexivity

3. If A = B and B = C, then A = C (Transitivity)

i.e., the relation of equality in the set of all matrices is an equivalence relation.

(j) *Diagonal Matrix:* A square matrix in which all elements except those in the main (or leading) diagonal are zero is know as a diagonal matrix.

For example $\begin{bmatrix} 2 & 0 & 0 \\ 0 & 3 & 0 \\ 0 & 0 & 7 \end{bmatrix}$ is a 3-rowed diagonal matrix.

The um of the diagonal elements of a square matrix A (say) is called the trace of the matrix A.

(k) *Sub-Matrix:* A matrix which is obtained from a given matrix by deleting any number of rows and number of columns is called a sub-matrix of the given matrix.

For example $\begin{bmatrix} 1 & 2 \\ 3 & 4 \end{bmatrix}$ is a sub-matrix of $\begin{bmatrix} 5 & 3 & 2 \\ 1 & 1 & 2 \\ 7 & 3 & 4 \end{bmatrix}$

(l) *Diagonal Element and Orinciple Diagonal:* Those elements a_{ij} of any matrix $[a_{ij}]$ are called diagonal elements for which i = j.

The line along which the above elements lie is called the Principal diagonal or the Diagonal of the matrix.

TRIANGULAR MATRICES

It every element above or below the leading diagonal is zero, then the matrix is called a *Triangular Matrix.*

(a) *Upper Triangular Matrix:* A square matrix A whose elements $a_{ij} = 0$ for $i < j$ is called an upper triangular matrix.

For example $\begin{bmatrix} a_{11} & a_{12} & a_{13} \ldots\ldots a_{1n} \\ 0 & a_{22} & a_{23} \ldots\ldots a_{2n} \\ 0 & 0 & a_{33} \ldots\ldots a_{3n} \\ \ldots & \ldots & \ldots\ldots\ldots \\ 0 & 0 & 0 \ldots\ldots a_{nn} \end{bmatrix}$

(b) *Lower Triangular Matrix:* A square matrix A whose element $a_{ij} = 0$ for $i < j$ is called a lower triangular matrix.

For example
$$\begin{bmatrix} a_{11} & 0 & 0 \ldots\ldots 0 \\ a_{21} & a_{22} & 0 \ldots\ldots 0 \\ a_{31} & a_{32} & a_{33} \ldots\ldots 0 \\ \ldots & \ldots & \ldots \ldots\ldots 0 \\ a_{n1} & a_{n2} & a_{n3} \ldots\ldots a_{nn} \end{bmatrix}$$

DIAGONALMATRIX

Defintion: A square matrix in which all element except those element there in the leading diagonal are zero is called a diagonal matrix. Thus, an n-rowed square matrix $[a_{ij}]$ is a diagonal matrix iff $a_{ij} = 0$ wherever $i \neq j$. If A $[a_{ij}]$ is a diagonal matrix of order n. It must be in the following form.

For eample
$$\begin{bmatrix} a_{11} & 0 & 0 \ldots\ldots 0 \\ 0 & a_{22} & 0 \ldots\ldots 0 \\ 0 & 0 & a_{33} \ldots\ldots \ldots \\ \ldots & \ldots & \ldots \ldots\ldots \ldots \\ 0 & 0 & 0 \ldots\ldots a_{nn} \end{bmatrix}$$

Theorem 1:

Any two diangonal matrices of the same order commute under mltiplication.

Proof:

Let any two diagonal matrices be

$$A = \begin{bmatrix} a_1 & 0 & 0 & \ldots & 0 \\ 0 & a_2 & 0 & \ldots & 0 \\ \ldots & \ldots & \ldots & \ldots & \ldots \\ 0 & 0 & 0 & \ldots & a_n \end{bmatrix} \text{ and } B = \begin{bmatrix} b_1 & 0 & 0 & \ldots & 0 \\ 0 & b_2 & 0 & \ldots & 0 \\ \ldots & \ldots & \ldots & \ldots & \ldots \\ 0 & 0 & 0 & \ldots & b_n \end{bmatrix}$$

Then we have

$$AB = \begin{bmatrix} a_1 & 0 & 0 & \ldots & 0 \\ 0 & a_2 & 0 & \ldots & 0 \\ \ldots & \ldots & \ldots & \ldots & \ldots \\ 0 & 0 & 0 & \ldots & a_n \end{bmatrix} \times \begin{bmatrix} b_1 & 0 & 0 & \ldots & 0 \\ 0 & b_2 & 0 & \ldots & 0 \\ \ldots & \ldots & \ldots & \ldots & \ldots \\ 0 & 0 & 0 & \ldots & b_n \end{bmatrix}$$

$$\Rightarrow \quad AB = \begin{bmatrix} a_1b_1 & 0 & 0 & \dots & 0 \\ 0 & a_2b_2 & 0 & \dots & 0 \\ \dots & 0 & \dots & \dots & \dots \\ 0 & 0 & 0 & \dots & a_nb_n \end{bmatrix} \quad \text{...(1)}$$

$$\text{and} \quad BA = \begin{bmatrix} b_1 & 0 & 0 & \dots & 0 \\ 0 & b_2 & 0 & \dots & 0 \\ \dots & \dots & \dots & \dots & 0 \\ 0 & 0 & 0 & \dots & b_n \end{bmatrix} \times \begin{bmatrix} a_1 & 0 & 0 & \dots & 0 \\ 0 & a_2 & 0 & \dots & 0 \\ \dots & \dots & \dots & \dots & \dots \\ 0 & 0 & 0 & \dots & a_n \end{bmatrix}$$

$$= \begin{bmatrix} b_1a_1 & 0 & 0 & \dots & 0 \\ 0 & b_2a_1 & 0 & \dots & 0 \\ \dots & \dots & \dots & \dots & 0 \\ 0 & 0 & 0 & \dots & b_na_n \end{bmatrix} \quad \text{...(ii)}$$

$\therefore$ From (1) and (2), we find that AB = BA and each one of them is a diagonal matrix. **Hence proved.**

Theorme 2:

Product of any two diagonal matrices of order n is a diagonal matrix of order n.

Proof:

The proof is same as of theorem 1 above. (Do yourself).

Theorem 3:

Sum of any two diagonal matrices of order n is a diagonal matrix of order n and commute under addition.

Proof:

Let any two diagonal matrices be

$$A = \begin{bmatrix} a_1 & 0 & 0 & \dots & 0 \\ 0 & a_2 & 0 & \dots & 0 \\ \dots & \dots & \dots & \dots & \dots \\ 0 & 0 & 0 & \dots & a_n \end{bmatrix} \text{ and } B = \begin{bmatrix} b_1 & 0 & 0 & \dots & 0 \\ 0 & b_2 & 0 & \dots & 0 \\ \dots & \dots & \dots & \dots & \dots \\ 0 & 0 & 0 & \dots & b_n \end{bmatrix}$$

$$\therefore A + B = \begin{bmatrix} a_1 + b_1 & 0 & 0 & \dots & 0 \\ 0 & a_2 + b_2 & 0 & \dots & \dots \\ \dots & \dots & \dots & \dots & 0 \\ 0 & 0 & 0 & \dots & a_n + b_n \end{bmatrix} \quad \dots(1)$$

$$\text{And } B + A = \begin{bmatrix} b_1 + a_1 & 0 & 0 & \dots & 0 \\ 0 & b_2 + a_2 & 0 & \dots & 0 \\ \dots & \dots & \dots & \dots & 0 \\ 0 & 0 & 0 & \dots & b_n + a_n \end{bmatrix} \quad \dots(2)$$

$\therefore$ From (1) and (2), we get A + B = B + A and each one of them is a diagonal matrix of odrer n.

SEALAR MATRIX

Definition: If in a square matrix A all the diagonal elements are equal to a (where $a \neq 0$) and all the remaining elements are equal to zero then it is called a scalar matrix. Thus, $n \times n$ square matrix $[a_{ij}]$ is called a sealar matrix iff for some member a.

For example $\begin{bmatrix} a & 0 & 0 & 0 \\ 0 & a & 0 & 0 \\ 0 & 0 & a & 0 \\ 0 & 0 & 0 & a \end{bmatrix}$ is a scalar matrix of order 4×4.

Commutative Matrices

Definition: If A and B are two square matrices such that AB = BA, then A and B are called *commutative* or are said to *commute.*

If **AB = – BA,** the matrices **A** and **B** are said to *anti-commute.*

Example 1:

Show that the matrices A and B anti-commute, where

$$A = \begin{bmatrix} 1 & -1 \\ 2 & -1 \end{bmatrix} \text{ and } B = \begin{bmatrix} 1 & 1 \\ 4 & -1 \end{bmatrix}$$

Solution:

$$\text{Here } AB = \begin{bmatrix} 1 & -1 \\ 2 & -1 \end{bmatrix} \times \begin{bmatrix} 1 & 1 \\ 4 & -1 \end{bmatrix}$$

$$= \begin{bmatrix} 1.1 - 1.4 & 1.1 + (-1).(-1) \\ 2.1 - 1.4 & 2.1 + (-1).(-1) \end{bmatrix} = \begin{bmatrix} -3 & 2 \\ -2 & 3 \end{bmatrix}$$

And $BA = \begin{bmatrix} 1 & 1 \\ 4 & -1 \end{bmatrix} \times \begin{bmatrix} 1 & -1 \\ 2 & -1 \end{bmatrix}$ ...(1)

$$= \begin{bmatrix} 1.1 + 1.2 & 1.(-1) + 1.(-1) \\ 4.1 + (-1).2 & 4.(-1) + (-1).(-1) \end{bmatrix} = \begin{bmatrix} 3 & -2 \\ 2 & -3 \end{bmatrix}$$

$= - \begin{bmatrix} -3 & 2 \\ -2 & 3 \end{bmatrix}$ ...(2)

∴ From (1) and (2) we find that **AB** = – **BA**.

Hence **A** and **B** anti-commute.

Example 2:

If $A = \begin{bmatrix} a & 0 & 0 \\ 0 & a & 0 \\ 0 & 0 & a \end{bmatrix}$ *and* $B = \begin{bmatrix} a_{11} & a_{12} & a_{13} \\ a_{21} & a_{22} & a_{23} \\ a_{31} & a_{32} & a_{33} \end{bmatrix}$

Then prove that AB = BA = aB.

Solution:

$$AB = \begin{bmatrix} a & 0 & 0 \\ 0 & a & 0 \\ 0 & 0 & a \end{bmatrix} \times \begin{bmatrix} a_{11} & a_{12} & a_{13} \\ a_{21} & a_{22} & a_{23} \\ a_{31} & a_{32} & a_{33} \end{bmatrix}$$

$$= \begin{bmatrix} aa_{11} & aa_{12} & aa_{13} \\ aa_{21} & aa_{22} & aa_{23} \\ aa_{31} & aa_{32} & aa_{33} \end{bmatrix} = a \begin{bmatrix} a_{11} & a_{12} & a_{13} \\ a_{21} & a_{22} & a_{23} \\ a_{31} & a_{32} & a_{33} \end{bmatrix}$$

= aB.

Similarly $BA = \begin{bmatrix} a_{11} & a_{12} & a_{13} \\ a_{21} & a_{22} & a_{23} \\ a_{31} & a_{32} & a_{33} \end{bmatrix} \times \begin{bmatrix} a & 0 & 0 \\ 0 & a & 0 \\ 0 & 0 & a \end{bmatrix}$

$$= \begin{bmatrix} aa_{11} & aa_{12} & aa_{13} \\ aa_{21} & aa_{22} & aa_{23} \\ aa_{31} & aa_{32} & aa_{33} \end{bmatrix} = a \begin{bmatrix} a_{11} & a_{12} & aa_{13} \\ a_{21} & a_{22} & a_{23} \\ a_{31} & a_{32} & a_{33} \end{bmatrix}$$

= aB.

Hence AB = BA = aB.

SCALAR MULTIPLE OF A MATRIX

If A is a matrix and λ is a number then λA is defined as the matrix each element of which λ times the corresponding element of the matrix. A.

For example: $2\begin{bmatrix}3 & 5 & 7\\2 & 3 & 4\end{bmatrix} = \begin{bmatrix}6 & 10 & 14\\4 & 6 & 8\end{bmatrix}$

⇒ if A $[a_{ij}]$, then $\lambda A = [\lambda a_{ij}]$, where λ is a number.

ADDITION OF MATRICES

If there be two m × n matrices given by $A = [a_{ij}]$ and $B = [b_{ij}]$, then the matrix A + B is defined as the matrix each element of which is the sum of the corresponding elements of A and B *i.e.*,

$$A + B = [a_{ij} + b_{ij}],$$

where i = 1, 2, 3,..., m and j = 1, 2, 3,..., n.

For example: If $A = \begin{bmatrix}a_1 & b_1 & c_1\\a_2 & b_2 & c_2\end{bmatrix}$ and $B = \begin{bmatrix}a_3 & b_3 & c_3\\a_4 & b_4 & c_4\end{bmatrix}$,

then $A = B = \begin{bmatrix}a_1 + a_3 & b_2 + b_3 & c_1 + c_3\\a_2 + a_4 & b_2 + b_4 & c_2 + c_4\end{bmatrix}$

SUBTRACTION OF MATRICES

If there be two m × n matrices given by $A = [a_{ij}]$ and $B = [b_{ij}]$, then the matrix A – B is defined as the matrix each element of which is obtained by subtracting the element of B from the corresponding element A *i.e.*, $A - B = [a_{ij} - b_{ij}]$,

where i = 1, 2,...,m and j = 1, 2,...,n.

For example: If $A = \begin{bmatrix}a_1 & b_1 & c_1\\a_2 & b_2 & c_2\end{bmatrix}$ and $B = \begin{bmatrix}a_3 & b_3 & c_3\\a_4 & b_4 & c_4\end{bmatrix}$

then $A - B = \begin{bmatrix}a_1 - a_3 & b_2 - b_3 & c_1 - c_3\\a_2 - a_4 & b_2 - b_4 & c_2 - c_4\end{bmatrix}$

Note: If the two matrices A and B are of the same order, then only their addition and subtraction is possible and these matrices are said to be conformable for addition or subtraction. On the other hand if the matrices A and B are of different orders, then their addition and subtraction is not possible and these matrices are called non-conformable for addition and subtraction.

INVOLUTORY MATRIX

Definition: *A square matrix A is called Involutory provided it satisfies the relation $A^2 = \mathbf{I}$, where **I** is the identity matrix.*

For example, the matrix $A = \begin{bmatrix} 1 & 0 \\ 1 & -1 \end{bmatrix}$ is involutory matrix,

$$\text{since} \quad A^2 = \begin{bmatrix} 1 & 0 \\ 1 & -1 \end{bmatrix} \times \begin{bmatrix} 1 & 0 \\ 1 & -1 \end{bmatrix}$$

$$= \begin{bmatrix} 1.1 + 0.0 & 1.0 + 0.(-1) \\ 0.1 + (-1).0 & 0.0 + (-1).(-1) \end{bmatrix} = \begin{bmatrix} 1 & 0 \\ 0 & 1 \end{bmatrix} = I$$

Example 1:

If A is any square martix of order n and I_n is the identity martix of order n, such that $(I_n - A) = O$, then show that A is an involutory martix.

Solution:

Given that $(I_n - A)(I_n + A) = O$

$\Rightarrow \quad I_n^2 + I_n \cdot A - A \cdot I_n - A^2 = O$

$\Rightarrow \quad I_n + A - A - A^2 = O, \qquad \because I_n^2 = I_n, I_n \cdot A = A = A \cdot I_n.$

$\Rightarrow \quad I_n - A^2 = O$

$\Rightarrow \quad A^2 = I_n$

i.e., **A** is involutory by definition.

Example 2:

Show that the matrix $A = \begin{bmatrix} -5 & -8 & 0 \\ 3 & 5 & 0 \\ 1 & 2 & -1 \end{bmatrix}$ *is involutory,*

Solution:

$$A^2 = \begin{bmatrix} -5 & -8 & 0 \\ 3 & 5 & 0 \\ 1 & 2 & -1 \end{bmatrix} \times \begin{bmatrix} -5 & -8 & 0 \\ 3 & 5 & 0 \\ 1 & 2 & -1 \end{bmatrix}$$

$$= \begin{bmatrix} (-5).(-5)+(-8).3+0.1 & (-5).(-8)+(-8).5+0.2 & (-5).0+(-8).0+0\,(-1) \\ 3.(-5)+5.3+0.1 & 3.(-8)+5.5+0.2 & 3.0+5.0+0.(-1) \\ 1.(-5)+2.3+(-1).1 & 1.(-8)+2.5+(-1).2 & 1.0+2.0+(-1)\,(-1) \end{bmatrix}$$

$$= \begin{bmatrix} 25-24+0 & -40-40+0 & 0+0+0 \\ -15+15+0 & -24+25+0 & 0+0+0 \\ -5+6-1 & -8+10-2 & 0+0+1 \end{bmatrix} = \begin{bmatrix} 1 & 0 & 0 \\ 0 & 1 & 0 \\ 0 & 0 & 1 \end{bmatrix} = I$$

Hence the given matrix a is involutory.

NILPOTENT MATRIX

Definition: *A square matrix **A** si called Nilpotent matrix of order **m**, provided it satisfies the relation* $A^m = O$ *and* $A^{m-1} \neq O$*, where m is a positive integer and **O** is the null matrix of order m.*

For example, the matrix A = $\begin{bmatrix} 0 & 1 \\ 0 & 0 \end{bmatrix}$ is a nilpotent matrix, since

$$A = \begin{bmatrix} 0 & 1 \\ 0 & 0 \end{bmatrix} \neq \mathbf{O},$$

$$A^2 = \begin{bmatrix} 0 & 1 \\ 0 & 0 \end{bmatrix} + \begin{bmatrix} 0 & 1 \\ 0 & 0 \end{bmatrix} = \begin{bmatrix} 0.0+1.0 & 0.1+1.0 \\ 0.0+0.0 & 0.1+0.0 \end{bmatrix}$$

$$= \begin{bmatrix} 0 & 0 \\ 0 & 0 \end{bmatrix} = \mathbf{O},$$

$$A^3 = \lambda^2 \cdot A = O \cdot A = O.$$

i.e., A is a matrix which is not itself a zero matrix though its powers are zero matrices and so it is a nilpotent matrix *(Another definition of nilpotent matrix).*

Example:

Show that $A = \begin{bmatrix} 1 & 2 & 3 \\ 1 & 2 & 3 \\ -1 & -2 & -3 \end{bmatrix}$ *is a nilpotent matrix of order 2.*

Solution:

Given $A = \begin{bmatrix} 1 & 2 & 3 \\ 1 & 2 & 3 \\ -1 & -2 & -3 \end{bmatrix} \neq O$

$$\therefore A^2 = \begin{bmatrix} 1 & 2 & 3 \\ 1 & 2 & 3 \\ -1 & -2 & -3 \end{bmatrix} \times \begin{bmatrix} 1 & 2 & 3 \\ 1 & 2 & 3 \\ -1 & -2 & -3 \end{bmatrix}$$

$$= \begin{bmatrix} 1.1 + 2.2 + 3(-1) & 1.2 + 2.2 + 3(-2) & 1.3 + 2.3 + 3(-3) \\ 1.1 + 2.1 + 3(-1) & 1.2 + 2.2 + 3(-2) & 1.3 + 2.3 + 3(-3) \\ -1.1 - 2.1 - 3(-1) & -1.2 - 2.2 - 3(-2) & -1.3 - 2.3 - 3(-3) \end{bmatrix}$$

$$= \begin{bmatrix} 0 & 0 & 0 \\ 0 & 0 & 0 \\ 0 & 0 & 0 \end{bmatrix} = \mathbf{O}, \text{ where } \mathbf{O} \text{ is the null matrix of order 3.}$$

i.e., $\mathbf{A}^2 = \mathbf{O}$ $\mathbf{A} \neq \mathbf{O}$. Hence **A** is a nilpotent matrix of order 2.

SOLVED EXAMPLES

Example 1:

I*if* $A = \begin{bmatrix} 1 & 1 & 3 \\ 2 & 2 & 6 \\ -1 & -1 & -3 \end{bmatrix}$, *show that* $A^2 = O$.

Solution:

$$A^2 = \begin{bmatrix} 1 & 1 & 3 \\ 2 & 2 & 6 \\ -1 & -1 & -3 \end{bmatrix} \times \begin{bmatrix} 1 & 1 & 3 \\ 2 & 2 & 6 \\ -1 & -1 & -3 \end{bmatrix}$$

$$= \begin{bmatrix} 1.1 + 1.2 + 3.(-1) & 1.1 + 1.2 + 3.(-1) & 1.3 + 1.6 + 3.(-3) \\ 2.1 + 2.2 + 6.(-1) & 2.1 + 2.2 + 6.(-1) & 2.3 + 2.6 + 6.(-3) \\ -1.1 - 1.2 - 3.(-1) & -1.1 + 1.2 - 3(-1) & -1.3 + 1.6 - 3.(-3) \end{bmatrix}$$

$$= \begin{bmatrix} 0 & 0 & 0 & 0 \\ 0 & 0 & 0 & 0 \\ 0 & 0 & 0 & 0 \end{bmatrix} = O, \text{ where O is } 3 \times 3 \text{ null matrix.}$$ **Hence proved**

Example 2:

If $A = \begin{bmatrix} 2 & 3 & 1 \\ 0 & -1 & 5 \end{bmatrix}$ and $B = \begin{bmatrix} 1 & 2 & -6 \\ 0 & -1 & 3 \end{bmatrix}$ *evaluate (a)* $A^2 - B^2$ *and (b) AB and BA.*

Solution:

(a) $A^2 = \begin{bmatrix} 2 & 3 & 1 \\ 0 & -1 & 5 \end{bmatrix} \times \begin{bmatrix} 2 & 3 & 1 \\ 0 & -1 & 5 \end{bmatrix}$, which does not exist as number of columns in the first matrix is not equal to number of rows in the second matrix.

Similarly B^2 does not exist.

(b) AB and BA both do not exist, the reason being the same as in part (a) above.

Example 3(a):

If $A = \begin{bmatrix} \cos\theta & -\sin\theta \\ \sin\theta & \cos\theta \end{bmatrix}$, $B = \begin{bmatrix} \cos\phi & -\sin\phi \\ \sin\phi & \cos\phi \end{bmatrix}$ *show that AB = BA.*

Solution:

$$AB = \begin{bmatrix} \cos\theta & -\sin\theta \\ \sin\theta & \cos\theta \end{bmatrix} \times \begin{bmatrix} \cos\phi & -\sin\phi \\ \sin\phi & \cos\phi \end{bmatrix}$$

$$= \begin{bmatrix} \cos\theta\cos\phi - \sin\theta\sin\phi & -\cos\theta\sin\phi - \sin\theta\cos\phi \\ \sin\theta\cos\phi + \cos\theta\sin\phi & -\sin\theta\sin\phi + \cos\theta\cos\phi \end{bmatrix}$$

$$= \begin{bmatrix} \cos(\theta+\phi) & -\sin(\theta+\phi) \\ \sin(\theta+\phi) & \cos(\theta+\phi) \end{bmatrix} \quad ...(1)$$

$$\text{And } BA = \begin{bmatrix} \cos\phi & -\sin\phi \\ \sin\phi & \cos\phi \end{bmatrix} \times \begin{bmatrix} \cos\theta & -\sin\theta \\ \sin\theta & \cos\theta \end{bmatrix}$$

$$= \begin{bmatrix} \cos\phi\cos\theta - \sin\phi\sin\theta & -\cos\phi\sin\theta - \sin\phi\cos\theta \\ \sin\phi\cos\theta + \cos\phi\sin\theta & -\sin\phi\sin\theta + \cos\phi\cos\theta \end{bmatrix}$$

$$= \begin{bmatrix} \cos(\theta+\phi) & -\sin(\theta+\phi) \\ \sin(\theta+\phi) & \cos(\theta+\phi) \end{bmatrix}$$

∴ From (1) and (2) we get AB = BA. **Hence proved.**

Example 3(b):

Calculate AB and BA if

$$A = \begin{bmatrix} 1 & -1 & 1 \\ -3 & 2 & -1 \\ -2 & 1 & 0 \end{bmatrix}, B = \begin{bmatrix} 1 & 2 & 3 \\ 2 & 4 & 6 \\ 1 & 2 & 3 \end{bmatrix}$$

Solution:

$$AB = \begin{bmatrix} 1 & -1 & 1 \\ -3 & 2 & -1 \\ -2 & 1 & 0 \end{bmatrix} \times \begin{bmatrix} 1 & 2 & 3 \\ 2 & 4 & 6 \\ 1 & 2 & 3 \end{bmatrix}$$

$$= \begin{bmatrix} 1.1 + (-1).2 + 1.1 & 1.2 \ (-1).4 + 1.2 & 1.3 + (-1).6 + 1.3 \\ (-3).1 + 2.2 + (-1).1 & (-3).2 + 2.4 + (-1).2 & (-3).3 + 2.6 + (-1).3 \\ (-2).1 + 1.2 + 0.1 & (-2).2 + 1.4 + 0.2 & (-2).3 + 1.6 + 0.3 \end{bmatrix}$$

$$= \begin{bmatrix} 0 & 0 & 0 \\ 0 & 0 & 0 \\ 0 & 0 & 0 \end{bmatrix}$$ = O, where O is 3 × 3 null matrix. **Ans.**

$$\text{And } BA = \begin{bmatrix} 1 & 2 & 3 \\ 2 & 4 & 6 \\ 1 & 2 & 3 \end{bmatrix} \times \begin{bmatrix} 1 & -1 & 1 \\ -3 & 2 & -1 \\ -2 & 1 & 0 \end{bmatrix}$$

$$= \begin{bmatrix} 1.1 + 2.(-3) + 3(-2) & 1(-1) + 2.2 + 3.1 & 1.1 + 2.(-1) + 3.0 \\ 2.1 + 4.(-3) + 6(-2) & 2(-1) + 4.2 + 6.1 & 2.1 + 4.(-1) + 6.0 \\ 1.1 + 2.(-3) + 3(-2) & 1(-1) + 2.2 + 3.1 & 1.1 + 2.(-1) + 30 \end{bmatrix}$$

$$= \begin{bmatrix} -11 & 6 & -1 \\ -22 & 12 & -2 \\ -11 & 6 & -1 \end{bmatrix}$$ **Ans.**

[Note: AB ≠ BA].

Example 3(c):

If $A = \begin{bmatrix} 2 & 3 & 4 \\ 1 & 2 & 3 \\ -1 & 1 & 2 \end{bmatrix}$ and $B = \begin{bmatrix} 1 & 3 & 0 \\ -1 & 2 & 1 \\ 0 & 0 & 2 \end{bmatrix}$ *evaluate AB, BA or which-ever exists.*

Solution:

$$AB = \begin{bmatrix} 2 & 3 & 4 \\ 1 & 2 & 3 \\ -1 & 1 & 2 \end{bmatrix} \times \begin{bmatrix} 1 & 3 & 0 \\ -1 & 2 & 1 \\ 0 & 0 & 2 \end{bmatrix}$$

$$= \begin{bmatrix} 2.1 + 3(-1) + 4.0 & 2.3 + 3.2 + 4.0 & 2.0 + 3.1 + 4.2 \\ 1.1 + 2(-1) + 3.0 & 1.3 + 2.2 + 3.0 & 1.0 + 2.1 + 3.2 \\ -1.1 + 1(-1) + 2.0 & 1.3 + 1.2 + 2.0 & -1.0 + 1.1 + 2.2 \end{bmatrix}$$

$$= \begin{bmatrix} -1 & 12 & 11 \\ -1 & 7 & 8 \\ -2 & -1 & 5 \end{bmatrix}$$

$$\text{And } BA = \begin{bmatrix} 1 & 3 & 0 \\ -1 & 2 & 1 \\ 0 & 0 & 2 \end{bmatrix} \times \begin{bmatrix} 2 & 3 & 4 \\ 1 & 2 & 3 \\ -1 & 1 & 2 \end{bmatrix}$$

$$= \begin{bmatrix} 1.2 + 3.1 + 0(-1) & 1.3 + 3.2 + 0.1 & 1.4 + 3.3 + 0.2 \\ -1.2 + 2.1 + 1(-1) & -1.3 + 2.2 + 1.1 & -1.4 + 2.3 + 1.2 \\ 0.2 + 0.1 + 2(-1) & 0.3 + 0.2 + 2.1 & 0.4 + 0.3 + 2.2 \end{bmatrix}$$

$$= \begin{bmatrix} 5 & 9 & 13 \\ -1 & 2 & 4 \\ -2 & 2 & 4 \end{bmatrix}$$

Example 4:

Find the product matrix of the matrices

$$\begin{bmatrix} 2 & 1 & 2 & 1 \\ 1 & 1 & 1 & 1 \end{bmatrix} \text{ and } \begin{bmatrix} 2 & -1 & 0 \\ 0 & 4 & 1 \\ -2 & 1 & 0 \\ 1 & -3 & 2 \end{bmatrix}$$

Solution:

The required matrix

$$= \begin{bmatrix} 2 & 1 & 2 & 1 \\ 1 & 1 & 1 & 1 \end{bmatrix} \times \begin{bmatrix} 2 & -1 & 0 \\ 0 & 4 & 1 \\ -2 & 1 & 0 \\ 1 & -3 & 2 \end{bmatrix}$$

$$= \begin{bmatrix} 2.2 + 1.0 + 2(-2) + 1.1 & 2(-1) + 1.4 + 2.1 + 1(-3) & 2.0 + 1.1 + 2.0 + 1.2 \\ 1.2 + 1.0 + 1(-2) + 1.1 & 1(-1) + 1.4 + 1.1 + 1(-3) & 1.0 + 1.1 + 1.0 + 1.2 \end{bmatrix}$$

$$= \begin{bmatrix} 1 & 1 & 3 \\ 1 & 1 & 3 \end{bmatrix}.$$ **Ans.**

Example 5(a):

$$A = \begin{bmatrix} 1 & 1 & -1 \\ 2 & -3 & 4 \\ 3 & -2 & 3 \end{bmatrix};\ B = \begin{bmatrix} -1 & -2 & -1 \\ 6 & 12 & 6 \\ 5 & 10 & 5 \end{bmatrix},\ C = \begin{bmatrix} -1 & -1 & 1 \\ 2 & 2 & -2 \\ -3 & -3 & 3 \end{bmatrix}$$

show that AB and CA are null matrices but AB ≠ O, AC ≠ O.

Solution:

$$AB = \begin{bmatrix} 1 & 1 & -1 \\ 2 & -3 & 4 \\ 3 & -2 & 3 \end{bmatrix} \times \begin{bmatrix} -1 & -2 & -1 \\ 6 & 12 & 6 \\ 5 & 10 & 5 \end{bmatrix}$$

$$= \begin{bmatrix} 1(-1) + 1.6 + (-1).5 & 1(-2) + 1.12 + (-1).10 & 1(-1) + 1.6 + (-1).5 \\ 2(-1) - 3.6 + 4.5 & 2(-2) - 3.12 + 4.10 & 2(-1) - 3.6 + 4.5 \\ 3(-1) - 2.6 + 3.5 & 3(-2) - 2.12 + 3.10 & 3(-1) - 2.6 + 3.5 \end{bmatrix}$$

$$= \begin{bmatrix} 0 & 0 & 0 \\ 0 & 0 & 0 \\ 0 & 0 & 0 \end{bmatrix},$$ which is null matrix.

This is known as 'unusual property' of matrix multiplication

$$CA = \begin{bmatrix} -1 & -1 & 1 \\ 2 & 2 & -2 \\ -3 & -3 & 3 \end{bmatrix} \times \begin{bmatrix} 1 & 1 & -1 \\ 2 & -3 & 4 \\ 3 & -2 & 3 \end{bmatrix}$$

$$= \begin{bmatrix} 1.1 - 1.2 + 1.3 & -1.1 - 1(-3) + 1(-2) & -1(-1) - 1.4 + 1.3 \\ 2.1 + 2.2 - 2.3 & 2.1 + 2(-3) - 2(-2) & 2(-1) + 2.4 - 2.3 \\ -3.1 - 3.2 + 3.3 & -3.1 - 3(-3) + 3(-2) & -3(-1) - 3.4 + 3.3 \end{bmatrix}$$

$= \begin{bmatrix} 0 & 0 & 0 \\ 0 & 0 & 0 \\ 0 & 0 & 0 \end{bmatrix}$, which is a null matrix. **Hence proved.**

We can prove in a similar way that BA ≠ O and AC ≠ O.

Example 5(b):

If $A = \begin{bmatrix} 1 & -2 & 3 \\ -4 & 2 & 5 \end{bmatrix}$ and $B = \begin{bmatrix} 2 & 3 \\ 4 & 5 \\ 2 & 1 \end{bmatrix}$ *find AB and show that AB ≠ BA.*

Solution:

$$AB = \begin{bmatrix} 1 & -2 & 3 \\ -4 & 2 & 5 \end{bmatrix} \times \begin{bmatrix} 2 & 3 \\ 4 & 5 \\ 2 & 1 \end{bmatrix}$$

$$= \begin{bmatrix} 1.2 + (-2).4 + 3.2 & 1.3 + (-2).5 + 3.1 \\ -4.2 + 2.4 + 5.2 & -4,3 + 2.5 + 5.1 \end{bmatrix}$$

$$= \begin{bmatrix} 0 & -4 \\ 10 & 3 \end{bmatrix}$$

and $BA = \begin{bmatrix} 2 & 3 \\ 4 & 5 \\ 2 & 1 \end{bmatrix} \times \begin{bmatrix} 1 & -2 & 3 \\ -4 & 2 & 5 \end{bmatrix}$

$$= \begin{bmatrix} 2.1 + 3(-4) & 2(-2) + 3(2) & 2(3) + 3(5) \\ 4.1 + 5(-4) & 4(-2) + 5(2) & 4(3) + 5(5) \\ 2.1 + 1(-4) & 2(-2) + 1(2) & 2(3) + 1(5) \end{bmatrix}$$

$$= \begin{bmatrix} -10 & 2 & 21 \\ -16 & 2 & 37 \\ -2 & -2 & 11 \end{bmatrix}$$

Hence AB ≠ BA.

Example 5(c):

If $A = \begin{bmatrix} 1 & 2 \\ 3 & 0 \\ 4 & 1 \end{bmatrix}$ and $B = \begin{bmatrix} 0 & 1 & 0 \\ 0 & 2 & 1 \\ 2 & 3 & 0 \end{bmatrix}$, *find BA.*

Solution:

$$BA = \begin{bmatrix} 0 & 1 & 0 \\ 0 & 2 & 1 \\ 2 & 3 & 0 \end{bmatrix} \times \begin{bmatrix} 1 & 2 \\ 3 & 0 \\ 4 & 1 \end{bmatrix}$$

$$= \begin{bmatrix} 0.1 + 1.3 + 0.4 & 0.2 + 1.0 + 0.1 \\ 0.1 + 2.3 + 1.4 & 0.2 + 2.0 + 1.1 \\ 2.1 + 3.3 + 0.4 & 2.2 + 3.0 + 0.1 \end{bmatrix}$$

$$= \begin{bmatrix} 3 & 0 \\ 10 & 1 \\ 11 & 4 \end{bmatrix}$$

Example 6:

If $A = \begin{bmatrix} 0 & 1 \\ 1 & 2 \end{bmatrix}$ and $B = \begin{bmatrix} 1 \\ 2 \end{bmatrix}$, *find AB and BA, if they exist.*

Solution:

$$AB = \begin{bmatrix} 0.1 + 1.2 \\ 1.1 + 2.2 \end{bmatrix} = \begin{bmatrix} 2 \\ 5 \end{bmatrix}$$ **Ans.**

Also BA does not exist as the number of columns in B and number of rows in A are not equal.

Example 7(a):

If $A = \begin{bmatrix} 1 & 4 & 0 \\ 2 & 5 & 0 \\ 3 & 6 & 0 \end{bmatrix}$, $B = \begin{bmatrix} 3 & 2 & 1 \\ 1 & 2 & 3 \\ 4 & 5 & 6 \end{bmatrix}$ and $C = \begin{bmatrix} 3 & 2 & 1 \\ 1 & 2 & 3 \\ 7 & 8 & 9 \end{bmatrix}$

then evaluate AB – AC.

Solution:

$$AB = \begin{bmatrix} 1 & 4 & 0 \\ 2 & 5 & 0 \\ 3 & 6 & 0 \end{bmatrix} \times \begin{bmatrix} 3 & 2 & 1 \\ 1 & 2 & 3 \\ 4 & 5 & 6 \end{bmatrix}$$

$$= \begin{bmatrix} 1.3 + 4.1 + 0.4 & 1.2 + 4.2 + 0.5 & 1.1 + 4.3 + 0.6 \\ 2.3 + 5.1 + 0.4 & 2.2 + 5.2 + 0.5 & 2.1 + 5.3 + 0.6 \\ 3.3 + 6.1 + 0.4 & 3.2 + 6.2 + 0.5 & 3.1 + 6.3 + 0.6 \end{bmatrix}$$

$$= \begin{bmatrix} 7 & 10 & 13 \\ 11 & 14 & 17 \\ 15 & 18 & 21 \end{bmatrix} \qquad ...(1)$$

And $AC = \begin{bmatrix} 1 & 4 & 0 \\ 2 & 5 & 0 \\ 3 & 6 & 0 \end{bmatrix} \times \begin{bmatrix} 3 & 2 & 1 \\ 1 & 2 & 3 \\ 7 & 8 & 9 \end{bmatrix}$

$$= \begin{bmatrix} 1.3 + 4.1 + 0.7 & 1.2 + 4.2 + 0.8 & 1.1 + 4.3 + 0.9 \\ 2.3 + 5.1 + 0.7 & 2.2 + 5.2 + 0.8 & 2.1 + 5.3 + 0.2 \\ 3.3 + 6.1 + 0.7 & 3.2 + 6.2 + 0.8 & 3.1 + 6.3 + 0.9 \end{bmatrix}$$

$$= \begin{bmatrix} 7 & 10 & 13 \\ 11 & 14 & 17 \\ 15 & 18 & 21 \end{bmatrix} \qquad ...(2)$$

$\therefore$ From (1) and (2) we get AB – AC

$$= \begin{bmatrix} 7 & 10 & 13 \\ 11 & 14 & 17 \\ 15 & 18 & 21 \end{bmatrix} - \begin{bmatrix} 7 & 10 & 13 \\ 11 & 14 & 17 \\ 15 & 18 & 21 \end{bmatrix} = \begin{bmatrix} 0 & 0 & 0 \\ 0 & 0 & 0 \\ 0 & 0 & 0 \end{bmatrix} \qquad \textbf{Ans.}$$

Example 7(b):

Find the square of the matrix

$$\begin{bmatrix} -1 & 1 & 1 & 1 \\ 1 & -1 & 1 & 1 \\ 1 & 1 & -1 & 1 \\ 1 & 1 & 1 & -1 \end{bmatrix}$$

Solution:

$$\begin{bmatrix} -1 & 1 & 1 & 1 \\ 1 & -1 & 1 & 1 \\ 1 & 1 & -1 & 1 \\ 1 & 1 & 1 & -1 \end{bmatrix}^2$$

$$= \begin{bmatrix} -1 & 1 & 1 & 1 \\ 1 & -1 & 1 & 1 \\ 1 & 1 & -1 & 1 \\ 1 & 1 & 1 & -1 \end{bmatrix} \times \begin{bmatrix} -1 & 1 & 1 & 1 \\ 1 & -1 & 1 & 1 \\ 1 & 1 & -1 & 1 \\ 1 & 1 & 1 & -1 \end{bmatrix}$$

$$= \begin{bmatrix} (-1)(-1) + 1.1 + 1.1 + 1.1 & (-1).1 + 1(-1) + 1,1 + 1.1 \\ 1.(-1) + (-1).1 + 1.1 + 1.1 & 1.1 + (-1)(-1) + 1.1 + 1.1 \\ 1.(-1) + 1.1 + (-1).1 + 1.1 & 1.1 + 1(-1) + (-1).1 + 1.1 \\ 1.(-1) + 1.1 + 1.1 + (-1).1 & 1.1 + 1.(-1) + 1.1 + (-1).1 \end{matrix}$$

$$\begin{bmatrix} (-1).1 + 1.1 + 1(-1) + 1.1 & (-1).1 + 1.1 + 1.1 + 1(-1) \\ 1.1 + (-1).1 + 1(-1) + 1.1 & 1.1 + (-1).1 + 1.1 + 1(-1) \\ 1.1 + 1.1 + (-1).(-1) + 1.1 & 1.1 + 1.1 + (-1).1 + 1(-1) \\ 1.1 + 1.1 + 1,(-1) + (-1).1 & 1.1 + 1.1 + 1.1 + (-1)(-1) \end{bmatrix}$$

$$= \begin{bmatrix} 4 & 0 & 0 & 0 \\ 0 & 4 & 0 & 0 \\ 0 & 0 & 4 & 0 \\ 0 & 0 & 0 & 4 \end{bmatrix} = 4 \begin{bmatrix} 1 & 0 & 0 & 0 \\ 0 & 1 & 0 & 0 \\ 0 & 0 & 1 & 0 \\ 0 & 0 & 0 & 1 \end{bmatrix}$$ **Ans.**

Example 8:

Evaluate A^3 *if* $A = \begin{bmatrix} \cosh\theta & \sinh\theta \\ \sinh\theta & \cosh\theta \end{bmatrix}$

Solution:

$$A^2 = \begin{bmatrix} \cosh\theta & \sinh\theta \\ \sinh\theta & \cosh\theta \end{bmatrix} \times \begin{bmatrix} \cosh\theta & \sinh\theta \\ \sinh\theta & \cosh\theta \end{bmatrix}$$

$$= \begin{bmatrix} \cosh^2\theta + \sinh^2\theta & \cosh\theta\sinh\theta + \sinh\theta\cosh\theta \\ \sinh\theta\cosh\theta + \cosh\theta\sinh\theta & \sinh^2\theta + \cosh\theta \end{bmatrix}$$

$$= \begin{bmatrix} \cosh 2\theta & \sinh 2\theta \\ \sinh 2\theta & \cosh 2\theta \end{bmatrix} \because \begin{matrix} \cosh^2\theta + \sinh^2\theta = \cosh 2\theta \\ 2\cosh\theta\cosh\theta = \sinh 2\theta \end{matrix}$$

$$\therefore \quad A^3 = A^2A = \begin{bmatrix} \cosh 2\theta & \sinh 2\theta \\ \sinh 2\theta & \cosh 2\theta \end{bmatrix} \begin{bmatrix} \cosh 2\theta & \sinh 2\theta \\ \sinh 2\theta & \cosh 2\theta \end{bmatrix}$$

$$= \begin{bmatrix} \cosh 2\theta\cosh\theta + \sinh 2\theta\sinh\theta & \cosh 2\theta\sinh\theta + \sinh 2\theta\cosh\theta \\ \sinh 2\theta\cosh\theta + \cosh 2\theta\sinh\theta & \sinh 2\theta\sinh\theta + \cosh 2\theta\cosh\theta \end{bmatrix}$$

$$= \begin{bmatrix} \cosh(2\theta + \theta) & \sinh(2\theta + \theta) \\ \sinh(2\theta + \theta) & \cosh(2\theta + \theta) \end{bmatrix},$$

$\because$ sinh (A + B) = sinh A cosh B + cosh A sin B

sinh (A + B) = cosh A cosh B + sinh A sin B

$$= \begin{bmatrix} \cosh 3\theta & \sinh 2\theta \\ \sinh 3\theta & \cosh 2\theta \end{bmatrix}$$

Example 9:

If A, B, C are three matrices such that

$A = [x, y, z]$, $B = \begin{bmatrix} a & h & g \\ h & b & f \\ g & f & c \end{bmatrix}$, $C = \begin{bmatrix} x \\ y \\ z \end{bmatrix}$, *evaluate ABC.*

Solution:

$$B = [x, y, z] \times \begin{bmatrix} a & h & g \\ h & b & f \\ g & f & c \end{bmatrix}$$

$$= [x.a + y.h + z.g \quad x.h + y.b + z.f \quad x.g + y.f + z.c]$$

$$\Rightarrow \quad ABC = [ax + hy + gz \quad hx + by + fz \quad gx + fy + cz \times \begin{bmatrix} x \\ y \\ z \end{bmatrix}$$

$$= [x (ax + hy + gz) + y (hx + by + fz) + z (gx + fy + cz)] \text{ (Note)}$$

$$= [ax^2 + by^2 + cz^2 + 2hxy + 2gzx + 2fyz]. \qquad \textbf{Ans.}$$

Example 10:

Find the product of the following two matrices

$$\begin{bmatrix} 0 & c & -b \\ -c & 0 & a \\ b & -a & 0 \end{bmatrix} \times \begin{bmatrix} a^2 & ab & ac \\ ab & b^2 & bc \\ ac & bc & c^2 \end{bmatrix}$$

Solution:

The required product

$$= \begin{bmatrix} 0 & c & -b \\ -c & 0 & a \\ b & -a & 0 \end{bmatrix} \times \begin{bmatrix} a^2 & ab & ac \\ ab & b^2 & bc \\ ac & bc & c^2 \end{bmatrix}$$

$$= \begin{bmatrix} 0.a^2 + c.ab - b.ac & 0.ab + c.b^2 - b.bc & 0ac + c.bc - b.c^2 \\ -ca^2 + 0.ab\ a.ac & -c.ab + 0b^2 + a.bc & -c.ac + 0.bc + a.c^2 \\ b.a^2 - a.ab + 0.ac & b.ab - a.b^2 + 0.bc & b.ac - a.bc + 0.c^2 \end{bmatrix}$$

$$= \begin{bmatrix} 0 & 0 & 0 \\ 0 & 0 & 0 \\ 0 & 0 & 0 \end{bmatrix}$$

Example 11(a):

Find the values of x, y, z in the following equation

$$\begin{bmatrix} 1 & 2 & 3 \\ 3 & 1 & 2 \\ 2 & 3 & 1 \end{bmatrix} \times \begin{bmatrix} x \\ y \\ z \end{bmatrix} = \begin{bmatrix} 4 & -2 \\ 0 & -6 \\ -1 & 2 \end{bmatrix} \times \begin{bmatrix} 2 \\ 1 \end{bmatrix}$$

Solution:

$$\begin{bmatrix} 1 & 2 & 3 \\ 3 & 1 & 2 \\ 2 & 3 & 1 \end{bmatrix} \times \begin{bmatrix} x \\ y \\ z \end{bmatrix} = \begin{bmatrix} 1.x + 2y + 2z \\ 3x + 1.y + 2z \\ 2x + 3y + 1.z \end{bmatrix} \quad ...(1)$$

And $$\begin{bmatrix} 4 & -2 \\ 0 & -6 \\ -1 & 2 \end{bmatrix} \times \begin{bmatrix} 2 \\ 1 \end{bmatrix} = \begin{bmatrix} 4.2 + (-2).1 \\ 0.2 + (-6).1 \\ -1.2 + 2.1 \end{bmatrix} = \begin{bmatrix} 6 \\ -6 \\ 0 \end{bmatrix} \quad .(2)$$

With the help of (1) and (2), the given equation reduces to

$$\begin{bmatrix} x + 2y + 3z \\ 3x + y + 2z \\ 2x + 3y + z \end{bmatrix} = \begin{bmatrix} 6 \\ -6 \\ 0 \end{bmatrix}$$

From this on comparing the corresponding elements on both sides we get $x + 2y + 3z = 6$; $3x + y + 2z = -6$ and $2x + 3y + z = 0$.

Solving these we get $x = -4$, $y = 2$, $z = 2$. **Ans.**

Example 11(b):

If $A = \begin{bmatrix} 1 & 2 & 3 \\ 0 & 1 & 2 \\ 0 & 0 & 1 \end{bmatrix}$; $B = \begin{bmatrix} x \\ y \\ z \end{bmatrix}$ and $AB = \begin{bmatrix} 6 \\ 3 \\ 1 \end{bmatrix}$ *find the values of x, y, z.*

Solution:

$$AB = \begin{bmatrix} 1 & 2 & 3 \\ 0 & 1 & 2 \\ 0 & 0 & 1 \end{bmatrix} \times \begin{bmatrix} x \\ y \\ z \end{bmatrix}$$

$$\Rightarrow \quad \begin{bmatrix} 6 \\ 3 \\ 1 \end{bmatrix} = \begin{bmatrix} 1.x + 2.y + 3z \\ 0.x + 1.y + 2z \\ 0.x + 0.y + 1.z \end{bmatrix}$$

$\Rightarrow \quad 6 = x + 2y + 3z,\ 3 = y + 2z,\ 1 = z,$ **(Note)**

comparing the corresponding elements of the matrices on both sides. Solving these we get x = 1, y = 1, z = 1. **Ans.**

Example 12:

If $A = \begin{bmatrix} i & 0 \\ 0 & -i \end{bmatrix}, B = \begin{bmatrix} 0 & -1 \\ 1 & 0 \end{bmatrix}, C = \begin{bmatrix} 0 & i \\ i & 0 \end{bmatrix}$, *prove that*

$A^2 = B^2 = C^2 = -I$ *and* $AB = -C = -BA$, *where* $I = \begin{bmatrix} 1 & 1 \\ 0 & 1 \end{bmatrix}$

Solution:

$$A^2 = \begin{bmatrix} i & 0 \\ 0 & -i \end{bmatrix} \times \begin{bmatrix} i & 0 \\ 0 & -i \end{bmatrix}$$

$$= \begin{bmatrix} i.i + 0.0 & i.0 + (-1) \\ 0.i - i.0 & 0.0 + (-i)(-i) \end{bmatrix} = \begin{bmatrix} -1 & 0 \\ 0 & -1 \end{bmatrix}$$

$$= \begin{bmatrix} 1 & 0 \\ 0 & 1 \end{bmatrix}$$

$$= -I.$$

$$\text{and } B^2 = \begin{bmatrix} 1 & -1 \\ 0 & 0 \end{bmatrix} \times \begin{bmatrix} 1 & -1 \\ 0 & 0 \end{bmatrix} = \begin{bmatrix} 0.0 - 1.1 & 0.(-1) + (-1).0 \\ 1.0 + 0.1 & 1(-1) + 0.0 \end{bmatrix}$$

$$= \begin{bmatrix} -1 & 0 \\ 0 & -1 \end{bmatrix} = -\begin{bmatrix} 1 & 0 \\ 0 & 1 \end{bmatrix} = -1$$

Similarly we can prove that $C^2 = -I$. Hence $A^2 = B^2 = C^2 = -I$.

$$\text{Again } AB = \begin{bmatrix} i & 0 \\ 0 & -i \end{bmatrix} \times \begin{bmatrix} 0 & -1 \\ 1 & 0 \end{bmatrix}$$

$$= \begin{bmatrix} i.0 + 0.1 & i(-1) + 0.0 \\ 0.0 - i(1) & 0(-1) - i.0 \end{bmatrix}$$

$$= \begin{bmatrix} 0 & -i \\ -i & 0 \end{bmatrix} = -\begin{bmatrix} 0 & i \\ i & 0 \end{bmatrix} = -C$$

and $BA = \begin{bmatrix} 0 & -1 \\ 1 & 0 \end{bmatrix} \times \begin{bmatrix} i & 0 \\ 0 & -i \end{bmatrix}$

$$= \begin{bmatrix} 0.i - i.o & 0.0 - 1(-i) \\ 1.i + 0.0 & 1.0 + 0(-1) \end{bmatrix}$$

$$= \begin{bmatrix} 0 & i \\ i & 0 \end{bmatrix} = C$$ Hence AB = – C = – BA.

Example 13:

Given $A_i = \begin{bmatrix} 0 & 0 & 0 & 1 \\ 0 & 0 & 1 & 0 \\ 0 & 1 & 0 & 0 \\ 1 & 0 & 0 & 0 \end{bmatrix}$; $A_2 = \begin{bmatrix} 0 & 0 & 0 & i \\ 0 & 0 & -i & 0 \\ 0 & i & 0 & 0 \\ -i & 0 & 0 & 0 \end{bmatrix}$

$A_2 = \begin{bmatrix} 0 & 0 & 1 & 0 \\ 0 & 0 & 0 & -1 \\ 1 & 0 & 0 & 0 \\ 0 & -1 & 0 & 0 \end{bmatrix}$ and $A_4 = \begin{bmatrix} 1 & 0 & 0 & 0 \\ 0 & 1 & 0 & 0 \\ 0 & 0 & -1 & 0 \\ 0 & 0 & 0 & -1 \end{bmatrix}$

Show that $A_i A_k + A_k A_i = 2I$ *or O according as* $i = k$ *or* $i \neq k$ *and I is the matrix of order 4 and i and k take the values 1, 2, 3 and 4.*

Solution:

Let i = k = 1 (say). Then

$A_iA_k = A_1A_2 = A_kA_i$

$$\therefore A_iA_k = A_1A_1 = \begin{bmatrix} 0 & 0 & 0 & 1 \\ 0 & 0 & 1 & 0 \\ 0 & 1 & 0 & 0 \\ 1 & 0 & 0 & 0 \end{bmatrix} \times \begin{bmatrix} 0 & 0 & 0 & 1 \\ 0 & 0 & 1 & 0 \\ 0 & 1 & 0 & 0 \\ 1 & 0 & 0 & 0 \end{bmatrix}$$

$$= \begin{bmatrix} 0+0+0+1 & 0+0+0+0 & 0+0+0+0 & 0+0+0+0 \\ 0+0+0+0 & 0+0+1+0 & 0+0+0+0 & 0+0+0+0 \\ 0+0+0+0 & 0+0+0+0 & 0+1+0+0 & 0+0+0+0 \\ 0+0+0+0 & 0+0+0+0 & 0+0+0+0 & 1+0+0+0 \end{bmatrix}$$

$$= \begin{bmatrix} 1 & 0 & 0 & 0 \\ 0 & 1 & 0 & 0 \\ 0 & 0 & 1 & 0 \\ 0 & 0 & 0 & 1 \end{bmatrix} = I$$

$\therefore \quad A_iA_k + A_kA_i = I + I = 2I$ **Hence proved.**

If i ≠ k, let i = 3 and k = 2

Then $A_iA_k = A_3A_2 = \begin{bmatrix} 0 & 0 & 1 & 0 \\ 0 & 1 & 0 & -1 \\ 1 & 0 & 0 & 0 \\ 0 & -1 & 0 & 0 \end{bmatrix} \times \begin{bmatrix} 0 & 0 & 0 & i \\ 0 & 0 & -i & 0 \\ 0 & i & 0 & 0 \\ -i & 0 & 0 & 0 \end{bmatrix}$

$$= \begin{bmatrix} 0+0+0+0 & 0+0+i+0 & 0+0+0+0 & 0+0+0+0 \\ 0+0+0+i & 0+0+0+0 & 0+0+0+0 & 0+0+0+0 \\ 0+0+0+0 & 0+0+0+0 & 0+0+0+0 & i+0+0+0 \\ 0+0+0+0 & 0+0+0+0 & 0+i+0+0 & 0+0+0+0 \end{bmatrix}$$

$$= \begin{bmatrix} 0 & i & 0 & 0 \\ i & 0 & 0 & 0 \\ 0 & 0 & 0 & i \\ 0 & 0 & i & 0 \end{bmatrix} = i \begin{bmatrix} 0 & 1 & 0 & 0 \\ 1 & 0 & 0 & 0 \\ 0 & 0 & 0 & 1 \\ 0 & 0 & 1 & 0 \end{bmatrix}$$

And $A_kA_i = A_2A_3 = \begin{bmatrix} 0 & 0 & 0 & i \\ 0 & 0 & -i & 0 \\ 0 & i & 0 & 0 \\ -i & 0 & 0 & 0 \end{bmatrix} \times \begin{bmatrix} 0 & 0 & 1 & 0 \\ 0 & 0 & 0 & -1 \\ 1 & 0 & 0 & 0 \\ 0 & -1 & 0 & 0 \end{bmatrix}$

$= \begin{bmatrix} 0 & -i & 0 & 0 \\ -i & 0 & 0 & 0 \\ 0 & 0 & 0 & -i \\ 0 & 0 & -i & 0 \end{bmatrix}$ Multiplying in the usual way

$$= -i \begin{bmatrix} 0 & 1 & 0 & 0 \\ 1 & 0 & 0 & 0 \\ 0 & 0 & 0 & 1 \\ 0 & 0 & 1 & 0 \end{bmatrix}$$

$\therefore A_iA_k + A_kA_i = i \begin{bmatrix} 0 & 1 & 0 & 0 \\ 1 & 0 & 0 & 0 \\ 0 & 0 & 0 & 1 \\ 0 & 0 & 1 & 0 \end{bmatrix} - i \begin{bmatrix} 0 & 1 & 0 & 0 \\ 1 & 0 & 0 & 0 \\ 0 & 0 & 0 & 1 \\ 0 & 0 & 1 & 0 \end{bmatrix}$

$= O$ **Hence proved.**

We can in a similar way prove the above result by giving i and k other values also.

EXERCISES

1. show that all positive integal powers of s symmetric matrix are symmertic.
2. If A be any square matrix, then show that

   $A + A^{\Theta}$ is Hermitian.
3. If A and B are symmetric and they commute, then A^{-1} B and $A^{-1}B^{1}$ are symmetic
4. If B is any square matrix, show that B' AB is symmetric provided B' AB is defined.
5. If A and B are idempotent, then A + B will be idempotent if AB = BA = O, where O is the null matrix.

   [Hint: $(A + B)^2 = A^2 + AB + BA + B^2 = A + O + O + B$]
6. Show that the matrix $\begin{bmatrix} ab & b^2 \\ -a^2 & -ab \end{bmatrix}$ is nilpotent.
7. Show that $\begin{bmatrix} 1 & 1 & 3 \\ 5 & 2 & 6 \\ -2 & -1 & -3 \end{bmatrix}$ is a nilpotent matrix of order 3.
8. If $A = \begin{bmatrix} 1 & -1 & 0 \\ 2 & 1 & 3 \\ 4 & 1 & 8 \end{bmatrix}$ *and* $B = \begin{bmatrix} 4 & 1 & 0 \\ 2 & -3 & 1 \\ 1 & 1 & -1 \end{bmatrix}$ then verify that (AB)' =B'A'
9. If A and B are symmetric (or skew-symmetric) matrices, then so is A + B.
10. Show that every square matrix can be expressed is one and only one way as P + iQ, where P and Q are Hermitian.
11. If A and B are symmetric matrices, then prove that AB + BA is symmertic and AB – BA is Skew Symmettic,
12. If A is any matrix, then show that A' A is a symmetric matrix.
13. Show that the matrices $\begin{bmatrix} 0 & 1 \\ 1 & 0 \end{bmatrix}$ *and* $\begin{bmatrix} 1 & 0 \\ 0 & -1 \end{bmatrix}$ anti-commute.

14. Show that the matrices $\begin{bmatrix} 1 & 2 \\ 2 & 1 \end{bmatrix}$ *and* $\begin{bmatrix} 5 & 7 \\ 7 & 5 \end{bmatrix}$ commute.

15. If A is symmetric matrix, then show that AA' = A'A and A^2 is symmetic.

16. Show that the matrix $A = \frac{1}{\sqrt{2}}\begin{bmatrix} 1 & i \\ -i & -1 \end{bmatrix}$ is unitary.

17. For any two orthogonal matrices A and B, show that BA is an orthogonal matrix.

18. For any two unitary matrices A and B, show that BA is an unitary matrix.

Ans. $XY = \begin{bmatrix} -2 & -4 \\ 3 & 2 \end{bmatrix}$

19. If $A = \begin{bmatrix} 1 & 2 \\ 3 & 4 \\ 5 & 6 \end{bmatrix}$ and $B = \begin{bmatrix} 9 & 8 & 7 \\ 6 & 5 & 4 \end{bmatrix}$, then find AB and BA.

Ans. $AB = \begin{bmatrix} 21 & 18 & 15 \\ 51 & 44 & 37 \\ 81 & 70 & 59 \end{bmatrix}$ and $BA = \begin{bmatrix} 68 & 92 \\ 41 & 56 \end{bmatrix}$

20. Find AB when $A = \begin{bmatrix} 2 & -1 & 0 \\ 0 & 2 & 1 \\ 1 & 0 & 1 \end{bmatrix}$ and $B = \begin{bmatrix} -2 & 1 & -1 \\ 1 & 2 & -2 \\ 2 & -1 & -4 \end{bmatrix}$

Ans. $\begin{bmatrix} -5 & 0 & 0 \\ 4 & 3 & -8 \\ 0 & 0 & -5 \end{bmatrix}$

21. Show that the matrix $A = \begin{bmatrix} 1 & 2 \\ 3 & 1 \end{bmatrix}$ satisfies the equation $A^2 - 2A - 5I = O$, where O is the 2 × 2 null matrix.

22. Evaluate $A^2 - 3A - 13I$, where I is the 2 × 2 unit matrix and $A = \begin{bmatrix} 2 & 5 \\ 3 & 1 \end{bmatrix}$

Ans. $\begin{bmatrix} 0 & 0 \\ 0 & 0 \end{bmatrix} = O$

23. Show that matrix $A = \begin{bmatrix} 1 & 0 & 0 \\ 2 & 1 & 0 \\ 3 & 2 & 1 \end{bmatrix}$

satisfies the equation $A^3 - 3A^2 + 3A - I = O$, where I is the unit matrix and O the null matrix of order 3.

24. If $A = \begin{bmatrix} 2 & 3 \\ 4 & -1 \end{bmatrix}$, $B = \begin{bmatrix} 3 & -2 \\ 2 & 1 \end{bmatrix}$, $C = \begin{bmatrix} 1 & 2 \\ 3 & 4 \end{bmatrix}$ verify (1) (AB) C = A (BC); (2) (A + B) C = AC + BC.

25. If $A = \begin{bmatrix} 1 & 2 \\ 3 & 4 \end{bmatrix}$, $B = \begin{bmatrix} 2 & 1 \\ 4 & 2 \end{bmatrix}$, $C = \begin{bmatrix} 5 & 1 \\ 7 & 4 \end{bmatrix}$,

show that A (B + C) = AB + AC.

26. If $A_\alpha = \begin{bmatrix} \cos\alpha & \sin\alpha \\ -\sin\alpha & \cos\alpha \end{bmatrix}$, then show that $A_\alpha^{\,n} = \begin{bmatrix} \cos n\alpha & \sin n\alpha \\ -\sin n\alpha & \cos n\alpha \end{bmatrix}$,

where n is any positive integer.

Also prove that A_α and A_β commute and $A_\alpha . A_\beta = A_{\alpha+\beta}$.

27. If $A = \begin{bmatrix} 1 & -2 & 3 \\ -4 & 2 & 5 \end{bmatrix}$; $B = \begin{bmatrix} 1 & 2 \\ -1 & 0 \\ 2 & 4 \end{bmatrix}$, then show that (AB)" = B'A'.

4

Rank of a Matrix

INTRODUCTION

Definition: *If in an $m \times n$ matrix A, at least one of its $r \times r$ minors is different from zero while all the minors of order $(r + 1)$ are zero, then r is defined as the rank of the matrix A.*

Or

A number r is defined as the rank of an $m \times n$ matrix A provide (i) A has at least one minor of order r which does not vanish and (ii) there is no minor of order $(r + 1)$ which is nor equal to zero.

Note 1: The rank of a matrix remains unaltered by the application of elementary row or column operations *i.e.,* all equivalent matrices have the same rank.

Note 2: From the definition of the rank of a matrix we conclude that :

(a) If a matrix A does not posses any minor of order (r+1) then $p(A) < r$.

(b) If at least one minor of order r of the matrix A is not equal to zero, then $p(A) < r$.

Note 3: If every minor of order p of a matrix A is zero, the every minor of order higher than p is definitely zero.

Note 4: The rank of a matrix A is also denoted by p (A).

Note 5: The rank of a zero matrix by definition is 0 *i.e.,* $p(O) = 0$.

Consider the matrix $A = \begin{bmatrix} 1 & 2 & 3 \\ 2 & 3 & 4 \\ 3 & 5 & 7 \end{bmatrix}$

This matrix A has only one three rowed minor *i.e.,* minor of order 3, *viz.* $\begin{vmatrix} 1 & 2 & 3 \\ 2 & 3 & 4 \\ 3 & 5 & 7 \end{vmatrix}$ and its value can easily be calculated to zero, by expanding with respect to first row.

The matrix A has 9 minors of order 2 (or two-rowed minors) and one of them is $\begin{vmatrix} 3 & 4 \\ 5 & 7 \end{vmatrix}$ which has the value

$$(3 \times 7) - (5 \times 4) = 21 - 20 = 1 \neq 0.$$

This fact that A is a matrix whose every minor of order 3 is zero and there is at least one minor of order 2 which is not equal to zero is also expressed as 'the rank of the matrix A is 2'.

Example 1:

Find the rank of the matrix $A = \begin{bmatrix} 6 & 1 & 3 & 8 \\ 4 & 2 & 6 & -1 \\ 10 & 3 & 9 & 7 \\ 16 & 4 & 12 & 15 \end{bmatrix}$

Solution:

The det. of order 4 formed by this matrix

$$= \begin{vmatrix} 6 & 1 & 3 & 8 \\ 4 & 2 & 6 & -1 \\ 10 & 3 & 9 & 7 \\ 16 & 4 & 12 & 15 \end{vmatrix}$$

$$= \begin{vmatrix} 6 & 1 & 3 & 8 \\ 4 & 2 & 6 & -1 \\ 6 & 1 & 3 & 8 \\ 6 & 1 & 3 & 5 \end{vmatrix},$$ replacing R_3 and R_4 by $R_3 - R_2$ and $R_4 - R_3$ respectively

= 0, as its three rows are identical

A minor of order 3

$$= \begin{vmatrix} 6 & 1 & 3 \\ 4 & 2 & 6 \\ 10 & 3 & 9 \end{vmatrix} = \begin{vmatrix} 6 & 1 & 3 \\ 4 & 2 & 6 \\ 6 & 1 & 2 \end{vmatrix},$$ repacing R_3 by $R_3 - R_2$

= 0, two rows being identical.

In a similar way we can prove that all the minors of order 3 are zero.

Now a minor of order 2 = $\begin{vmatrix} 6 & 1 \\ 4 & 2 \end{vmatrix}$ = 12 – 4 = 8 ≠ 0

Hence, the rank of the given matrix = 2. **Ans.**

Example 2(a):

Find the rank of the matrix $\begin{bmatrix} 3 & 4 & 5 & 6 & 7 \\ 4 & 5 & 6 & 7 & 8 \\ 5 & 6 & 7 & 8 & 9 \\ 15 & 16 & 17 & 18 & 19 \end{bmatrix}$

Solution:

One minor of order 3 of A

$= \begin{vmatrix} 5 & 7 & 8 \\ 6 & 8 & 9 \\ 6 & 18 & 19 \end{vmatrix} = \begin{vmatrix} 5 & 7 & 8 \\ 1 & 1 & 1 \\ 11 & 11 & 11 \end{vmatrix}$, replacing R_2 and R_3 by $R_2 - R_1$ and $R_3 - R_1$ respectively.

$= \begin{vmatrix} 5 & 7 & 8 \\ 1 & 1 & 1 \\ 0 & 0 & 0 \end{vmatrix}$, replacing R_3 by $R_3 - 11R_2$

= 0.

In a similar way we can prove that all the minors of order 3 of A are zero.

This shows that all minors of order 4 of A are automatically zero.

Now one minor of order 2 of A

$= \begin{vmatrix} 7 & 8 \\ 8 & 9 \end{vmatrix}$ = (7 × 9) – (8 × 8) = 63 – 64 == – 1 ≠ 0.

Hence the rank of A is 2. **Ans.**

Example 2(b):

Find the rank of A = $\begin{bmatrix} 1 & 1 & 1 \\ b + c & c + a & a + b \\ bc & ca & ab \end{bmatrix}$

Solution:

$$|A| = \begin{vmatrix} 1 & 1 & 1 \\ b+c & c+a & a+b \\ bc & ca & ab \end{vmatrix}$$

$= -(a-b)(b-c)(c-a)$, on evaluating. ...(i)

Now following cases arises :

Case I: $a = b = c$.

If a = b = c, then $A = \begin{vmatrix} 1 & 1 & 1 \\ 2a & 2a & 2a \\ a^2 & a^2 & a^2 \end{vmatrix}$

Therefore all minors of order 2 and 3 of A vanish.

Also A has non-zero minor of order 1, since no element of A is zero.

Hence the rank of A in this case is 1. **Ans.**

Case II: *Two of numbers a, b, c are equal but are different from the third.*

Let $a = b \neq c$.

Then $|A| = \begin{vmatrix} 1 & 1 & 1 \\ a+c & c+a & 2a \\ ac & ca & a^2 \end{vmatrix} = 0$, as C_1, C_2 are identical.

Also A has a minor order 2 viz. $\begin{vmatrix} 1 & 1 \\ a+c & 2a \end{vmatrix}$

$= 2a - (a+c) = a - c \neq 0$, $\because a \neq c$,

Hence the rank of A in this case is 2.

Similarly we can discuss the cases $b = c \neq a$, $c = a \neq b$. **Ans.**

Case III: a, b, c are all different.

In this case $|A| \neq 0$, as is evident from (i) above.

i.e., A has a non-zero minor of order 3 and there exists no minor of order greater than 3.

Example 3:

Find the rank of the matrix $A = \begin{bmatrix} 1 & a & b & 0 \\ 0 & c & d & 1 \\ 1 & a & b & 0 \\ 0 & c & d & 1 \end{bmatrix}$

Solution:

$$|A| = \begin{vmatrix} 1 & a & b & 0 \\ 0 & c & d & 1 \\ 1 & a & b & 0 \\ 0 & c & d & 1 \end{vmatrix}$$

= 0, as R_1, R_3 are identical.

A minor of order 3 of A

$$= \begin{vmatrix} a & b & 0 \\ c & d & 1 \\ a & b & 0 \end{vmatrix} = 0, \text{ as } R_1, R_3 \text{ are identical.}$$

In a similar way we can show that all the minors of order 3 are zero in value.

A minor order 2 of A

$$= \begin{vmatrix} a & b \\ c & d \end{vmatrix} = ad - bc \neq 0$$

Hence the rank of the matrix A is 2. **Ans.**

Example 4: .

Find the rank of the matrix

$$A = \begin{bmatrix} 1 & 1 & 1 \\ a & b & c \\ a^3 & b^3 & c^3 \end{bmatrix}, \text{ where } a, b, c \text{ are all real.}$$

Solution:

$$|A| = \begin{vmatrix} 1 & 1 & 1 \\ a & b & c \\ a^3 & b^3 & c^3 \end{vmatrix} = \begin{vmatrix} 1 & 0 & 0 \\ a & b-a & c-a \\ a^3 & b^3-a^3 & c^3-a^3 \end{vmatrix}, \text{ replacing } C_2, C_3 \text{ by } C_2 - C_1, C_3 - C_1$$

$$= \begin{vmatrix} b-a & c-a \\ b^3-a^3 & c^3-a^3 \end{vmatrix}, \text{ expanding with respect to } R_1$$

$$= (b-a)(c-a) \begin{vmatrix} 1 & 1 \\ b^2+ab+a^2 & c^2+ca+a^2 \end{vmatrix},$$

taking (b – a), (c – a) common from C_1 and C_2

$$= (b-a)(c-a)\begin{vmatrix} 1 & 1 \\ b^2+ab+a^2 & c^2+ca-b^2-ab \end{vmatrix},$$

replacing C_2 by $C_2 - C_1$

$$= (b-a)(c-a)[(c^2+ca-b^2-ab)-0]$$

$$= (b-a)(c-a)[(c^2-b^2)+a(c-b)] \quad \textbf{(Note)}$$

$$= (b-a)(c-a)[(c-b)(c+b+a)]$$

$$\Rightarrow \quad |A| = (a-b)(b-c)(c-a)(a+b+c) \quad \ldots(i)$$

Now following cases arise :

Case I: *$a = b = c$.*

If a = b = c, then $A = \begin{bmatrix} 1 & 1 & 1 \\ a & a & a \\ a^3 & a^3 & a^3 \end{bmatrix}$

Therefore all minors of order 3 and 2 of A are zero.

Also as no element of A is zero, so A has non zero minors of order 1.

Hence in this case, the rank of A is 1. **Ans.**

Case II: *Two of the numbers a, b, c are equal but are different from the third.*

let $a = b \neq c$

Then $= |A| = \begin{vmatrix} 1 & 1 & 1 \\ a & a & a \\ a^3 & a^3 & a^3 \end{vmatrix} = 0$, as C_1 and C_2 are identical.

Also A has a minor of order 2, viz. $\begin{vmatrix} 1 & 1 \\ a & c \end{vmatrix} = c - a \neq 0. \quad \because a \neq c$

Hence in this case the rank of A is 2. **Ans.**

Similarly we can discuss the cases $b = c \neq a$, $c = a \neq b$.

Case III: *a, b, c are all different but $a + b + c = 0$.*

In this case from (i), it is evident that $|A| = 0$. **(Note)**

Also A has a minor of order 2, viz. $\begin{vmatrix} 1 & 1 \\ a & b \end{vmatrix} = b - a \neq 0, \quad \because a \neq b$

Hence in this case the rank of A is 2. **Ans.**

Case IV: *a, b, c are all different but $a + b + c \neq 0$.*

In this case from (i), it is evident that $|A| \neq 0$. **(Note)**

i.e., A has a non-zero minor of order 3.

Also A has no minor of order greater than 3.

Hence in this case the rank of A is 3. **Ans.**

SWEEP OUT METHOD OF FINDING THE RANK OF A MATRIX

In the process of evaluation of the rank of a matrix by means of elementary row and column transformations, if certain rows of columns are zero-rows or zero-columns (*i.e.*, each element of these rows or columns are zero) then we can remove these rows or columns without any effect on the rank of the matrix. This method is generally called the *Sweep out* method.

NORMAL FORM OF A MATRIX

Every non-zero matrix A of order m × n can be reduced by application of elementary row and column operations into equivalent matrix of one of the following forms:

(i) $\begin{bmatrix} I_r & 0 \\ 0 & 0 \end{bmatrix}$, (ii) $\begin{bmatrix} I_r \\ 0 \end{bmatrix}$, (iii) $[I_r\ 0]$, (iv) $[I_r]$,

where I_r is r × r identity matrix and 0 is null matrix of any order.

These four forms are called **Normal** or **canonical form of A.**

Important Theorems (Without Proof)

Theorem 1:

If m × n matrix A is reduced to the canonical form or normal form $\begin{bmatrix} I_r & 0 \\ 0 & 0 \end{bmatrix}$ *by application of elementary row or column operations, then r, the order to the identity sub-matrix I_r is the rank of the matrix A.*

Theorem 2:

If a non-singular matrix of order n × n is reduce to the identity matrix I_n (which is its canonical or normal form), then the rank of the matrix is n.

Example 1:

Find the rank of the matrix $A = \begin{bmatrix} 1 & 2 & 3 \\ 4 & 5 & 6 \\ 2 & 1 & 2 \end{bmatrix}$

Solution:

$$A \sim \begin{bmatrix} 1 & 0 & 0 \\ 4 & -3 & -6 \\ 2 & -3 & -4 \end{bmatrix}, \text{ replacing } C_2, C_3 \text{ by } C_2 - 2C_1, \; C_3 - 3C_1 \text{ respectively.}$$

$$\sim \begin{bmatrix} 1 & 0 & 0 \\ 4 & -3 & -6 \\ 2 & -3 & -4 \end{bmatrix}, \text{ replacing } R_2, R_3 \text{ by } R_2 - 4R_1 \; R_3 - 2R_1 \text{ respectively}$$

$$\sim \begin{bmatrix} 1 & 0 & 0 \\ 0 & 0 & -2 \\ 0 & -3 & 4 \end{bmatrix}, \text{ replacing } R_3 \text{ by } R_2 - R_3$$

$$\sim \begin{bmatrix} 1 & 0 & 0 \\ 0 & 0 & 1 \\ 0 & 1 & 2 \end{bmatrix}, \text{ replacing } C_2, C_3 \text{ by } -\frac{1}{3}C_2, \; -\frac{1}{2}C_3 \text{ respectively}$$

$$\sim \begin{bmatrix} 1 & 0 & 0 \\ 0 & 0 & 1 \\ 0 & 1 & 0 \end{bmatrix}, \text{ replacing } R_3 \text{ by } R_3 - 2R_2$$

$$\sim \begin{bmatrix} 1 & 0 & 0 \\ 0 & 1 & 0 \\ 0 & 0 & 1 \end{bmatrix}, \text{ interchanging } C_2 \text{ and } C_3$$

$$\sim [\, I_3 \,]$$

Hence the rank of A is 3. **Ans.**

Example 2(a):

Find the rank of the matrix $A = \begin{bmatrix} -2 & -1 & -3 & -1 \\ 1 & 2 & 3 & -1 \\ 1 & 0 & 1 & 1 \\ 0 & 1 & 1 & -1 \end{bmatrix}$

Solution:

$$A \sim \begin{bmatrix} 0 & -1 & -1 & 1 \\ 0 & 2 & 2 & -2 \\ 1 & 0 & 1 & 1 \\ 0 & 1 & 1 & -1 \end{bmatrix}, \text{ replacing } R_1, R_2 \text{ by } R_1 + 2R_3, R_2 - R_3 \text{ respectively}$$

$$\sim \begin{bmatrix} 0 & 0 & 0 & 1 \\ 0 & 0 & 0 & -2 \\ 1 & 1 & 2 & 1 \\ 0 & 0 & 0 & -1 \end{bmatrix},$$ replacing C_2, C_3 by $C_2 + C_4, C_3 + C_4$ respectively

$$\Rightarrow \quad A \sim \begin{bmatrix} 0 & 0 & 0 & 1 \\ 0 & 0 & 0 & 0 \\ 1 & 1 & 2 & 0 \\ 0 & 0 & 0 & 0 \end{bmatrix},$$ replacing R_2, R_3, R_4 by $R_2 + 2R_1, R_3 - R_1, R_4 + R_1$ respectively

$$\sim \begin{bmatrix} 0 & 0 & 0 & 1 \\ 0 & 0 & 0 & 0 \\ 1 & 0 & 0 & 0 \\ 0 & 0 & 0 & 0 \end{bmatrix},$$ replacing C_2, C_3 by $C_2 - C_1, C_3 - 2C_1$ respectively.

$$\sim \begin{bmatrix} 1 & 0 & 0 & 0 \\ 0 & 0 & 0 & 0 \\ 0 & 0 & 0 & 1 \\ 0 & 0 & 0 & 0 \end{bmatrix},$$ interchanging R_1 and R_4

$$\sim \begin{bmatrix} 1 & 0 & 0 & 0 \\ 0 & 0 & 0 & 0 \\ 0 & 1 & 0 & 0 \\ 0 & 0 & 0 & 0 \end{bmatrix},$$ interchanging C_2 and C_4

$$\sim \begin{bmatrix} 1 & 0 & 0 & 0 \\ 0 & 1 & 0 & 0 \\ 0 & 0 & 0 & 0 \\ 0 & 0 & 0 & 0 \end{bmatrix},$$ interchanging R_2 and R_3

$$\sim \begin{bmatrix} I_2 & 0 \\ 0 & 0 \end{bmatrix}$$

Hence the rank of A is 2. **Ans.**

Example 2(b):

Find the rank of the matrix $A = \begin{bmatrix} 1 & -1 & 2 & -3 \\ 4 & 1 & 0 & 2 \\ 0 & 3 & 0 & 4 \\ 0 & 1 & 0 & 2 \end{bmatrix}$

Solution:

$$A \sim \begin{bmatrix} 1 & 0 & 2 & 0 \\ 4 & 5 & 0 & 14 \\ 0 & 3 & 0 & 4 \\ 0 & 1 & 0 & 2 \end{bmatrix}, \text{ replacing } C_2, C_4 \text{ by } C_2 + C_1, C_4 + 3C_1 \text{ respectively}$$

$$\sim \begin{bmatrix} 1 & 0 & 1 & 0 \\ 4 & 5 & 0 & 7 \\ 0 & 3 & 0 & 2 \\ 0 & 1 & 0 & 1 \end{bmatrix}, \text{ replacing } C_3, C_4 \text{ by } \frac{1}{2}C_3, \frac{1}{2}C_4 \text{ respectively}$$

$$\Rightarrow \quad A \sim \begin{bmatrix} 0 & 0 & 1 & 0 \\ 4 & 5 & 0 & 2 \\ 0 & 3 & 0 & -1 \\ 0 & 1 & 0 & 0 \end{bmatrix}, \text{ replacing } C_1, C_4 \text{ by } C_1 - C_3, C_4 - C_2 \text{ respectively}$$

$$\sim \begin{bmatrix} 0 & 0 & 1 & 0 \\ 1 & 5 & 0 & 2 \\ 0 & 3 & 0 & -1 \\ 0 & 1 & 0 & 0 \end{bmatrix}, \text{ replacing } C_1 \text{ by } \frac{1}{2} C_1$$

$$\sim \begin{bmatrix} 0 & 0 & 1 & 0 \\ 1 & 0 & 0 & 0 \\ 0 & 3 & 0 & -1 \\ 0 & 1 & 0 & 0 \end{bmatrix}, \text{ replacing } C_2, C_4 \text{ by } C_2 - 5C_1, C_4 - 2C_1 \text{ respectively}$$

$$\sim \begin{bmatrix} 0 & 0 & 1 & 0 \\ 1 & 0 & 0 & 0 \\ 0 & 0 & 0 & 1 \\ 0 & 1 & 0 & 0 \end{bmatrix}, \text{ replacing } C_2, C_4 \text{ by } C_2 + 3C_4, -C_4 \text{ respectively}$$

$$\sim \begin{bmatrix} 1 & 0 & 0 & 0 \\ 0 & 1 & 0 & 0 \\ 0 & 0 & 1 & 0 \\ 0 & 0 & 0 & 1 \end{bmatrix}, \text{ rearranging columns}$$

$\sim [\ I_4\ ]$.

$\therefore$ Rank of A is 4.

SOME IMPORTANT THEOREMS

Theorem 1:

The rank of the product matrix AB of two matrices A and B is less than the rank of either of the matrices A and B.

Proof:

Let r_1 and r_2 be the rank of the matrices A and B.

$\because$ r_1 is the rank of A, therefore $A \sim \begin{bmatrix} M \\ O \end{bmatrix}$, where M is a submatrix of ran r_1 and contains r_1 rows.

Post-Multiplying it by B, we get

$$AB \sim \begin{bmatrix} M \\ O \end{bmatrix} B.$$

But $\begin{bmatrix} M \\ O \end{bmatrix}$ B can have r_1 non-zero rows at the most which are obtained on multiplying r_1 non-zero rows of M with columns of B.

$\therefore$ Rank of AB = Rank of $\begin{bmatrix} M \\ O \end{bmatrix} B \leq r_1$

i.e., Rank of AB £ rank of A ...(i)

In a similar way we get B ~ [N O], where N is a submatrix of ran r_2 and contains r_2 columns.

Pre-multiplying it by A, we get

$$AB \sim A\ [N\ O].$$

But [N O] can have r_2 non-zero columns at the most which are obtained on multiplying the rows of A with r_2 non-zero columns of [N O].

∴ Rank of AB = Rank of A[N O] $\leq r_2$

i.e., Rank of AB ≤ rank of B ...(ii)

Hence the theorem from (i) and (ii).

Theorem 2:

The rank of a matrix is equal to the rank of the transposed matrix.

Proof:

Let A = $[a_{ij}]$ be any m × n matrix.

The transposed matrix A' = $[A_{ij}]$ is an n × m matrix.

Let the rank of A be r and let B be the r × r sub-matrix of A such that | B | ≠ 0.

Also we know that the value of a determinant remains unaltered if its rows and columns are interchanged.

i.e., | B' | = | B | ≠ 0, where B' is evidently a r × r sub-matrix of A'

∴ The rank of A' ≥ r.

Again if C be a (r + 1) × (r + 1) submatrix of A, then by definition of rank we must e all | C | = 0.

Also C' is a (r + 1) × (r + 1) submatrix of A' so we have

| C' | = | C | = 0, as explained above.

∴ We conclude that there cannot be any (r + 1) × (r + 1) submatrix of A' with non-zero determinant.

∴ The rank of A' ≥ r and it cannot be greater than r as above.

∴ The rank of A' is r which is also the rank of A. **Hence proved.**

Example 1:

Show that AA', has the same rank as A, where A' is the transpose of A.

Solution:

Let B = AA', then rank of B ≤ rank of A ...(i)

Also A^{-1} B = A', and so we have

rank A = rank A' ≤ rank of B ...(ii)

∴ From (i) and (ii), rank of A = rank of B.

Example 2:

Show that the rank of a matrix (A does not alter by pre or post) multiplying it with any non-singular matrix R.

Solution:

Let B = RA

Then rank of B = rank of RA $\leq$ rank A ...(i)

Also A = R^{-1} B, where R^{-1} is the inverse matrix of R.

$\therefore$ rank of A = rank of (R^{-1} B) $\leq$ rank of B. ...(ii)

$\therefore$ From (i) and (ii) we conclude that

rank of A = rank of B. **Hence proved.**

Example 3:

Prove that if A is a matrix of order m × n and if B is a non singular matrix of order n, then the product P = AB has the same rank as A.

Solution:

Here A ~ (m × n), B ~ (n × n)

P = AB ~ (m × n)

If m < n, rank of A $\leq$ m but rank of B = n

$\therefore$ rank of A < rank of B.

Now rank (P) = rank (AB $\leq$ rank A ...(i)

But we can write A = PB^{-1}

$\therefore$ rank of A = rank of (PB^{-1}) $\leq$ rank of P ...(ii)

$\therefore$ From (i) and (ii) we get rank of P = rank of A.

ECHELON FORM OF A MATRIX

Definition: *If in a matrix,*

(i) all the non-zero rows, if any, precede the zero rows,

(ii) the number of zero proceeding the first non-zero element in a row is less than the number of such zero in the succeeding row,

(iii) the first non-zero element in a row is unity, then it is in the Echelon form.

Note: The number of non-zero rows of a matrix given in the Echelon form is its rank. **(Remember)**

Examples of a matrix in the Echelon Form :

$$\begin{bmatrix} 1 & 3 & 4 & 5 & 6 \\ 0 & 1 & 2 & 3 & 4 \\ 0 & 0 & 1 & 2 & 3 \\ 0 & 0 & 0 & 0 & 0 \end{bmatrix}$$

In this matrix we observe that

(i) the first three non-zero rows precede the fourth row which is a zero row,

(ii) the number of zero is R_1, R_2 and R_3 are 5, 2 and 1 respectively which are in descending order,

(iii) the first non-zero term in each row is unity.

Hence all the three conditions of the Echelon form are satisfied.

Also there being three non-zero rows in this matrix, its rank is 3. This fact can be proved by actually finding the rank of this matrix.

In this matrix, a minor of order 4.

$$= \begin{vmatrix} 1 & 3 & 4 & 5 \\ 0 & 1 & 2 & 3 \\ 0 & 0 & 1 & 2 \\ 0 & 0 & 0 & 0 \end{vmatrix} = 0, \text{ one row being of zero.}$$

In a similar way we can show that all minors of order 4 are zero.

Now a minor of order 3 $= \begin{vmatrix} 1 & 3 & 4 \\ 0 & 1 & 2 \\ 0 & 1 & 0 \end{vmatrix}$

$$= \begin{vmatrix} 1 & 2 \\ 0 & 1 \end{vmatrix}, \text{ expanding w.r. to } C_1$$

$$= 1 \neq 0$$

Hence the rank of this matrix = 3.

Example :

Find the rank of the matrix $A = \begin{bmatrix} 0 & 1 & 2 & 3 \\ 0 & 0 & 1 & -1 \\ 0 & 0 & 0 & 0 \end{bmatrix}$

Solution:

In the given matrix we observe that

(i) the first two non-zero rows precede the third row which is a zero row,

(ii) the number of zero in R_2, R_3 and R_1 are 4, 2 and 1 respectively which are in descending order, and

(iii) the first non-zero term in each row is unity.

Hence all three conditions of the Echelon form are satisfied.

Also there being two non-zero rows in this matrix, its rank is 2. **Ans.**

INVARIANCE OF RANK UNDER ELEMENTARY OPERATIONS

Theorem:

All equivalent matrices have the same ranks i.e. the rank of a matrix remains unaltered by the application of elementary row and column operations.

Proof:

Let r be the rank of an m × n matrix $A = [a_{ij}]$.

Case I: *If ith and jth rows are interchanged (which may be written symbolically as R_{ij} or $R_I \leftrightarrow R_j$) then it does not effect the rank.*

Let B denote the matrix obtained from the matrix A by the elementary operation $R_i \leftrightarrow R_j$ and let p be the rank of B.

Also if D be any (r + 1) rowed square sub-matrix of B, then $|D| = \pm |C|$, where C is a particular (r + 1) rowed submatrix of A.

As r is the rank of the matrix A so every (r + 1) rowed minor of A vanishes and therefore p, the rank osw *are multiplied by a non-zero number λ (which may be written symbolically, as $R_i \rightarrow \lambda R_j$, $\lambda \neq 0$) it does not effect the rank.*

Let B denote the matrix obtained from the matrix A by the elementary operation $R_i \rightarrow \lambda R_j$ and let p be the rank of B.

Let D be any (r + 1) rowed square submatrix of B and let C be the submatrix of A having the same position as D. Then either $|D| = |C|$ or $|D| = \lambda |C|$.

[Here $|D| = |C|$ happens if the ith row of A is one of those rows which are removed to obtain D from B and $|D| = \lambda |C|$ happens when the ith row is not removed while obtaining C from A].

Also as r is the rank of the matrix A so every (r + 1) rowed minor of A vanishes and therefore in particular $|C| = 0$ and consequently in both the above cases $|D| = 0$.

$\therefore$ p, the rank of B, cannot exceed r, the rank of A.

i.e., $p \leq r$

Also we can obtain A from B by the elementary operation $R_i \to \lambda^{-1} R_j$, therefore in that case interchanging the roles of A and B we shall gets $r \leq p$.

Hence r = p.

Case III: *If to the elements of the ith row are added the products by any non-zero numbers λ of the corresponding elements of jth row (which may be written symbolically as $R_i \to R_j + \lambda R_j$, $\lambda \neq 0$)*

Let B denote the matrix obtained from the matrix A by the elementary operation $R_i \to R_j + \lambda R_i$ and let p be the rank of B.

Let D be any (r + 1) rowed square submatrix of B and let C be the submatrix of A having the same position as D.

Now three sub-cases arise:-

(i) If A and B differ only in the ith row *i.e.*, if ith row of B is one of those rows which have been removed while obtaining C.

In this case D = C and therefore | D | = | C |.

$\therefore$ The rank of A is r, so | C | = 0 and consequently | D | = 0

(ii) If ith row of B has not been removed but jth row has been removed while obtaining D.

In this case $| D | = | C | + \lambda | C_0 |$, where C_0 is an (r + 1) rowed matrix which is obtained from C by replacing A_{ik} by a_{jk} *i.e.*, C_0 is obtained from C by performing the elementary operation R_{ij} or $R_i \leftrightarrow R_j$ and then removing those rows and columns of the new matrix which are removed to obtain D from B.

$\therefore$ $| C_0 |$ is negative of some (r + 1) rowed minor of A and as the rank of A is r so every (r + 1) rowed minor of A is zero *i.e.*, $| C | = 0, | C_0 | = 0$ and consequently | D | = 0.

(iii) If neither the ith row nor the jth row of B has been removed while obtained D.

Here | D | = | C | and so as before | D | = 0.

$\therefore$ Every (r + 1) rowed minor of B vanishes so p, the rank of B, cannot exceed r, the rank of A *i.e.*, $p \leq r$.

Also we can obtain A from B by the elementary operation $R_i \to R_j - \lambda R_j$, therefore in that case interchanging the roles of A and B we shall get $r \leq p$.

Hence r = p.

Thus we have observed that the rank of a matrix remains invariant under elementary row operations. Similarly it can be shown that the rank of a matrix remains invariant under elementary column operations too.

Note: By the application of the above theorem we can easily obtain the rank of a matrix for if we can obtain a matrix B by elementary operations on a matrix A and of the rank of B can be easily determined by inspection or simple calculations as given in previous articles in this chapter, then we can determine the rank of A.

SOLVED EXAMPLES

Example 1:

Find the rank of the matrix $A = \begin{bmatrix} 1 & 3 & 4 & 5 \\ 1 & 2 & 6 & 7 \\ 1 & 5 & 0 & 1 \end{bmatrix}$

Solution:

One minor of order three of A

$$= \begin{vmatrix} 1 & 4 & 5 \\ 1 & 6 & 7 \\ 1 & 0 & 1 \end{vmatrix} = \begin{vmatrix} 1 & 0 & 0 \\ 1 & 2 & 2 \\ 1 & -4 & -4 \end{vmatrix}$$, replacing C_2, C_3 by $C_2 - 4C_1$ and $C_3 - 5\,C_1$ respectively.

$$= \begin{vmatrix} 2 & 2 \\ -4 & -4 \end{vmatrix}$$, expanding with respect to R_1

$= 2\,(-4) - 2\,(-4) = 0.$

In a similar way we can prove that all the four minors of order three are zero.

Now a minor or order 2 is $\begin{vmatrix} 2 & 6 \\ 5 & 0 \end{vmatrix} = 2.0 - 6.5 \neq 0.$

Hence the rank of A is 2.

Example 2:

Under what condition the rank of the following matrix A is 3? Is it possible for the rank to be 1? Why?

$$A = \begin{bmatrix} 2 & 4 & 3 \\ 3 & 1 & 2 \\ 1 & 0 & x \end{bmatrix}$$

Solution:

If the rank of the matrix A is 3, then the minor of order 3 of A should be non-zero

i.e.,, $\begin{vmatrix} 2 & 4 & 3 \\ 3 & 1 & 2 \\ 1 & 0 & x \end{vmatrix} \neq 0$, which is the required condition.

Also the rank of A can not be 1 as at least one minor of order 1 of A *i.e.*, one element of A is zero.

[If we are to find the condition under which the rank of A is 2, then the same is | A | = 0 *i.e.*, minor of order 3 of A must be zero.]

i.e., $\begin{vmatrix} 2 & 4 & 2 \\ 3 & 1 & 2 \\ 1 & 0 & x \end{vmatrix} = 0$ *i.e.*, $\begin{vmatrix} 2 & 4 & 2 \\ 1 & -3 & 0 \\ 1 & 0 & x \end{vmatrix} = 0$,

replacing R_2 by $R_2 - R_1$

i.e., $\begin{vmatrix} 0 & 10 & 2 \\ 1 & -3 & 0 \\ 1 & 0 & x \end{vmatrix} = 0$, replacing R_1 by $R_1 - 2R_2$

i.e., $\begin{vmatrix} 10 & 2 \\ -3 & 0 \end{vmatrix} - 0 + x \begin{vmatrix} 0 & 10 \\ 1 & -3 \end{vmatrix} = 0$, expanding with respect to R_2

i.e., $6 - 10x = 0$ *i.e.*, $x = 6/10 = 3/5$. **Ans.**

Example 3:

Prove that the points (x_1, y_1), (x_2, y_2), (x_3, y_3) *are collinear if the rank of the matrix* $\begin{bmatrix} x_1 & y_1 & 1 \\ x_2 & y_2 & 1 \\ x_3 & y_3 & 1 \end{bmatrix}$ *is less than 3.*

Solution:

If the rank of the given matrix is less than 3, then the minor of order 3 of this matrix must be zero.

i.e., $\begin{vmatrix} x_1 & y_1 & 1 \\ x_2 & y_2 & 1 \\ x_3 & y_3 & 1 \end{vmatrix} = 0$ ...(i)

Now the area of the triangle whose vertices are (x_1, y_1), (x_2, y_2) and

$$(x_3, y_3) = \frac{1}{2}\begin{vmatrix} x_1 & y_1 & 1 \\ x_2 & y_2 & 1 \\ x_3 & y_3 & 1 \end{vmatrix}$$

$= 0$, from (i)

Since the area of this triangle is zero, so its vertices (x_1, y_1) (x_2, y_2) and (x_3, y_3) are collinear. **Hence proved.**

Example 4(a):

Are the matrices $A = \begin{bmatrix} 1 & 2 & 3 \\ 2 & 5 & 4 \\ 3 & 7 & 9 \end{bmatrix}$ *and*

$$B = \begin{bmatrix} 1 & 0 & -5 & 6 \\ 3 & -2 & 1 & 2 \\ 5 & -2 & -9 & 14 \\ 4 & -2 & -4 & 8 \end{bmatrix}$$ *equivalent?*

Example 4(b):

Find the rank of the matrix $A = \begin{bmatrix} 1 & 3 & 2 \\ 1 & 2 & 3 \\ 1 & 5 & 4 \end{bmatrix}$

Solution:

The determinant of order 3 formed by A

$$= \begin{vmatrix} 1 & 3 & 2 \\ 1 & 2 & 3 \\ 1 & 5 & 4 \end{vmatrix} = \begin{vmatrix} 1 & 3 & 2 \\ 0 & -1 & 1 \\ 0 & 2 & 2 \end{vmatrix},$$ replacing R_2, R_3 by $R_2 - R_1, R_3 - R_1$ respectively.

$$= \begin{vmatrix} -1 & 1 \\ 2 & 2 \end{vmatrix} = -2 - 2 = -4 \neq 0.$$

$\therefore \quad p(A) \geq 3$ ...(i)

Also the matrix A does not possess any matrix of order 4 *i.e.*, 3 + 1, so

$p(A) \leq 3.$ ...(ii)

∴ From (i) and (ii) we get p (A) = 3. **Ans.**

Example 4(c):

Find the rank of the matrix $A = \begin{bmatrix} 0 & 1 & 2 \\ 1 & 2 & 3 \\ 3 & 1 & 1 \end{bmatrix}$

Solution:

The determinant of order 3 formed by A

$$= \begin{vmatrix} 0 & 1 & 2 \\ 1 & 2 & 3 \\ 3 & 1 & 1 \end{vmatrix} = \begin{vmatrix} 0 & 1 & 0 \\ 1 & 2 & -1 \\ 3 & 1 & -1 \end{vmatrix}, \text{ replacing } C_3 \text{ by } C_3 - 2C_3$$

$$= - \begin{vmatrix} 1 & -1 \\ 3 & -1 \end{vmatrix} = -[-1 + 3] = -2 \neq 0.$$

∴ p (A) ≥ 3 ...(i)

Also the matrix A does not possesses any matrix of order 4 *i.e.*, 3 + 1, so p (A) ≤ 3. ...(ii)

∴ From (i) and (ii) we get p (A) = 3. **Ans.**

Example 5:

Find the rank of the matrix $A = \begin{bmatrix} 1 & 2 & 3 \\ 2 & 5 & 8 \\ 4 & 10 & 18 \end{bmatrix}$

Solution:

The determinant of order 2 formed by A

$$= \begin{vmatrix} 1 & 2 & 3 \\ 2 & 5 & 8 \\ 4 & 10 & 18 \end{vmatrix}$$

$$= \begin{vmatrix} 1 & 0 & 0 \\ 2 & 1 & 2 \\ 4 & 2 & 6 \end{vmatrix}, \text{replacing } C_2, C_3 \text{ by } C_2 - 2C_1, C_3 - 3C_1 \text{ respectively}$$

$$= \begin{vmatrix} 1 & 2 \\ 2 & 6 \end{vmatrix} = (1 \times 6) - (2 \times 2) = 6 - 4 = 2 \neq 0$$

$\therefore$ $p(A) \geq 3$ ...(i)

Also matrix A does not possesses any matrix of order 4 *i.e.*, 3 + 1, so

$p(A) \leq 3.$ ...(ii)

$\therefore$ From (i) and (ii) we get $p(A) = 3$. **Ans.**

Example 6:

Find the rank of the matrix $A = \begin{bmatrix} 1 & 2 & 3 \\ 2 & 3 & 4 \\ 4 & 10 & 18 \end{bmatrix}$

Solution:

The determinant of order 3 formed by A

$$= \begin{vmatrix} 1 & 2 & 3 \\ 2 & 3 & 4 \\ 4 & 10 & 18 \end{vmatrix} = \begin{vmatrix} 1 & 0 & 0 \\ 2 & -1 & -2 \\ 4 & 2 & 6 \end{vmatrix}$$, replacing C_2, C_3 by $C_2 - 2C_1, C_3 - 3C_1$ respectively.

$$= \begin{vmatrix} -1 & -2 \\ 2 & 6 \end{vmatrix} = -6 + 4 = -2 \neq 0$$

$\therefore$ $p(A) \geq 3$ ...(i)

Also the matrix A does not possesses any matrix of order 4 *i.e.*, 3 + 1, so

$p(A) \leq 3.$ ...(ii)

$\therefore$ From (i) and (ii) we get $p(A) = 3$. **Ans.**

Example 7(a):

Find the rank of the matrix $A = \begin{bmatrix} 1 & -3 & 2 \\ 3 & -9 & 6 \\ -2 & 6 & -4 \end{bmatrix}$

Solution:

The determinant of order 3 formed by this matrix A

$$= \begin{vmatrix} 1 & -3 & 2 \\ 3 & -9 & 6 \\ -2 & 6 & -4 \end{vmatrix} = \begin{vmatrix} 1 & 0 & 0 \\ 3 & 0 & 0 \\ -2 & 0 & 0 \end{vmatrix}$$, replacing C_2, C_3 by $C_2 + 3C_1$ and $C_3 - 2C_1 = 0$ respectively

Also there exists no minor of order 2 of A which is not equal to zero. (Students can verify for themselves).

Finally all minors of order 1 of the matrix A are non-zero, as no element of the matrix A is zero.

Hence the rank of A is 1. **Ans.**

Example 7(b):

Find the rank of the matrix $\begin{bmatrix} 1 & 2 & 3 & 1 \\ 2 & 4 & 6 & 2 \\ 1 & 2 & 3 & 2 \end{bmatrix}$

Solution:

In this matrix, a minor of order

$$= \begin{vmatrix} 1 & 2 & 3 \\ 2 & 4 & 6 \\ 1 & 2 & 3 \end{vmatrix} = 0, \because R_1, R_2 \text{ are identical}$$

In a similar way we prove that all the minors of order 3 are zero.

Now a minor order 2 = $\begin{vmatrix} 1 & 2 \\ 2 & 4 \end{vmatrix} = 0.$

But another minor order 2 = $\begin{vmatrix} 3 & 1 \\ 3 & 2 \end{vmatrix} \neq 0$

Hence rank of the given matrix is 2. **Ans.**

Example 7(c):

Find the rank of the matrix $\begin{bmatrix} 13 & 16 & 19 \\ 14 & 17 & 20 \\ 15 & 18 & 21 \end{bmatrix}$

Solution:

The determinant of order 3 formed by this matrix

$$= \begin{vmatrix} 13 & 16 & 19 \\ 14 & 17 & 20 \\ 15 & 18 & 21 \end{vmatrix}$$

$= \begin{vmatrix} 13 & 3 & 3 \\ 14 & 3 & 3 \\ 15 & 3 & 3 \end{vmatrix}$, replacing C_2 and C_3 by $C_2 - C_1$ and $C_3 - C_2$ respectively.

= 0, since two columns viz. C_2 and C_3 are identical.

A minor of order 2 = $\begin{vmatrix} 13 & 16 \\ 14 & 17 \end{vmatrix} \neq 0$.

Hence the rank of the given matrix is 2. **Ans.**

Example 8:

Find the rank of the matrix A = $\begin{bmatrix} 1 & 2 & 3 \\ 2 & 3 & 1 \\ -2 & -3 & -1 \end{bmatrix}$

Solution:

The determinant of order 3 formed by this matrix A

$= \begin{vmatrix} 1 & 2 & 3 \\ 2 & 3 & 1 \\ -2 & -3 & -1 \end{vmatrix} = \begin{vmatrix} 1 & 2 & 3 \\ 2 & 3 & 1 \\ 0 & 0 & 0 \end{vmatrix}$, replacing R_3 by $R_3 + R_2$

= 0

Also there exists a minor of order 2 of A

viz. $\begin{vmatrix} 1 & 2 \\ 2 & 3 \end{vmatrix} = 3 - 4 = -1 \neq 0$

Hence the rank of the given matrix A is 2. **Ans.**

Example 9:

Find the rank of the matrix A = $\begin{bmatrix} 1 & 2 & 3 \\ 2 & 3 & 4 \\ 4 & 5 & 6 \end{bmatrix}$

Solution:

The determinant of order 3 formed by A

$$= \begin{vmatrix} 1 & 2 & 3 \\ 2 & 3 & 4 \\ 4 & 5 & 6 \end{vmatrix} = \begin{vmatrix} 1 & 0 & 0 \\ 2 & -1 & -2 \\ 4 & -3 & -6 \end{vmatrix}$$, replacing C_2, C_3 by $C_2 - 2C_1, C_3 - 3C_1$ respectively.

$$= \begin{vmatrix} -1 & -2 \\ - & -6 \end{vmatrix} = 6 - 6 = 0$$

Also there exists a minor of order 2 of A

viz. $\begin{vmatrix} 3 & 4 \\ 5 & 6 \end{vmatrix} = 18 - 20 = -2 \neq 0.$

Hence the rank of the given matrix A is 2. **Ans.**

Example 10:

Find the rank of the matrix $\begin{bmatrix} 1 & -1 & -2 & -4 \\ 3 & 1 & 3 & -2 \\ 6 & 3 & 0 & -7 \\ 2 & 3 & -1 & -1 \end{bmatrix}$

Solution:

The determinant of order 4 formed by the given matrix

$$= \begin{vmatrix} 1 & -1 & -2 & -4 \\ 3 & 1 & 3 & -2 \\ 6 & 3 & 0 & -7 \\ 2 & 3 & -1 & -1 \end{vmatrix}$$

$$= \begin{vmatrix} 1 & 0 & 0 & 0 \\ 3 & 4 & 9 & 10 \\ 6 & 9 & 12 & 17 \\ 2 & 5 & 3 & 7 \end{vmatrix}$$, replacing C_2, C_3, C_4 by $C_2 + C_1, C_3 + 2C_1, C_4 + 4C_1$ respectively.

$$= \begin{vmatrix} 4 & 9 & 10 \\ 9 & 12 & 17 \\ 5 & 3 & 7 \end{vmatrix} = \begin{vmatrix} 4 & 9 & 10 \\ 5 & 3 & 7 \\ 5 & 3 & 7 \end{vmatrix}$$, replacing R_2 by $R_2 - R_1$

= 0, as its two row and identical.

A minor of order 3

$$= \begin{vmatrix} 1 & -1 & -2 \\ 3 & 1 & 3 \\ 6 & 3 & 0 \end{vmatrix} = \begin{vmatrix} 1 & 0 & 0 \\ 3 & 4 & 9 \\ 6 & 9 & 12 \end{vmatrix},$$ replacing C_2, C_3 by $C_2 + C_1, C_3 + 2C_1$ respectively.

$$= \begin{vmatrix} 4 & 9 \\ 9 & 12 \end{vmatrix} = 48 - 81 = -33 \neq 0.$$

Hence the rank of the given matrix is 3. **Ans.**

Example 11:

Find the rank of the matrix $A = \begin{bmatrix} 1 & 1 & 1 & -1 \\ 1 & 2 & 3 & 4 \\ 3 & 4 & 5 & 2 \end{bmatrix}$

Solution:

In this matrix there are four minors of order 3 of a Viz.

$$\begin{vmatrix} 1 & 1 & 1 \\ 1 & 2 & 3 \\ 3 & 4 & 5 \end{vmatrix}; \begin{vmatrix} 1 & 1 & -1 \\ 1 & 2 & 4 \\ 3 & 4 & 2 \end{vmatrix}; \begin{vmatrix} 1 & 1 & -1 \\ 1 & 3 & 4 \\ 3 & 5 & 2 \end{vmatrix}; \begin{vmatrix} 1 & 1 & -1 \\ 2 & 3 & 4 \\ 4 & 5 & 2 \end{vmatrix}$$

and each one of them is zero in value (prove it).

Also therexsts a minor of or 2 of A viz.

$$\begin{vmatrix} 2 & 3 \\ 4 & 5 \end{vmatrix} = 10 - 12 = -2 \neq 0$$

Hence the rank of the given matrix A is 2. **Ans.**

Example 12:

Fine the rank of the matrix $\begin{bmatrix} 1 & 3 & 4 & 3 \\ 3 & 9 & 12 & 9 \\ -1 & -3 & -4 & -3 \end{bmatrix}$

Solution:

In this matrix, a minor of order 3

$$= \begin{vmatrix} 1 & 3 & 4 \\ 3 & 9 & 12 \\ -1 & -3 & -4 \end{vmatrix} = 3 \begin{vmatrix} 1 & 3 & 4 \\ 1 & 3 & 4 \\ -1 & -3 & -4 \end{vmatrix},$$ taking common from R_3

$= 0$, as R_1 and R_2 are identical.

In a similar way we can prove that all minors of order 3 are zero.

Now a minor of order 2.

$$= \begin{vmatrix} 1 & 3 \\ 3 & 9 \end{vmatrix} = 3 \begin{vmatrix} 1 & 3 \\ 1 & 3 \end{vmatrix}, \text{ taking out 3 common from } R_3$$

$= 0$, as rows are identical.

Similarly, all the minors of order 2 are zero.

Hence, we are left with minors of order unity, viz. the elements of the given matrix, which are not equal to zero.

Hence rank of the given matrix = 1. **Ans.**

Example 13:

Find the rank of the matrix $A = \begin{bmatrix} 1 & -1 & 3 & 6 \\ 1 & 3 & -3 & -4 \\ 5 & 3 & 3 & 11 \end{bmatrix}$

Solution:

The given matrix A possesses a minor of order 3 viz.

$$\begin{vmatrix} 1 & 3 & 6 \\ 1 & -3 & -4 \\ 5 & 3 & 11 \end{vmatrix} = \begin{vmatrix} 2 & 0 & 2 \\ 1 & -3 & -4 \\ 6 & 0 & 7 \end{vmatrix}, \text{ replacing } R_1, R_2 \text{ by } R_2 + R_1, R_3 + R_2$$

$$= -3 \begin{vmatrix} 2 & 2 \\ 6 & 7 \end{vmatrix} = -3\,(14 - 12) = -6 \neq 0$$

$\therefore$ $\quad p\,(A) \geq 3$ ...(i)

Also A does not posses any minor of order 4 *i.e.*,, 3 + 1, so

$p\,(A) \leq 3$ ...(ii)

$\therefore$ From (i) and (ii) we get $p\,(A) = 3$. **Ans.**

Example 14:

Find the rank of the matrix $A = \begin{bmatrix} 1 & 3 & 5 & 1 \\ 2 & 4 & 8 & 0 \\ 3 & 1 & 7 & 5 \end{bmatrix}$

Solution:

The given matrix A possesses a minor of order 3 viz.

$$\begin{vmatrix} 1 & 3 & 1 \\ 2 & 4 & 0 \\ 3 & 1 & 5 \end{vmatrix} = \begin{vmatrix} 0 & 0 & 1 \\ 2 & 4 & 0 \\ -2 & -14 & 5 \end{vmatrix}, \text{ repacing } C_1 \text{ and } C_2 \text{ by } C_1 - C_3 \text{ and } C_2 - 3C_2$$

$$= \begin{vmatrix} 2 & 4 \\ -2 & -14 \end{vmatrix}, \text{expanding with respect to } R_1$$

$$= 2\,(-14) - (4)\,(-2) = -28 + 8 \neq 0$$

$\therefore$ $p\,(A) \geq 3$ ...(i)

Also A does not posses any minor of order 4 *i.e.,*, 3 + 1, so

$p\,(A) \leq 3$ ...(ii)

From (i) and (ii) we get p (A) = 3 *i.e.*, the rank of A is 3. **Ans.**

Example 15:

Determinant the rank of A $= \begin{bmatrix} 6 & 1 & 8 & 3 \\ 2 & 1 & 0 & 2 \\ 4 & -1 & -8 & -3 \end{bmatrix}$

Solution:

The given matrix A possesses a minor of order 3 viz.

$$\begin{vmatrix} 6 & 1 & 8 \\ 2 & 1 & 0 \\ 4 & -1 & -8 \end{vmatrix} = \begin{vmatrix} 10 & 0 & 0 \\ 6 & 0 & -8 \\ 4 & -1 & -8 \end{vmatrix}, \text{ replacing } R_1, R_2 \text{ by } R_1 + R_3, R_2 + R_3$$

$$= 10 \begin{vmatrix} 0 & -8 \\ -1 & -8 \end{vmatrix} = 10\,(0 - 8) = -80 \neq 0$$

$\therefore$ $p\,(A) \geq 3$ ...(i)

Also A does not posses any minor of order 4 *i.e.,*, 3 + 1, so

$p\,(A) \leq 3$ ...(ii)

$\because$ From (i) and (ii) we get p (A) = 3. **Ans.**

Example 16(a):

Reduce the matrix A by the normal form and hence find the rank of the matrix A, where

$$A = \begin{bmatrix} 1 & -1 & 1 & -1 \\ 4 & 2 & -1 & 2 \\ 2 & 2 & -2 & 2 \end{bmatrix}$$

Solution:

$$A \sim \begin{bmatrix} 1 & 0 & 1 & 0 \\ 4 & 6 & -1 & 1 \\ 2 & 4 & -2 & 0 \end{bmatrix},$$ replacing C_2, C_4 by $C_2 + C_1, C_4 + C_3$ respectively.

$$\sim \begin{bmatrix} 1 & 0 & 1 & 0 \\ 5 & 6 & 0 & 1 \\ 4 & 4 & 0 & 0 \end{bmatrix},$$ replacing R_2, R_3 by $R_2 + R_1, R_3 + 2R_1$ respectively.

$$\Rightarrow \quad A \sim \begin{bmatrix} 0 & 0 & 1 & 0 \\ 5 & 3 & 0 & 1 \\ 4 & 2 & 0 & 0 \end{bmatrix},$$ replacing C_1, C_2 by $C_1 - C_3, \frac{1}{2}C_2$ respectively

$$\sim \begin{bmatrix} 0 & 0 & 1 & 0 \\ -1 & 3 & 0 & 1 \\ 0 & 2 & 0 & 0 \end{bmatrix},$$ replacing C_1 by $C_1 - 2C_2$

$$\sim \begin{bmatrix} 0 & 0 & 1 & 0 \\ 0 & 0 & 0 & 1 \\ 0 & 2 & 0 & 0 \end{bmatrix},$$ replacing C_1, C_2 by $C_1 + C_4, C_2 - 3C_1$ respectively.

$$\sim \begin{bmatrix} 1 & 0 & 0 & 0 \\ 0 & 0 & 0 & 1 \\ 0 & 1 & 0 & 0 \end{bmatrix},$$ replacing C_2 by $\frac{1}{2}C_2$ and interchanging C_1 and C_2

$$\sim \begin{bmatrix} 1 & 0 & 0 & 0 \\ 0 & 1 & 0 & 0 \\ 0 & 0 & 1 & 0 \end{bmatrix},$$ rearranging columns

$$\sim [\, I_3 \; 0 \,]$$

Hence the rank of A is 3. **Ans.**

Example 16(b):

Find the rank of the matrix $A = \begin{bmatrix} 1 & -1 & 2 & -3 \\ 4 & 1 & 0 & 2 \\ 0 & 3 & 0 & 4 \\ 0 & 1 & 0 & 2 \end{bmatrix}$

Solution:

$$A \sim \begin{bmatrix} 1 & 0 & 1 & 0 \\ 4 & 5 & 0 & 14 \\ 0 & 3 & 0 & 4 \\ 0 & 1 & 0 & 2 \end{bmatrix},$$ replacing C_2, C_3, C_4 by $C_2 + C_1, \frac{1}{2}C_3, C_4 + 3C_1$ respectively.

$$\sim \begin{bmatrix} 0 & 0 & 1 & 0 \\ 4 & 5 & 0 & 7 \\ 0 & 3 & 0 & 2 \\ 0 & 1 & 0 & 1 \end{bmatrix},$$ replacing C_1, C_4 by $C_1 - C_3, \frac{1}{2}C_4$ respectively.

$$\sim \begin{bmatrix} 0 & 0 & 1 & 0 \\ 1 & 5 & 0 & 2 \\ 0 & 3 & 0 & -1 \\ 0 & 1 & 0 & 0 \end{bmatrix},$$ replacing C_1, C_4 by $\frac{1}{4}C_1$, $C_4 - C_2$ respectively.

$$\sim \begin{bmatrix} 0 & 0 & 1 & 0 \\ 1 & 0 & 0 & 0 \\ 0 & 3 & 0 & -1 \\ 0 & 1 & 0 & 0 \end{bmatrix},$$ replacing C_2, C_4 by $C_2 - 5C_1, C_4 - 2C_1$ respectively.

$$\Rightarrow \quad A \sim \begin{bmatrix} 1 & 0 & 0 & 0 \\ 0 & 0 & 1 & 0 \\ 0 & 0 & 0 & 1 \\ 0 & 1 & 0 & 0 \end{bmatrix},$$ replacing C_2, C_4 by $C_2 + 3C_4, -C_4$ respectively and interchanging C_1, C_2

$$\sim \begin{bmatrix} 1 & 0 & 0 & 0 \\ 0 & 1 & 0 & 0 \\ 0 & 0 & 1 & 0 \\ 0 & 0 & 0 & 1 \end{bmatrix},$$ rearranging columns

$$\sim [\, I_4 \,]$$

Hence the rank of A is 4. **Ans.**

Example 17:

Find the rank of the matrix $A = \begin{bmatrix} 1 & 3 & 4 & 7 \\ 2 & 4 & 5 & 8 \\ 3 & 1 & 2 & 4 \end{bmatrix}$

Solution:

$$A \sim \begin{bmatrix} 1 & 3 & 4 & 7 \\ 1 & 1 & 1 & 1 \\ 3 & 1 & 2 & 4 \end{bmatrix}, \text{ replacing } R_2 \text{ by } R_2 - R_1$$

$$\sim \begin{bmatrix} 1 & 2 & 3 & 6 \\ 1 & 0 & 0 & 0 \\ 3 & -2 & -1 & 1 \end{bmatrix}, \text{ replacing } C_2, C_3, C_4 \text{ by } C_2 - C_1, C_3 - C_1, C_4 - C_1 \text{ respectively.}$$

$$\sim \begin{bmatrix} 0 & 2 & 3 & 6 \\ 1 & 0 & 0 & 0 \\ 3 & -2 & -1 & 1 \end{bmatrix}, \text{ replacing } R_1, R_3 \text{ by } R_1 - R_2, R_3 - 3R_2$$

$$\sim \begin{bmatrix} 0 & 1 & 3 & 0 \\ 1 & 0 & 0 & 0 \\ 3 & -1 & -1 & 3 \end{bmatrix}, \text{ replacing } C_2, C_4 \text{ by } \frac{1}{2}C_2, C_4 - 2C_3 \text{ respectively.}$$

$$\Rightarrow \quad A \sim \begin{bmatrix} 0 & 1 & 0 & 0 \\ 1 & 0 & 0 & 0 \\ 0 & -1 & 2 & 3 \end{bmatrix}, \text{ replacing } C_3 \text{ by } C_3 - 3C_2$$

$$\sim \begin{bmatrix} 0 & 1 & 0 & 0 \\ 1 & 0 & 0 & 0 \\ 0 & 0 & 2 & 3 \end{bmatrix}, \text{ replacing } R_3 \text{ by } R_3 + R_1$$

$$\sim \begin{bmatrix} 0 & 1 & 0 & 0 \\ 1 & 0 & 0 & 0 \\ 0 & 0 & 1 & 1 \end{bmatrix}, \text{ replacing } C_3, C_4 \text{ by } \frac{1}{2}C_3 \text{ and } \frac{1}{3}C_4 \text{ respectively.}$$

$$\sim \begin{bmatrix} 0 & 1 & 0 & 0 \\ 1 & 0 & 0 & 0 \\ 0 & 0 & 1 & 0 \end{bmatrix}, \text{ replacing } C_4 \text{ by } C_4 - C_3$$

$$\sim \begin{bmatrix} 1 & 0 & 0 & 0 \\ 0 & 1 & 0 & 0 \\ 0 & 0 & 1 & 0 \end{bmatrix}, \text{ interchanging } R_1 \text{ and } R_2$$

$$\sim [\, I_3 \; 0 \,]$$

Hence the rank of A is 3. **Ans.**

Example 18:

Find the rank of $A = \begin{bmatrix} 0 & 1 & -3 & -1 \\ 1 & 0 & 1 & 1 \\ 3 & 1 & 0 & 2 \\ 1 & 1 & -2 & 0 \end{bmatrix}$

Solution:

$$A \sim \begin{bmatrix} 0 & 1 & -3 & -1 \\ 1 & 0 & 0 & 0 \\ 3 & 1 & -3 & -1 \\ 1 & 1 & -3 & -1 \end{bmatrix}, \text{ replacing } C_4, C_4 \text{ by } C_3 - C_1 \text{ and } C_4 - C_1 \text{ respectively}$$

$$\sim \begin{bmatrix} 0 & 1 & -3 & -1 \\ 1 & 0 & 0 & 0 \\ 3 & 0 & 0 & 0 \\ 1 & 0 & 0 & 0 \end{bmatrix}, \text{ replacing } R_3, R_4 \text{ by } R_3 - R_1 \text{ and } R_4 - R_1 \text{ respectively.}$$

$$\sim \begin{bmatrix} 0 & 1 & 0 & 0 \\ 1 & 0 & 0 & 0 \\ 3 & 0 & 0 & 0 \\ 1 & 0 & 0 & 0 \end{bmatrix}, \text{ replacing } C_3, C_4 \text{ by } C_3 + 3C_2, C_4 + C_2 \text{ respectively}$$

$$\sim \begin{bmatrix} 0 & 1 & 0 & 0 \\ 1 & 0 & 0 & 0 \\ 0 & 0 & 0 & 0 \\ 0 & 0 & 0 & 0 \end{bmatrix}, \text{ replacing } R_3, R_4 \text{ by } R_3 - 3R_2, R_4 - R_2 \text{ respectively.}$$

$$\sim \begin{bmatrix} 1 & 0 & 0 & 0 \\ 0 & 1 & 0 & 0 \\ 0 & 0 & 0 & 0 \\ 0 & 0 & 0 & 0 \end{bmatrix}, \text{ interchanging } C_1, C_2$$

$$\sim \begin{bmatrix} I_2 & 0 \\ 0 & 0 \end{bmatrix}$$

Hence the rank of A is 2. **Ans.**

Example 19:

Find the rank of the matrix $A = \begin{bmatrix} 1 & 0 & 2 & 1 \\ 0 & 1 & -2 & 1 \\ 1 & -1 & 4 & 0 \\ -2 & 2 & 8 & 0 \end{bmatrix}$

Solution:

$$A \sim \begin{bmatrix} 1 & 0 & 2 & 1 \\ 0 & 1 & -2 & 1 \\ 0 & -1 & 2 & -1 \\ 0 & 2 & 12 & 2 \end{bmatrix}, \text{ replacing } R_3, R_4 \text{ by } R_3 - R_2 \text{ and } R_4 + 2R_1 \text{ respectively}$$

$$\sim \begin{bmatrix} 1 & 0 & 2 & 1 \\ 0 & 1 & -2 & 1 \\ 0 & 0 & 0 & 0 \\ 0 & 0 & 16 & 0 \end{bmatrix}, \text{ replacing } R_3 \text{ and } R_4 \text{ by } R_3 + R_2 \text{ and } R_4 - 2R_2 \text{ respectively.}$$

$$\sim \begin{bmatrix} 1 & 0 & 2 & 1 \\ 0 & 1 & 0 & 1 \\ 0 & 0 & 0 & 0 \\ 0 & 0 & 16 & 0 \end{bmatrix}, \text{ replacing } C_2 \text{ by } C_3 + 2C_2$$

$$\sim \begin{bmatrix} 1 & 0 & 0 & 0 \\ 0 & 1 & 0 & 0 \\ 0 & 0 & 0 & 0 \\ 0 & 0 & 16 & 0 \end{bmatrix}, \text{ replacing } C_3, C_4 \text{ by } C_3 - 2C_1, C_4 - C_1 - C_2 \text{ respectively}$$

$$\sim \begin{bmatrix} 1 & 0 & 0 & 0 \\ 0 & 1 & 0 & 0 \\ 0 & 0 & 0 & 0 \\ 0 & 0 & 1 & 0 \end{bmatrix}, \text{ replacing } C_3 \text{ by } \frac{1}{16} C_2$$

$$\sim \begin{bmatrix} 1 & 0 & 0 & 0 \\ 0 & 1 & 0 & 0 \\ 0 & 0 & 1 & 0 \\ 0 & 0 & 0 & 0 \end{bmatrix}, \text{ interchanging } R_3 \text{ and } R_4$$

$$\sim \begin{bmatrix} I_3 & 0 \\ 0 & 0 \end{bmatrix}$$

Hence the rank of the given matrix = 3. **Ans.**

Example 20(a):

Use elementary transformation to reduce the following matrix A to triangular form and hence find the rank of A.

$$A = \begin{bmatrix} 5 & 3 & 14 & 4 \\ 0 & 1 & 2 & 1 \\ 1 & -1 & 2 & 0 \end{bmatrix}$$

Solution:

$$A \sim \begin{bmatrix} 5 & 3 & 8 & 1 \\ 0 & 1 & 0 & 0 \\ 1 & -1 & 4 & 1 \end{bmatrix}, \text{ replacing } C_3, C_4 \text{ by } C_3 - 2C_2, C_4 - C_2 \text{ respectively}$$

$$\Rightarrow \quad A \sim \begin{bmatrix} 5 & 8 & -12 & -4 \\ 0 & 1 & 0 & 0 \\ 1 & 0 & 0 & 0 \end{bmatrix}, \text{ replacing } C_2, C_3, C_4 \text{ by } C_2 + C_1, C_3 - 4C_1, C_4 - C_1$$

$$\sim \begin{bmatrix} 5 & 0 & 0 & -4 \\ 0 & 1 & 0 & 0 \\ 1 & 0 & 0 & 0 \end{bmatrix}, \text{ replacing } C_2, C_3 \text{ by } C_2 + 2C_4, C_3 - 3C_4 \text{ respectively}$$

$$\sim \begin{bmatrix} 5 & 0 & 0 & 1 \\ 0 & 1 & 0 & 0 \\ 1 & 0 & 0 & 0 \end{bmatrix}, \text{ replacing } C_4 \text{ by } -\frac{1}{4} C_4$$

$$\sim \begin{bmatrix} 0 & 0 & 0 & 1 \\ 0 & 1 & 0 & 0 \\ 1 & 0 & 0 & 0 \end{bmatrix}, \text{ replacing } C_1 \text{ by } C_1 - 5C_4$$

$$\sim \begin{bmatrix} 0 & 1 & 0 & 0 \\ 0 & 0 & 0 & 1 \\ 1 & 0 & 0 & 0 \end{bmatrix}, \text{ interchanging } R_1, R_2$$

$$\sim \begin{bmatrix} 0 & 1 & 0 & 0 \\ 1 & 0 & 0 & 0 \\ 0 & 0 & 0 & 1 \end{bmatrix}, \text{ interchanging } R_2, R_3$$

$$\sim \begin{bmatrix} 1 & 0 & 0 & 0 \\ 0 & 1 & 0 & 0 \\ 0 & 0 & 0 & 1 \end{bmatrix}, \text{ interchanging } R_1, R_2$$

$$\sim \begin{bmatrix} 1 & 0 & 0 & 0 \\ 0 & 1 & 0 & 0 \\ 0 & 0 & 1 & 0 \end{bmatrix}, \text{ interchanging } C_3, C_4$$

$$\sim [\, I_3 \; 0 \,]$$

Hence the rank of A is 3. **Ans.**

Example 20(b):

Find the rank of the matrix $A = \begin{bmatrix} 1 & 2 & -1 & 3 \\ 2 & 4 & -4 & 7 \\ -1 & -2 & -1 & 1 \end{bmatrix}$

Solution:

$$A \sim \begin{bmatrix} 1 & 0 & 0 & 0 \\ 2 & 0 & -2 & 1 \\ -1 & 0 & -2 & 5 \end{bmatrix}, \text{ replacing } C_2, C_3, C_4 \text{ by } C_2 - 2C_1,\ C_3 + C_1 \text{ and } C_4 - 3C_1 \text{ respectively}$$

$$\sim \begin{bmatrix} 1 & 0 & 0 & 0 \\ 2 & 0 & -2 & 1 \\ -3 & 0 & 0 & 4 \end{bmatrix}, \text{ replacing } R_3 \text{ by } R_3 - R_2$$

$$\sim \begin{bmatrix} 1 & 0 & 0 & 0 \\ 0 & 0 & -2 & 1 \\ 0 & 0 & 0 & 4 \end{bmatrix}, \text{ replacing } R_2, R_3 \text{ by } R_2 - 2R_1 \text{ and } R_3 + 3R_1 \text{ respectively}$$

$$\sim \begin{bmatrix} 1 & 0 & 0 & 0 \\ 0 & 0 & 1 & 1 \\ 0 & 0 & 0 & 1 \end{bmatrix}, \text{ replacing } C_3 \text{ by } -\frac{1}{2}C_3 \text{ and } R_3 \text{ by } \frac{1}{4}R_3$$

$$\sim \begin{bmatrix} 1 & 0 & 0 & 0 \\ 0 & 0 & 1 & 0 \\ 0 & 0 & 0 & 1 \end{bmatrix}, \text{ replacing } C_4 \text{ by } C_4 - C_3$$

$$\sim \begin{bmatrix} 1 & 0 & 0 & 0 \\ 0 & 1 & 0 & 0 \\ 0 & 0 & 0 & 1 \end{bmatrix}, \text{ interchanging } C_2 \text{ and } C_3$$

$$\sim \begin{bmatrix} 1 & 0 & 0 & 0 \\ 0 & 1 & 0 & 0 \\ 0 & 0 & 1 & 0 \end{bmatrix}, \text{ interchanging } C_3 \text{ and } C_4$$

$\sim [\ I_3\ 0\ ]$ **(Note)**

Hence the rank of A is . **Ans.**

Example 20(c):

Determine by reducing to normal form the rank of the matrix

$$A = \begin{bmatrix} 8 & 1 & 3 & 6 \\ 0 & 3 & 2 & 2 \\ -8 & -1 & -3 & 4 \end{bmatrix}$$

Solution:

$$A \sim \begin{bmatrix} 1 & 1 & 3 & 3 \\ 0 & 3 & 2 & 1 \\ -1 & -1 & -3 & 2 \end{bmatrix}, \text{ replacing } C_1 \text{ by } \frac{1}{8}C_1 \text{ and } C_4 \text{ by } \frac{1}{2}C_4$$

$$\sim \begin{bmatrix} 1 & 0 & 0 & 0 \\ 0 & 3 & 2 & 1 \\ -1 & 0 & 0 & 5 \end{bmatrix}, \text{ replacing } C_2, C_3 \text{ and } C_4 \text{ by } C_2 - C_1, C_3 - 3C_1, C_4 - 3C_1 \text{ respectively.}$$

$$\sim \begin{bmatrix} 1 & 0 & 0 & 0 \\ 0 & 3 & 2 & 1 \\ 0 & 0 & 0 & 5 \end{bmatrix}, \text{ replacing } R_3 \text{ by } R_3 + R_1$$

$$\sim \begin{bmatrix} 1 & 0 & 0 & 0 \\ 0 & 1 & 1 & 1 \\ 0 & 0 & 0 & 5 \end{bmatrix}, \text{ repacing } C_2 \text{ by } \frac{1}{3} C_2 \text{ and } C_2 \text{ by } \frac{1}{2} C_3$$

$$\sim \begin{bmatrix} 1 & 0 & 0 & 0 \\ 0 & 1 & 0 & 0 \\ 0 & 0 & 0 & 5 \end{bmatrix}, \text{ repacing } C_3 \text{ and } C_4 \text{ by } C_3 - C_2 \text{ and } C_4 - C_2 \text{ respectively.}$$

$$\sim \begin{bmatrix} 1 & 0 & 0 & 0 \\ 0 & 1 & 0 & 0 \\ 0 & 0 & 5 & 0 \end{bmatrix}, \text{ int erchanging } C_3 \text{ and } C_4$$

$$\Rightarrow \quad A \sim \begin{bmatrix} 1 & 0 & 0 & 0 \\ 0 & 1 & 0 & 0 \\ 0 & 0 & 1 & 0 \end{bmatrix}, \text{ replacing } C_3 \text{ by } \frac{1}{5} C_3$$

$$\sim [\, I_3 \;\, 0 \,]$$ **(Note)**

Hence the rank of A is . **Ans.**

Example 21:

Find the rank of the matrix $A = \begin{bmatrix} 1 & 1 & 1 \\ 2 & 2 & 2 \\ 3 & 3 & 3 \end{bmatrix}$

Solution:

$$A \sim \begin{bmatrix} 1 & 0 & 0 \\ 2 & 0 & 0 \\ 3 & 0 & 0 \end{bmatrix}, \text{ replacing } C_2, C_3 \text{ by } C_3 - C_1, C_3 - C_1 \text{ respectively}$$

$$\sim \begin{bmatrix} 1 & 0 & 0 \\ 0 & 0 & 0 \\ 0 & 0 & 0 \end{bmatrix}, \text{ replacing } R_2, R_3 \text{ by } R_2 - 2R_1 \;\; R_3 - 3R_1 \text{ respectively}$$

$$\sim \begin{bmatrix} I_1 & 0 \\ 0 & 0 \end{bmatrix}$$

Hence the rank of the matrix A is 1. **Ans.**

Example 22:

Find the rank of the matrix $A = \begin{bmatrix} 0 & 2 & 3 \\ 0 & 4 & 6 \\ 0 & 6 & 9 \end{bmatrix}$

Solution:

$$A \sim \begin{bmatrix} 0 & 2 & 1 \\ 0 & 4 & 2 \\ 0 & 6 & 3 \end{bmatrix}, \text{ replacing } C_3 \text{ by } C_3 - C_2$$

$$\sim \begin{bmatrix} 0 & 0 & 1 \\ 0 & 0 & 2 \\ 0 & 0 & 3 \end{bmatrix}, \text{ replacing } C_2 \text{ by } C_2 - 2C_3$$

$$\sim \begin{bmatrix} 0 & 0 & 1 \\ 0 & 0 & 0 \\ 0 & 0 & 0 \end{bmatrix}, \text{ replacing } R_2, R_3 \text{ by } R_2 - 2R_1, R_3 - 3R_1$$

$$\sim \begin{bmatrix} 1 & 0 & 0 \\ 0 & 0 & 0 \\ 0 & 0 & 0 \end{bmatrix}, \text{ interchanging } C_1 \text{ and } C_2$$

$$\sim \begin{bmatrix} I_1 & 0 \\ 0 & 0 \end{bmatrix}$$

Hence the rank of A is 1. **Ans.**

Example 23:

Reduce the matrix A to its normal form where

$A = \begin{bmatrix} 0 & 1 & 2 & -2 \\ 4 & 0 & 2 & 6 \\ 2 & 1 & 3 & 1 \end{bmatrix}$ *and hence determine its rank.*

Solution:

$$A \sim \begin{bmatrix} 0 & 1 & 0 & 0 \\ 4 & 0 & 2 & 8 \\ 2 & 1 & 1 & 4 \end{bmatrix}, \text{ replacing } C_3, C_4 \text{ by } C_3 - 2C_2 \text{ and } C_4 + C_3 \text{ respectively}$$

$$\sim \begin{bmatrix} 0 & 1 & 0 & 0 \\ 4 & 0 & 2 & 8 \\ 2 & 1 & 1 & 4 \end{bmatrix}, \text{ replacing } R_2 \text{ by } R_2 - 2R_3$$

$$\sim \begin{bmatrix} 0 & 1 & 0 & 0 \\ 0 & 0 & 0 & 0 \\ 2 & 0 & 1 & 4 \end{bmatrix}, \text{ replacing } R_2, R_3 \text{ by } R_2 + 2R_1 \text{ and } R_3 - R_1 \text{ respectively.}$$

$$\sim \begin{bmatrix} 0 & 1 & 0 & 0 \\ 0 & 0 & 0 & 0 \\ 0 & 0 & 1 & 0 \end{bmatrix}, \text{ replacing } C_1, C_4 \text{ by } C_1 - 3C_3, C_4 - 4C_3 \text{ respectively.}$$

$$\sim \begin{bmatrix} 1 & 0 & 0 & 0 \\ 0 & 0 & 0 & 0 \\ 0 & 0 & 1 & 0 \end{bmatrix}, \text{ int erchanging } C_1 \text{ and } C_2$$

$$\sim \begin{bmatrix} 1 & 0 & 0 & 0 \\ 0 & 0 & 0 & 0 \\ 0 & 1 & 0 & 0 \end{bmatrix}, \text{ int erchanging } C_2, C_3$$

$$\sim \begin{bmatrix} 1 & 0 & 0 & 0 \\ 0 & 1 & 0 & 0 \\ 0 & 0 & 0 & 0 \end{bmatrix}, \text{ int erchanging } R_2, R_3$$

$$\sim \begin{bmatrix} I_2 & 0 \\ 0 & 0 \end{bmatrix}$$

Hence the rank of A is 2. **Ans.**

Example 24:

Find the rank of the matrix $A = \begin{bmatrix} -1 & -2 & -1 \\ 6 & 12 & 6 \\ 5 & 10 & 5 \end{bmatrix}$

Solution:

$$A \sim \begin{bmatrix} -1 & 0 & 0 \\ 6 & 0 & 0 \\ 5 & 0 & 0 \end{bmatrix}, \text{ replcaing } C_2, C_3 \text{ by } C_2 - 2C_1, C_3 - C_1 \text{ respectively.}$$

$$\sim \begin{bmatrix} -1 & 0 & 0 \\ 0 & 0 & 0 \\ 0 & 0 & 0 \end{bmatrix},$$ replacing R_2, R_3 by $R_2 + 6R_1$, $R_3 + 5R_1$ respectively

$$\sim \begin{bmatrix} 1 & 0 & 0 \\ 0 & 0 & 0 \\ 0 & 0 & 0 \end{bmatrix},$$ replacing R_1 by $-R_1$

$$\sim \begin{bmatrix} I_1 & 0 \\ 0 & 0 \end{bmatrix}$$

Hence the rank of the matrix A is 2. **Ans.**

5

Transpose Matrix

INTRODUCTION

If $A = [a_{ij}]$ be a matrix of order $m \times n$, then the matrix $B = [b_{ij}]$ of order $n \times m$, such that $b_{ij} = a_{ij}$ is known as transposed matrix of A or the transpose of the matrix A and is denoted by A' or A^t.

If $A = \begin{bmatrix} 1 & 3 & 5 \\ 2 & 4 & 6 \end{bmatrix}$ then $A' = \begin{bmatrix} 1 & 2 \\ 3 & 4 \\ 5 & 6 \end{bmatrix}$

Or

The matrix of order n ´m obtained by interchanging the rows and columns of a matrix A of order m ´n is called the transpose matrix of A or transpose of the matrix A and is denoted by A' or A^t.

SOME IMPORTANT THEOREMS ON TRANSPOSED MATRICES

Theorem 1:

The transpose of the sum of two matrices is the sum of their transpose i.e., **(A + B)' = A' + B.**

Proof:

Let $A = [a_{ij}]$ and $B = [b_{ij}]$.

Then $\mathbf{A + B} = [a_{ij} + b_{ij}] = [c_{ij}]$, say

then $c_{ij} = a_{ij} + b_{ij}$

$\therefore$ $\mathbf{(A + B)'}\ [d_{ji}]$, where $d_{ji} = c_{ij}$ for all $1 \le i \le m$, $1 \le j \le n$

i.e., $d_{ji} = a_{ij} + b_{ij}$, for all $1 \le i \le m$, $1 \le j \le n$

$\Rightarrow$ $\mathbf{(A + B)'} = [c_{ij}] = [a_{ij} + b_{ij}]$

Also $\mathbf{A'} = [f_{ji}]$, where $f_{ji} = a_{ij}$ for all $1 \le i \le m$, $1 \le j \le n$

and $\quad$ **B'** $= [f_{ji}]$, where $g_{ji} = b_{ij}$ for all $1 \le i \le m,\ 1 \le j \le n$

$\therefore \quad$ **A' + B'** $= [f_{ji}] + [g_{ji}] = [f_{ji} + g_{ji}]$

$= [a_{ij} + b_{ij}]$ ...(2)

$\therefore \quad$ From (1) and (2) we get **(A + B)' = A' + B'**ss

Theorem 2(a):

The transpose of the transpose of a matrix is the matrix itself i.e., (A') = **A**.

Proof:

Let **A** = $[a_{ij}]$ be an $m \times n$ matrix. The **A'** *i.e.,* the transpose of **A** is an $n \times m$ matrix and **(A')** *i.e.,* The transpose of **A'** (or the transpose of **A**) is an $m \times n$ matrix.

Therefore, the matrices **A** and **(A')** are both $m \times n$ matrices and hence comparable. ...(1)

Also, the element in the ith row and jth column of **(A')'**

= the element in the jth row and ith column of **A'**

= the element in the jth row and ith column of **A**

i.e.,, the corresponding elements of **(A')** and **A** are equal ...(2)

$\therefore$ From (1) and (2), we conclude that **(A') = A.** **Hence proved.**

Theorem 2(b):

The transpose of the product of two matrices is the product in reverse order of their transpose i.e. **(AB)' = B'A'.**

Proof:

Let A = $[a_{ij}]$ and B = $[b_{ij}]$ be the two matrices of orderes $m \times n$ and $n \times p$ respectively.

Let **C** = **AB** = $[a_{ij}] \times [b_{ij}] = [c_{ij}]$, say,

where **C** is a matrix of order $m \times p$.

$\therefore$ The element in the ith row and jth column of **AB** is $c_{ij} = \sum_{k=1}^{n} a_{jk}\ b_{kj}$.

This is also the element in the ith row and jth column of **(AB)'**. ...(1)

The elements in the jth rwo of **B'**are $b_{1j}, b_{2j}, b_{3j}, \ldots, b_{nj}$ and the elements in the ith column of **A'** are $a_{i1}, a_{i2}, a_{i3}, \ldots, a_{in}$. Then the elements in the jth row and ith column of **B' A'** is

$$\sum_{k=1}^{n} b_{kj}\, a_{ik} = \sum_{k=1}^{n} a_{ik}\, b_{kj} = c_{ij} \qquad ...(2)$$

Hence from (1) and (2) we conclude the **(AB)'** = **B'A'**.

Theorem 3:

*if A is any m × n matrix, then (k**A**)' = k**A'**, where k is any number*

Proof:

Let A = $[a_{ij}]$ be any m × n matrix. Then k**A** is also an m × n matrix and therefore (k**A**)' *i.e.,* the transpose of the matrix k**A** is an n × m matrix.

Also A', the transpose of the matrix A, is an n × m matrix and kA' is also an n × m matrix.

Thus, we find that the matrices (kA)' and kA' are both n × m matrices and hence comparable. ...(1)

Again the element in ith row and jth column of (kA)'

= the element in jth row and ith column of kA

= k times the element in jth row and ith column of A

= k times the element in ith row and jth column of A

= k a_{ji}

= the element in ith row and jth column of kA'

i.e., the corresponding elements of (kA)' and kA' are equal ...(2)

∴ From (1) and (2), we conclude that (kA)' = kA'. **Hence proved.**

Example 1:

If $A = \begin{bmatrix} 2 & 3 \\ 0 & 1 \end{bmatrix}$, $B = \begin{bmatrix} 3 & 4 \\ 2 & 1 \end{bmatrix}$ *then verify that [AB]' = B'A'.*

Solution:

$$AB = \begin{bmatrix} 2 & 3 \\ 0 & 1 \end{bmatrix} \times \begin{bmatrix} 3 & 4 \\ 2 & 1 \end{bmatrix}$$

$$= \begin{bmatrix} 2.3 + 3.2 & 2.4 + 3.1 \\ 0.3 + 1.2 & 0.4 + 1.1 \end{bmatrix} = \begin{bmatrix} 12 & 11 \\ 2 & 1 \end{bmatrix}$$

∴ **[AB]'** = transposed matrix of **AB**

$\Rightarrow \quad [\mathbf{AB}]' = \begin{bmatrix} 12 & 2 \\ 11 & 1 \end{bmatrix}$, by defintion ...(1)

Again $B' = \begin{bmatrix} 3 & 2 \\ 4 & 1 \end{bmatrix}$ and $A' = \begin{bmatrix} 2 & 0 \\ 3 & 1 \end{bmatrix}$

$\therefore \quad \mathbf{B'\,A'} = \begin{bmatrix} 3 & 2 \\ 4 & 1 \end{bmatrix} \times \begin{bmatrix} 2 & 0 \\ 3 & 1 \end{bmatrix}$

$= \begin{bmatrix} 3.2 + 2.3 & 3.0 + 2.1 \\ 4.2 + 1.3 & 4.0 + 1.1 \end{bmatrix} = \begin{bmatrix} 12 & 2 \\ 11 & 1 \end{bmatrix}$

$= [\mathbf{AB}]'$, from (1). **Hence proved.**

Example 2:

If $A = \begin{bmatrix} \cos\alpha & -\sin\alpha \\ -\sin\alpha & \cos\alpha \end{bmatrix}$, *verify that* $AA' = I_2 = A'A$.

Solution:

Here $A' = \begin{bmatrix} \cos\alpha & -\sin\alpha \\ \sin\alpha & \cos\alpha \end{bmatrix}$

$\therefore \quad \mathbf{AA'} = \begin{bmatrix} \cos\alpha & \sin\alpha \\ -\sin\alpha & \cos\alpha \end{bmatrix} \times \begin{bmatrix} \cos\alpha & -\sin\alpha \\ \sin\alpha & \cos\alpha \end{bmatrix}$

$= \begin{bmatrix} \cos^2\alpha + \sin^2\alpha & -\cos\alpha\sin\alpha + \sin\alpha\cos\alpha \\ -\sin\alpha\cos\alpha + \cos\alpha\sin\alpha & \sin^2\alpha + \cos^2\alpha \end{bmatrix}$

$= \begin{bmatrix} 1 & 0 \\ 0 & 1 \end{bmatrix} = \mathbf{I_2}.$

Similarly we can prove that

$\mathbf{AA'} = \begin{bmatrix} \cos\alpha & -\sin\alpha \\ \sin\alpha & \cos\alpha \end{bmatrix} \times \begin{bmatrix} \cos\alpha & \sin\alpha \\ -\sin\alpha & \cos\alpha \end{bmatrix}$

$= \begin{bmatrix} \cos^2\alpha + \sin^2\alpha & \cos\alpha\sin\alpha - \sin\alpha\cos\alpha \\ \sin\alpha\cos\alpha - \cos\alpha\sin\alpha & \sin^2\alpha + \cos^2\alpha \end{bmatrix}$

$= \begin{bmatrix} 1 & 0 \\ 0 & 1 \end{bmatrix} = \mathbf{I_2}.$

Hence $\quad \mathbf{AA' = I_2 = A'A}.$

TRIANGULAR MATRIX

Definition: *A matrix $[a_{ij}]$ is called a triangular matrix if $a_{ij} = 0$ for $i > j$.*

$$\begin{bmatrix} 2 & 3 & 1 & 4 \\ 0 & 1 & 2 & 3 \\ 0 & 0 & 5 & 7 \end{bmatrix} \text{ or } \begin{bmatrix} 2 & 3 & 4 & 5 \\ 0 & 1 & 3 & 4 \\ 0 & 0 & 2 & 5 \\ 0 & 0 & 0 & 7 \end{bmatrix}$$

Note:

1. The elements a_{ij} for which $i \leq j$ are not necessarily zero.
2. Triangular matrix need not be square. If it is square, then it is called upper triangular matrix.

Theorem:

Every matrix can be reduced to triangle form by elementary row operations.

Proof:

We shall prove this theorem by Mathematical Induction.

Assume that this theorem holds for all matrices containing n – 1 rows and let $A = [a_{ij}]$ be an $n \times m$ matrix given below:

$$A = \begin{bmatrix} a_{11} & a_{12} & a_{13} & \cdots & \cdots & \cdots & a_{1m} \\ a_{21} & a_{22} & a_{23} & \cdots & \cdots & \cdots & a_{2m} \\ \cdots & \cdots & \cdots & \cdots & \cdots & \cdots & \cdots \\ \cdots & \cdots & \cdots & \cdots & \cdots & \cdots & \cdots \\ a_{n1} & a_{n2} & a_{n3} & \cdots & \cdots & \cdots & a_{nm} \end{bmatrix}$$

Now the following cases arise:

Case I: If $a_{11} \neq 0$, then replacing R_1 by $(1/a_{11})\, R_1$ (*i.e.,* by applying elementary row operation) the matrix A reduces to an $n \times m$ matrix.

$$B = [b_{ij}] = \begin{bmatrix} b_{11} & b_{12} & \cdots & \cdots & b_{1m} \\ b_{21} & b_{22} & \cdots & \cdots & b_{2m} \\ \cdots & \cdots & \cdots & \cdots & \cdots \\ b_{n1} & b_{n2} & \cdots & \cdots & b_{nm} \end{bmatrix}$$

where $b_{11} = 1$.

Now apply elementary row-operation $R_k - R_{ji} R_1$ to R_k where k = 1, 2,..., n *i.e.*, subtract b_{ki} times R_1 from R_k, where k takes value from 1 to n. This reduces the matrix B to matrix $C = [c_{ij}]$ where $c_{k1} = 0$ whenever k > 1 and we have

$$C = \begin{bmatrix} 1 & c_{12} & c_{13} & \dots & c_{1m} \\ 0 & c_{22} & c_{23} & \dots & c_{2m} \\ \dots & \dots & \dots & \dots & \dots \\ 0 & c_{n2} & c_{n3} & \dots & c_{nm} \end{bmatrix} \qquad \dots(1)$$

Now by our assumption that the theorem which we are going to prove holds for matrices containing (n – 1) rows we find that (n – 1) rowed matrix

$$\begin{bmatrix} 0 & c_{22} & c_{23} & \dots & c_{2m} \\ 0 & c_{32} & c_{33} & \dots & c_{3m} \\ \dots & \dots & \dots & \dots & \dots \\ 0 & c_{n2} & c_{n3} & \dots & c_{nm} \end{bmatrix}$$

can always be reduced to triangular form by elementary row operations and hence from (1) the matrix C will reduce to triangular form when the same elementary row operations are applied to c.

Case II: If $a_{11} = 0$ but $a_{k1} \neq 0$ for some value of k then interchanging R_1 and R_k the matrix **A** reduces to the matrix $\mathbf{D} = [d_{ij}]$ where $d_{11} \neq 0$.

Then the matrix D can always be reduced to the triangular from as in case **I** above.

Case III: If $a_{k1} = 0$ for all values of k, then we have

$$A = \begin{bmatrix} 0 & a_{12} & a_{13} & \dots & a_{1m} \\ 0 & a_{22} & a_{23} & \dots & a_{2m} \\ \dots & \dots & \dots & \dots & \dots \\ 0 & a_{n2} & a_{n3} & \dots & a_{nm} \end{bmatrix}$$

By hypothesis (inductive) then (n – 1) rowed matrix

$$\begin{bmatrix} 0 & a_{22} & a_{23} & \dots & a_{2m} \\ \dots & \dots & \dots & \dots & \dots \\ 0 & a_{n2} & a_{n3} & \dots & a_{nm} \end{bmatrix}$$

as in case I above can be reduced to triangular form by elementary row operations and the same elementary operations when applied on A will reduce A to triangular form.

Hence the matrix A can always be reduced to triangular form and the proof is complete by mathematical induction.

Example 1:

Reduce $\begin{bmatrix} -1 & 2 & 1 & 8 \\ 2 & 1 & -1 & 0 \\ 3 & 2 & 1 & 7 \end{bmatrix}$ *to triangular form.*

Solution:

Let A = $\begin{bmatrix} -1 & 2 & 1 & 8 \\ 2 & 1 & -1 & 0 \\ 3 & 2 & 1 & 7 \end{bmatrix}$

$\sim \begin{bmatrix} 1 & -2 & -1 & -8 \\ 2 & 1 & -1 & 0 \\ 3 & 2 & 1 & 7 \end{bmatrix}$, replacing R_1 by $-R_1$

$\sim \begin{bmatrix} 1 & -2 & -1 & -8 \\ 0 & 5 & 1 & 16 \\ 0 & 8 & 4 & 31 \end{bmatrix}$, replacing R_1 by $R_2 - 2R_1$ and R_3 by $R_3 - 3R_1$

$\sim \begin{bmatrix} 1 & -2 & -1 & -8 \\ 0 & 5 & 1 & 16 \\ 0 & 40 & 20 & 155 \end{bmatrix}$, replacing R_3 by $5R_3$

$\sim \begin{bmatrix} 1 & -2 & -1 & -8 \\ 0 & 5 & 1 & 16 \\ 0 & 0 & 12 & 27 \end{bmatrix}$, replacing R_3 by $R_3 - 8R_2$

This is the required triangular form.

Example 2:

Reduce the matrix $\begin{bmatrix} 3 & 1 & 4 \\ 1 & 2 & -5 \\ 0 & 1 & 2 \end{bmatrix}$ *to triangular form.*

Solution:

Let A = $\begin{bmatrix} 3 & 1 & 4 \\ 1 & 2 & -5 \\ 0 & 1 & 2 \end{bmatrix}$

$$\sim \begin{bmatrix} 1 & \frac{1}{3} & \frac{4}{3} \\ 1 & 2 & -5 \\ 0 & 1 & 2 \end{bmatrix}, \text{ replacing } R_1 \text{ by } \frac{1}{3} R_1$$

$$\sim \begin{bmatrix} 1 & \frac{1}{3} & \frac{4}{8} \\ 1 & \frac{5}{8} & -\frac{19}{8} \\ 0 & 1 & 2 \end{bmatrix}, \text{ replacing } R_1 \text{ by } \frac{1}{3} R_1$$

$$\sim \begin{bmatrix} 0 & \frac{1}{3} & \frac{4}{8} \\ 0 & \frac{5}{8} & -\frac{19}{8} \\ 0 & 0 & \frac{29}{5} \end{bmatrix}, \text{ replacing } R_3 \text{ by } R_3 - \frac{3}{5} R_2$$

This is the required triangular form.

Aliter. $A = \begin{bmatrix} 3 & 1 & 4 \\ 1 & 2 & -5 \\ 0 & 1 & 2 \end{bmatrix}$

$$\sim \begin{bmatrix} 1 & 2 & -5 \\ 3 & 1 & 4 \\ 0 & 1 & 2 \end{bmatrix}, \text{ int erchanging } R_1 \text{ and } R_2$$

$$\sim \begin{bmatrix} 1 & 2 & -5 \\ 0 & -5 & 19 \\ 0 & 1 & 2 \end{bmatrix}, \text{ replacing } R_2 \text{ by } R_2 - 3R_1$$

$$\sim \begin{bmatrix} 1 & 2 & -5 \\ 0 & -5 & 19 \\ 0 & 5 & 10 \end{bmatrix}, \text{ replacing } R_3 \text{ by } 5R_3$$

$$\sim \begin{bmatrix} 1 & 2 & -5 \\ 0 & -5 & 19 \\ 0 & 0 & 29 \end{bmatrix}, \text{ replacing } R_3 \text{ by } R_3 + R_2$$

This is also a triangular matrix

ORDER OF A MINOR

Definition: *If any r rows and any r columns from an m × n matrix A are retained and remaining (m – r) rows and (n – r) columns removed, then the determinant of the remaining r × r sub-matrix of A is called minor of A of order r.*

In the matrix $\begin{bmatrix} a_{11} & a_{12} & a_{13} & a_{14} \\ a_{21} & a_{22} & a_{23} & a_{24} \\ a_{31} & a_{32} & a_{33} & a_{34} \\ a_{41} & a_{42} & a_{43} & a_{44} \\ a_{51} & a_{52} & a_{53} & a_{54} \end{bmatrix}$

element a_{11}, a_{13}, a_{31} etc. are minors of order unity;

$$\begin{vmatrix} a_{11} & a_{12} \\ a_{21} & a_{22} \end{vmatrix}, \begin{vmatrix} a_{11} & a_{13} \\ a_{21} & a_{23} \end{vmatrix}, \begin{vmatrix} a_{33} & a_{34} \\ a_{53} & a_{54} \end{vmatrix} \text{ etc.}$$

are minors of order 2;

$$\begin{vmatrix} a_{11} & a_{12} & a_{13} \\ a_{21} & a_{22} & a_{23} \\ a_{31} & a_{32} & a_{33} \end{vmatrix}, \begin{vmatrix} a_{21} & a_{23} & a_{24} \\ a_{41} & a_{43} & a_{44} \\ a_{51} & a_{53} & a_{54} \end{vmatrix} \text{ etc.}$$

$$\begin{vmatrix} a_{11} & a_{12} & a_{13} & a_{14} \\ a_{21} & a_{22} & a_{23} & a_{24} \\ a_{41} & a_{42} & a_{43} & a_{44} \\ a_{51} & a_{52} & a_{53} & a_{54} \end{vmatrix}, \begin{vmatrix} a_{21} & a_{22} & a_{23} & a_{24} \\ a_{31} & a_{32} & a_{33} & a_{34} \\ a_{41} & a_{42} & a_{43} & a_{44} \\ a_{51} & a_{52} & a_{53} & a_{54} \end{vmatrix} \text{ etc.}$$

are minors of order 4.

COMPLEX CONJUGATE (OR CONJUGATE) OF A MATRIX

Definition: The matrix obtained from any given matrix A of order m × n with complex elements a_{ij} by replaclng its elements by the corresponding conjugate complex numbers is called the complex conjugate or conjugate of **A** and is denoted by $\overline{A}$ and read as '**A** conjugate'.

⇒ if A = $[a_{ij}]$ and a_{ij} is the complex conjugate of the element a_{ij} then $\overline{A}$ = $[a_{ij}]$, for all $1 \le i \le m$, $1 \le j \le n$.

For exmaple: If **A** = $\begin{bmatrix} 1+i & 2+3i \\ 2 & 3i \end{bmatrix}$, then $$\overline{A} = \begin{bmatrix} 1+i & 2+3i \\ 2 & 3i \end{bmatrix}$$

Rea Matrix

Definition: **A** matrix **A** is called real provided it satisfles the relation

$$A = \overline{A}.$$

Imaginary Matrix

Definition: *A m,atrix* **A** *is called imaginary provided it satisfies the relation*

$$A = -\overline{A}.$$

THEOREMS ON COMPLEX CONJUGATE OF A MATRIX

Theorem 1:

If **A** = $[a_{ij}]$ *be any m × n matrix and* **B** = $[b_{ij}]$ *be any n × p matrix i.e., if* **A** *and* **B** *are conformable to the product* **AB**, *then* $\overline{AB} = \overline{A} + \overline{B}$.

Proof:

Since **A** and **B** are conformable to the product **AB,** so **AB** = $[a_{ij}] \times [b_{jk}]$ = $[c_{ij}]$, where $c_{ik} = a_{ij}\, b_{jk}$, for all $1 \le i \le m$, $1 \le k \le p$ and there is summation on j where j = 1, 2, 3,...,n.

Also $\overline{A} = [\overline{a}_{ij}]$, for all $1 \le i \le m$, $1 \le j \le n$

and $\overline{B} = [\overline{b}_{jk}]$, for all $1 \le j \le n$, $1 \le k \le p$

$\overline{A}\,\overline{B}$ is defined and we have

$\overline{A}\,\overline{B}\,[\overline{a}_{ij}] \times [\overline{b}_{jk}] = [d_{ik}]$ ...(1)

where $d_{ik} = \overline{a}_{ij}\,\overline{b}_{jk}$ for all $1 \le i \le m$, $1 \le k \le p$ and j = 1, 2,...,n.

Again $\overline{AB}$ = complex conjugate of **AB** *i.e.,* $[c_{ik}]$

$= [\overline{c}_{ik}]$, where $c_{ik} = a_{ij}\, b_{jk}$

$= [\overline{a_{ij}\, b_{jk}}] = [\overline{a}_{ij}\,\overline{b}_{jk}]$ $\because \overline{z_1 z_2} = \overline{z}_1\,\overline{z}_2$, for any complex numbers z_1 and z_2

$= [d_{ik}]$, since $d_{ik} = \overline{a}_{ij}\,\overline{b}_{jk}$ for all $1 \le i \le m$,

$1 \le k \le p$ and j = 1, 2,..., n ...(2)

$\therefore$ From (1) and (2), we conclude that $\overline{\mathbf{A}\,\mathbf{B}} = \overline{\mathbf{A}}\,\overline{\mathbf{B}}$.

Theorem 2:

If $\mathbf{A} = [a_{ij}]$ *be any* $m \times n$ *matrix with complex elements* a_{ij}, *then* $\overline{\lambda\,\mathbf{A}} = \lambda\overline{\mathbf{A}}$.

Proof:

By defintion, we know

$\overline{\mathbf{A}}\,[\bar{a}_{ij}]$, for all $1 \le i \le m$, $1 \le j \le n$ and $\bar{a}_{ij}$ is the complex conjugate of a_{ij}.

Also $\lambda A = [\lambda a_{ij}]$, for all $1 \le i \le m$, $1 \le j \le n$

$\therefore\ \overline{\lambda\,\mathbf{A}} = [\overline{\lambda\ a_{ij}}] = [\overline{\lambda\ a_{ij}}]$ for all $1 \le i \le m$, $1 \le j \le n$ ...(1)

and we know that $\overline{z_1\ z_2} = \bar{z}_1\ \bar{z}_2$, where z_2, z_2 are any two complex numbers.

Again $\bar{\lambda}\,\overline{\mathbf{A}} = [b_{ij}]$, where b_{ij}

$= \overline{\lambda\ a_{ij}}$ for all $1 \le i \le m$, $1 \le j \le n$

$= [\bar{\lambda}\,\bar{a}_{ij}]$, for all $1 \le i \le m$, $1 \le j \le n$. ...(2)

$\therefore$ From (1) and (2) we conclude that the corresponding elements of $\overline{\lambda\mathbf{A}}$ and $\bar{\lambda}\,\overline{\mathbf{A}}$ are equal. Also it is evident that $\overline{\lambda\mathbf{A}}$ and $\bar{\lambda}\,\overline{\mathbf{A}}$ are matrices of the same order. Hence, we conclude that

$$\overline{\lambda\mathbf{A}} = \bar{\lambda}\,\overline{\mathbf{A}}.$$

Theorem 3:

If $A = [a_{ij}]$ *be any* $m \times n$ *matrix with complex elements* a_{ij}, *then the complex conjugate of* $\overline{\mathbf{A}}$ *is the matrix A itself.*

Proof:

We know that $\overline{\mathbf{A}} = [a_{ij}]$, for all $1 \le i \le m$, $1 \le j \le n$ and a_{ij} is the complex conjugate of a_{ij}.

i.e., the element in the ith row and jth column of complex conjugate of a *i.e.,* $\overline{\mathbf{A}}$.

= the complex conjugate of the element in ith row and jth column of A

∴ The element in ith row and jth column of the complex conjugate of $\overline{\mathbf{A}}$ *i.e.,* $\overline{\overline{\mathbf{A}}}$

= the complex conjugate of the element in ith row and jth column of $\overline{\mathbf{A}}$

= the complex conjuate of a_{ij}

= a_{ij} *i.e.,* the element in ith row and jth column of **A**

i.e., the corresponding elements of **A** and the complex conjugate of $\overline{\mathbf{A}}$ are equal. ...(1)

Also it is evident that **A**, $\overline{\mathbf{A}}$ and its complex conjugate are m × n matrices and hence comparable. ...(2)

∴ From (1) and (2), we conclude that the complex conjugate of $\overline{\mathbf{A}}$ is equal to **A** or $\overline{\overline{\mathbf{A}}} = \mathbf{A}$.

Theorem 4:

If **A** *and* **B** *are two matrices conformable to addition, then* $\overline{\mathbf{A}+\mathbf{B}} = \overline{\mathbf{A}} + \overline{\mathbf{B}}$.

Proof:

Let A = $[a_{ij}]$ and B = $[b_{ij}]$ be any two matrices of order m × n. Then as these matrices are given as conformable to addition, so we have

$\mathbf{A} + \mathbf{B} = [a_{ij} + b_{ij}]$, for all $1 \le i \le m$, $1 \le j \le n$. ...(1)

Also $\overline{\mathbf{A}} = [a_{ij}]$ and $\overline{\mathbf{B}} = [b_{ij}]$, by definition.

$\therefore\ \overline{\mathbf{A}} + \overline{\mathbf{B}} = \left[\bar{a}_{ij} + \bar{b}_{ij}\right] = \left[\overline{a_{ij}} + \overline{b_{ij}}\right]$, ...(2)

for all $1 \le i \le m$, $1 \le j \le n$.

and also a $\bar{z}_j + \bar{z}_2 = \overline{z + z_2}$, where z_1, z_2

are any two complex numbers.

Again from (1), we have

$\overline{\mathbf{A}+\mathbf{B}}$ = complex conjugate of $[a_{ij} + b_{ij}]$

= complex conjugate of $[c_{ij}]$, were $c_{ij} = a_{ij} + b_{ij}$

= $[\bar{c}_{ij}] = [\overline{a_{ij} + b_{ij}}]$, for all $1 \le i \le m$, $1 \le j \le n$

i.e., $\overline{\mathbf{A}+\mathbf{B}} = [\overline{a_{ij} + b_{ij}}]$, for all $1 \le i \le m$, $1 \le j \le n$ ...(3)

∴ From (2) and (3) we conclude that the corresponding elements of $\overline{\mathbf{A}} + \overline{\mathbf{B}}$ and $\overline{\mathbf{A} + \mathbf{B}}$ are matrices of order m × **n** as **A** and **B** are given as conformable to addition. Hence we conclude that $\overline{\mathbf{A} + \mathbf{B}} = \overline{\mathbf{A}} + \overline{\mathbf{B}}$.

EQUIVALENT MATRICES (GENERAL DEFINITION)

Definition: *Two n × n matrices **A** and **B** are called equivalent if one can be obtained from the other by a finite number of row and column operations (or elementary operation) and written as **B** ~ **A**.*

TRIANGULAR ROW OPERATIONS

Let us consider the matrices

$$A = \begin{bmatrix} 1 & 2 & 3 \\ 4 & 5 & 6 \\ 7 & 8 & 9 \end{bmatrix}, B = \begin{bmatrix} 4 & 5 & 6 \\ 1 & 2 & 3 \\ 7 & 8 & 9 \end{bmatrix},$$

$$C = \begin{bmatrix} 3 & 6 & 9 \\ 4 & 5 & 6 \\ 7 & 8 & 9 \end{bmatrix}, D = \begin{bmatrix} 1 & 2 & 3 \\ 6 & 9 & 12 \\ 7 & 8 & 9 \end{bmatrix}$$

Here we observe that the matrices B, C, D are related to the matrix A in as much as:

(a) **B** can be obtained from **A** by interchanging first and second rows of **A**;

(b) **C** can be obtained from **A** by multiplying the first row of **A** by 3 and

(c) **D** can be obtained from **A** by adding two times the first row to the second row of **A**.

Such operations on the rows of a matrix are known as elementary row operations. Formal definition is given below:

Definition: *Let A_i denote the ith row of the matrix A - $[a_{ij}]$, then the elementary row operations on the matrix A are defined as:*

1. the interchanging of any two rows A_i and A_j (*i.e.,* ith and jth rows). The symbols R_{ij} or $R_i \leftrightarrow R_j$ are generally employed for this operation.
2. the multiplication of every element of A_i by a non-zero scalar c *i.e.,* replacing the ith rwo A_i by cA_i. The symbols R_i (c) pr $R_i \to cR_j$ are employed for this operation.
3. the addition to the elements of row A_i of c (a scalar) times the corresponding elements of the row A_k *i.e.,* replacing the row A_i by $A_i + cA_k$.

The symbols R_{ik} (c) or $R_i \rightarrow R_j + cR_k$ are used for this operation.

Example:

Let $A = \begin{bmatrix} 1 & 2 & 3 \\ 3 & 4 & 5 \\ 5 & 6 & 7 \end{bmatrix}$

The effect of the elementary rwo operation $A_2 - A_1$ or R_{21} (-1) is to product the matrix.

$$B = \begin{bmatrix} 1 & 2 & 3 \\ 3-1 & 4-2 & 5-3 \\ 5 & 6 & 7 \end{bmatrix} = \begin{bmatrix} 1 & 2 & 3 \\ 2 & 2 & 2 \\ 5 & 6 & 7 \end{bmatrix}$$

Again the effect of elementary row operation $B_2 + B_1$ or R_{21} (1) is to produce the matrix

$$\begin{bmatrix} 1 & 3 & 3 \\ 2+1 & 2+2 & 2+3 \\ 5 & 6 & 7 \end{bmatrix} = \begin{bmatrix} 1 & 2 & 3 \\ 3 & 4 & 5 \\ 5 & 6 & 7 \end{bmatrix} \text{ i.e. the matrix A}$$

Thus the above two operations are the inverse elementary row operations.

ROW EQUIVALENT MATRICES

Definition: *If an m × n matrix* ***B*** *can be obtained from and m × n matrix A by a finite number of elementary row operations, then B is called the row equivalent to* ***A*** *and is written as*

$$\mathbf{B} \overset{\text{row}}{\sim} \mathbf{A}.$$

Note: Equivalent matrices have the same order.

Example:

$$\begin{bmatrix} 1 & 3 & 4 & 7 \\ 2 & -3 & 5 & 6 \\ 1 & 0 & 3 & 2 \end{bmatrix} \overset{\text{row}}{\sim} \begin{bmatrix} 2 & -3 & 5 & 6 \\ 1 & 3 & 4 & 7 \\ 1 & 0 & 3 & 2 \end{bmatrix}$$

(interchanging first and second rows).

ELEMENTARY ROW MATRIX

Definition: *The matrix obtained by the application of one elementary row operation to the identity matrix I_n is called an elementary row matrix.*

Example:

Examples of elementary matrices obtained from I_3, where

$$I_3 = \begin{bmatrix} 1 & 0 & 0 \\ 0 & 1 & 0 \\ 0 & 0 & 1 \end{bmatrix}$$

1. $$I_3 \sim \begin{bmatrix} 0 & 1 & 0 \\ 1 & 0 & 0 \\ 0 & 0 & 1 \end{bmatrix} = E_a \text{ (say)},$$

obtained by interchanging first two rows.

2. $$I_3 \sim \begin{bmatrix} 1 & 0 & 0 \\ 0 & c & 0 \\ 0 & 0 & 1 \end{bmatrix} = E_b \text{ (say)},$$

obtained by multiplying the elements of second row by c.

3. $$I_3 \sim \begin{bmatrix} 1 & 2 & 0 \\ 0 & 1 & 0 \\ 0 & 0 & 1 \end{bmatrix} = E_c \text{ (say)},$$

obtained by adding two times the elements of second row to the corresponding elements of first row *i.e.*, replacing Λ_1 by Λ_1 + 2A2 *i.e.*,, R_{12} (2).

Theorem:

Each elementary row operation on an m × n matrix can be effected by premultiplying it by the corresponding elementary matrix.

Example:

$$\text{Let } A = \begin{bmatrix} a_{11} & a_{12} & a_{13} & a_{14} \\ a_{21} & a_{22} & a_{23} & a_{24} \\ a_{31} & a_{32} & a_{33} & a_{34} \end{bmatrix}$$

1. Interchanging the first and third rows, we have

$$A \sim \begin{bmatrix} a_{31} & a_{32} & a_{33} & a_{34} \\ a_{21} & a_{22} & a_{23} & a_{24} \\ a_{11} & a_{12} & a_{13} & a_{14} \end{bmatrix}$$

The corresponding elementary matrix (obtained by interchanging first and third rows of I_3) is given by

$$E_{13} = \begin{bmatrix} 0 & 0 & 1 \\ 0 & 1 & 0 \\ 1 & 0 & 0 \end{bmatrix}$$

[Here students should not that as we are to premultiply A therefore the number of columns of E_{13} should be 3, the number of rows of A].

$$\text{Now } E_{13}A = \begin{bmatrix} 0 & 0 & 1 \\ 0 & 1 & 0 \\ 1 & 0 & 0 \end{bmatrix} \times \begin{bmatrix} a_{11} & a_{12} & a_{13} & a_{14} \\ a_{21} & a_{22} & a_{23} & a_{24} \\ a_{31} & a_{32} & a_{33} & a_{34} \end{bmatrix}$$

$$= \begin{bmatrix} a_{31} & a_{32} & a_{33} & a_{34} \\ a_{21} & a_{22} & a_{23} & a_{24} \\ a_{11} & a_{12} & a_{13} & a_{14} \end{bmatrix}$$

This shows that **B** can be obtained from **A** by pre-multiplying it by E_{13}, the corresponding elementary matrix.

2. Let $A = \begin{bmatrix} 1 & 2 & 3 \\ 4 & 5 & 6 \\ 7 & 8 & 9 \end{bmatrix}$

Multiplying the elements of second row by 2, we get

$$A \sim \begin{bmatrix} 1 & 2 & 3 \\ 8 & 10 & 12 \\ 7 & 8 & 9 \end{bmatrix} = B \text{ (say)}$$

The corresponding elementary matrix (obtained by multiplying the elements of second row of I_3 by)2 is given by

$$E_2(2) = \begin{bmatrix} 1 & 0 & 0 \\ 0 & 2 & 0 \\ 0 & 0 & 1 \end{bmatrix}$$

$$\begin{bmatrix} 1 & 0 & 0 \\ 0 & 2 & 0 \\ 0 & 0 & 1 \end{bmatrix} \times \begin{bmatrix} 1 & 2 & 3 \\ 4 & 5 & 6 \\ 7 & 8 & 9 \end{bmatrix}$$

$$= \begin{bmatrix} 1.1 + 0.4 + 0.7 & 1.2 + 0.5 + 0.8 & 1.3 + 0.6 + 0.9 \\ 0.1 + 2.4 + 0.7 & 0.2 + 2.5 + 0.8 & 0.3 + 2.6 + 0.9 \\ 0.1 + 0.4 + 1.7 & 0.2 + 0.5 + 1.8 & 0.3 + 0.6 + 1.9 \end{bmatrix}$$

$$= \begin{bmatrix} 1 & 2 & 3 \\ 8 & 10 & 12 \\ 7 & 8 & 9 \end{bmatrix} = B$$

i.e.,, B can be obtained from A by pre-multiplying it by E_3 (2).

3. Let $A = \begin{bmatrix} 1 & -2 & 3 \\ -3 & 4 & 5 \\ 5 & 6 & -7 \end{bmatrix}$

Replacing A_1 by $A_1 + 2A_2$ i.e, adding two times the elements of second row to the corresponding elements of first row, we get

$$A \sim \begin{bmatrix} -5 & 2 & 13 \\ -3 & 4 & 5 \\ 5 & 6 & -7 \end{bmatrix} = B \text{ (say)}$$

The corresponding elementary matrix (obtained by adding two times the elements of second row of I_3 to the corresponding elements of the first row) is given by

$$E_{12}(2) = \begin{bmatrix} 1 & 2 & 0 \\ 0 & 1 & 0 \\ 0 & 0 & 1 \end{bmatrix}$$

Then $E_{12}(2) \times A = \begin{bmatrix} 1 & 2 & 0 \\ 0 & 1 & 0 \\ 0 & 0 & 1 \end{bmatrix} \times \begin{bmatrix} 1 & -2 & 3 \\ -3 & 4 & 5 \\ 5 & 6 & -7 \end{bmatrix}$

$$= \begin{bmatrix} 1.1 + 2(-3) + 0.5 & 1.(-2) + 2.4 + 0.6 & 1.3 + 2.5 + 0.(-7) \\ 0.1 + 1(-3) + 0.5 & 0.(-2) + 1.4 + 0.6 & 0.3 + 1.5 + 0.(-7) \\ 0.1 + 0(-3) + 1.5 & 0.(-2) + 0.4 + 1.6 & 0.3 + 0.5 + 1.(-7) \end{bmatrix}$$

$$= \begin{bmatrix} -5 & 6 & 13 \\ -3 & 4 & 5 \\ 5 & 6 & -7 \end{bmatrix} = B$$

i.e., **B** can be obtained from A by pre-multiplying it by E_{12} (2).

TYPES OF ELEMENTARY ROW MATRICES AND THEIR SYMBOLS

1. E_{ij} denotes the elementary matrix obtained by interchanging the ith and jth rows (or columns) of an identity (or unit) matrix.
2. E_i (c) denotes the elementary matrix obtained by multiplying the ith row (or column) of the identity matrix by c.
3. E_{ik} (c) denotes the elementary matrix obtained by adding to the elements of the ith row of the identity matrix c times the corresponding elements of the kth row.
4. E'_{ik} (c) denotes the transpose of E_{ik} (c) and can be obtained by adding to the elements of the ith column of the identity matrix c times the corresponding elements of the kth column.

Theorem 1:

The elementary matrix E_{ij} E_i (c), E_{jk} (1) are non-singular.

Proof:

1. The elementary matrix $\mathbf{E}_{ij}$ is obtained by interchanging the ith and jth rows of **I**. We shall get back **I** if we now apply the same row operation upon E_{ij} which can also be effected by pro-multiplying E_{ij} by E_{ij}

   $\therefore \quad \mathbf{E}_{ij}\ \mathbf{E}_{ij} = \mathbf{I}.$

   i.e., $\mathbf{E}_{ij}$ is its own inverse *i.e.,*, $\mathbf{E}_{ij}$ is non-singular.

2. The elementary matrix $\mathbf{E}_i$ (c) is obtained by multiplying the ith row of the identity matrix by c (where $c \neq 0$). We shall get back **I** if we now multiply the elements of ith row of $\mathbf{E}_i$ (c) by 1/c which can also be effected by pre-multiplying $\mathbf{E}_i$ (c) with the corresponding elementary matrix which is obtained from **I** by multiplying its ith row by 1/c, which is therefore the inverse of $\mathbf{E}_i$ (c).

   For example, let $I = \begin{bmatrix} 1 & 0 & 0 \\ 0 & 1 & 0 \\ 0 & 0 & 1 \end{bmatrix}$ and $E_3\ (c) = \begin{bmatrix} 1 & 0 & 0 \\ 0 & 1 & 0 \\ 0 & 0 & c \end{bmatrix}$

   Then $\{E_3\ (c)\}^{-1} = \begin{bmatrix} 1 & 0 & 0 \\ 0 & 1 & 0 \\ 0 & 0 & 1/c \end{bmatrix}$, where $\{E_3\ (c)\}^{-1}$ is the inverse of E_3 (c)

3. The elementary matrix E_{jk} (1) obtained from **I** by replacing its jth row by (jth row + kth row).

We shall get back **I** if we now replace the jth row of E_{ji} (1) by (jth row –kth row).

Hence the inverse of $\mathbf{E}_{jk}$ (1) is the elementary matrix obtained from **I** by replacing its jth row by (jth row –kth row)

Let $\mathbf{I} = \begin{bmatrix} 1 & 0 & 0 \\ 0 & 1 & 0 \\ 0 & 0 & 1 \end{bmatrix}$ and $E_{13}\ (1) = \begin{bmatrix} 1 & 0 & 1 \\ 0 & 1 & 0 \\ 0 & 0 & 1 \end{bmatrix}$ obtained from **I** by replacing its lst row by (lst row + 3rd row).

Then $\{E_{13}\ (1)\}^{-1} = \begin{bmatrix} 1 & 0 & -1 \\ 0 & 0 & 1 \\ 0 & 0 & 1 \end{bmatrix}$, obtained from I by replacing its first row by (lst row – 3rd row).

Theorem 2:

If the matrix **B** *is row equivalent to the matrix* **A** *the* **B** = **SA**, *where* **S** *is non-singular.*

We know that

$$\text{if } \mathbf{B} \overset{\text{row}}{\sim} \mathbf{A} \text{ then } \mathbf{B} = \mathbf{SA}$$

where **S** is the product of the elementary matrices and we have proved that elementary matrices are non-singular and hence their product is also non-singular.

This proves the above theorem.

Theorem 3:

If **a** *square matrix* **A** *of order n is row equivalent to the identity matrix* $\mathbf{I}_n$ *then* **A** *is non-singular.*

Proof:

We know that

$\mathbf{A} \overset{\text{row}}{\sim} \mathbf{I}_n$ then $\mathbf{A} = \mathbf{S}.\mathbf{I}_n$, where **S** is non-singular.

Now S I_n being the product of two non-singular matrices is non-singular. Therefore A is non-singular.

Note. The converse of this theorem is also true.

Theorem 4:

If a sequence of row operations applied to a square matrix **A** *reduces it to the identity matrix I, then the same sequence of row operations applied to the identity matrix gives the inverse of* **A** *(i.e.* $\mathbf{A^{-1}}$*).*

Proof:

We know that **SA** = **I**, where **S** is the product of the elementary matrices.

i.e., $(E_k \ldots E3.\ E2.\ E1)\ A = I$, where E_i denotes the elementary matrices.

$\Rightarrow$ $(E_k \ldots E_3.E_2.E_1)\ AA^{-1} = IA^{-1}$

$\Rightarrow$ $(E_k \ldots E_3.E_2.E_1)\ I = A^{-1}$, since $A^{-1} = I$ and $IA^{-1} = A^{-1}$

Hence the theorem.

In the following examples we shall show the successive matrices row equivalent to **A** and **I** in the left hand and right hand columns respectively. When ultimately A is reduced to **I** in the left hand column, I is reduced to A^{-1} in the right hand column.

Also R_1, R_2, R_3,... etc, stand for first row, second row, third row etc.

Example 1:

Find the inverse of the matrix $A = \begin{bmatrix} 1 & 2 & 2 \\ -1 & 3 & 0 \\ 0 & -2 & 1 \end{bmatrix}$

Solution:

$$\overset{A}{\begin{bmatrix} 1 & 2 & -2 \\ -1 & 3 & 0 \\ 0 & -2 & 1 \end{bmatrix}} \sim \overset{I}{\begin{bmatrix} 1 & 0 & 0 \\ 0 & 1 & 0 \\ 0 & 0 & 1 \end{bmatrix}}$$

$$\sim \begin{bmatrix} 1 & -2 & 0 \\ -1 & 3 & 0 \\ 0 & -2 & 1 \end{bmatrix} \sim \begin{bmatrix} 1 & 0 & 2 \\ 0 & 1 & 0 \\ 0 & 0 & 1 \end{bmatrix}$$

(Replacing R_1 by $R_1 + 2R_3$)

$$\sim \begin{bmatrix} 1 & -2 & 0 \\ -1 & 1 & 0 \\ 0 & -2 & 1 \end{bmatrix} \sim \begin{bmatrix} 1 & 0 & 2 \\ 1 & 1 & 2 \\ 0 & 0 & 1 \end{bmatrix}$$

(Replacing R_2 by $R_2 + R_1$)

$$\sim \begin{bmatrix} 1 & 0 & 0 \\ 0 & 1 & 0 \\ 0 & 0 & 1 \end{bmatrix} \Bigg\| \sim \begin{bmatrix} 3 & 2 & 6 \\ 1 & 1 & 2 \\ 2 & 2 & 5 \end{bmatrix}$$

(Replacing R_1 by $R_1 + 2R_2$ and R_3 by $R_3 + 2R_2$)

$= I \quad | \quad = A^{-1}$

$$\therefore A^{-1} = \begin{bmatrix} 3 & 2 & 6 \\ 1 & 1 & 2 \\ 2 & 2 & 5 \end{bmatrix}$$ **Ans.**

Example 2(a):

$$A = \begin{bmatrix} 1 & 2 & 1 \\ 3 & 2 & 3 \\ 1 & 1 & 2 \end{bmatrix}, \textit{evaluate } A^{-1}.$$

Solution:

$$\overset{A}{\begin{bmatrix} 1 & 2 & 1 \\ 3 & 2 & 3 \\ 1 & 1 & 2 \end{bmatrix}} \Bigg\| \sim \overset{I}{\begin{bmatrix} 1 & 0 & 0 \\ 0 & 1 & 0 \\ 0 & 0 & 1 \end{bmatrix}}$$

$$\sim \begin{bmatrix} 1 & 2 & 1 \\ 0 & -4 & 0 \\ 0 & -1 & 1 \end{bmatrix} \Bigg\| \sim \begin{bmatrix} 1 & 0 & 0 \\ -3 & 1 & 0 \\ -1 & 0 & 1 \end{bmatrix}$$

(Replacing R_2 by $R_2 - 3R_1$ and R_4 by $R_3 - R_1$)

$$\sim \begin{bmatrix} 1 & 3 & 0 \\ 0 & 1 & 0 \\ 0 & -1 & 1 \end{bmatrix} \Bigg\| \sim \begin{bmatrix} 2 & 0 & -1 \\ \frac{3}{4} & -\frac{1}{4} & 0 \\ -1 & 0 & 1 \end{bmatrix}$$

(Replacing R_1 by $R_1 - R_3$ and R_2 by $- 1/4R_2$)

$$\sim \begin{bmatrix} 1 & 0 & 0 \\ 0 & 1 & 0 \\ 0 & 0 & 1 \end{bmatrix} \Bigg\| \sim \begin{bmatrix} -\frac{1}{4} & \frac{3}{4} & -1 \\ \frac{3}{4} & -\frac{1}{4} & 0 \\ -\frac{1}{4} & \frac{1}{4} & 1 \end{bmatrix}$$

(Replacing R_1 by $R_1 - 2R_2$ and R_3 by $R_3 + R_2$)

$= I \quad | \quad = A^{-1}$

$$\therefore A^{-1} \begin{bmatrix} -\frac{1}{4} & \frac{3}{4} & -1 \\ \frac{3}{4} & -\frac{1}{4} & 0 \\ -\frac{1}{4} & \frac{1}{4} & 1 \end{bmatrix}$$

Example 2(b):

Find the inverse of the matrix $A = \begin{bmatrix} 1 & -3 & 2 \\ 2 & 0 & 0 \\ 1 & 4 & 1 \end{bmatrix}$

Solution:

$$\overset{A}{\begin{bmatrix} 1 & -3 & 2 \\ 2 & 0 & 0 \\ 1 & 4 & 1 \end{bmatrix}} \Bigg\| \sim \overset{I}{\begin{bmatrix} 0 & 0 & 0 \\ 0 & 1 & 0 \\ 0 & 0 & 1 \end{bmatrix}}$$

$$\sim \begin{bmatrix} 1 & -3 & 2 \\ 1 & 0 & 0 \\ 1 & 4 & 1 \end{bmatrix} \Bigg\| \sim \begin{bmatrix} 1 & 0 & 0 \\ 0 & \frac{1}{2} & 0 \\ 0 & 0 & 1 \end{bmatrix}$$

(Replacing R_2 by $\frac{1}{2}$ R_2)

$$\sim \begin{bmatrix} 1 & 0 & 0 \\ 1 & 0 & 2 \\ 1 & 4 & 1 \end{bmatrix} \Bigg\| \sim \begin{bmatrix} 0 & \frac{1}{2} & 0 \\ 1 & 0 & 0 \\ 0 & 0 & 1 \end{bmatrix}$$

(Interchanging R_1 and R_2)

$$\sim \begin{bmatrix} 1 & 0 & 0 \\ 0 & -3 & 2 \\ 0 & 4 & 1 \end{bmatrix} \Bigg\| \sim \begin{bmatrix} 0 & \frac{1}{2} & 0 \\ 1 & -\frac{1}{2} & 0 \\ 0 & -\frac{1}{2} & 1 \end{bmatrix}$$

(Replacing R_2 by $R_2 - R_1$ and R_3 by $R_3 - R_1$)

$$\sim \left[\begin{array}{ccc|ccc} 1 & 0 & 0 & 0 & \frac{1}{2} & 0 \\ 0 & -11 & 0 & 1 & \frac{1}{2} & -2 \\ 0 & 4 & 1 & 0 & -\frac{1}{2} & 1 \end{array}\right]$$

(Replacing R_2 by $R_2 - 2 R_2$)

$$\sim \left[\begin{array}{ccc|ccc} 1 & 0 & 0 & 0 & \frac{1}{2} & 0 \\ 0 & 1 & 0 & -\frac{1}{11} & -\frac{1}{22} & \frac{2}{11} \\ 0 & 4 & 1 & 0 & -\frac{1}{2} & 1 \end{array}\right]$$

(Replacing R_2 by $-\frac{1}{11} R_2$)

$$\sim \left[\begin{array}{ccc|ccc} 1 & 0 & 0 & 0 & \frac{1}{2} & 0 \\ 0 & 1 & 0 & -\frac{1}{11} & -\frac{1}{22} & \frac{2}{11} \\ 0 & 0 & 1 & \frac{4}{11} & -\frac{1}{2} & 1 \end{array}\right]$$

(Replacing R_3 by $R_3 - 4 R_2$)

$= I \quad | \quad = A^{-2}$

$$\therefore A^{-1} = \begin{bmatrix} 0 & \frac{1}{2} & 0 \\ -\frac{1}{11} & -\frac{1}{22} & \frac{2}{11} \\ \frac{4}{11} & -\frac{7}{22} & \frac{2}{11} \end{bmatrix}$$

Ans.

Example 3:

Find the inverse of the matrix $A = \begin{bmatrix} 1 & 2 & 3 \\ 3 & 1 & 2 \\ 0 & 1 & 2 \end{bmatrix}$

Solution:

$$\begin{matrix} & A & & I \\ \sim & \begin{bmatrix} 1 & 2 & 1 \\ 3 & 1 & 2 \\ 0 & 1 & 2 \end{bmatrix} & \sim & \begin{bmatrix} 1 & 0 & 0 \\ 0 & 1 & 0 \\ 0 & 0 & 1 \end{bmatrix} \end{matrix}$$

$$\sim \begin{bmatrix} 1 & 2 & 1 \\ 3 & -5 & -1 \\ 0 & 1 & 2 \end{bmatrix} \sim \begin{bmatrix} 1 & 0 & 0 \\ -3 & 1 & 0 \\ 0 & 0 & 1 \end{bmatrix}$$

(Replacing R_2 by $R_2 - 3\,R_1$)

$$\sim \begin{bmatrix} 1 & 2 & 1 \\ 0 & 1 & 11 \\ 0 & 1 & 2 \end{bmatrix} \sim \begin{bmatrix} 1 & 0 & 0 \\ -3 & 1 & 6 \\ 0 & 0 & 1 \end{bmatrix}$$

(Replacing R_2 by $R_2 + 6\,R_3$)

$$\sim \begin{bmatrix} 1 & 0 & -21 \\ 0 & 1 & 11 \\ 0 & 1 & 2 \end{bmatrix} \sim \begin{bmatrix} 7 & -2 & -12 \\ -3 & 1 & 6 \\ 0 & 0 & 1 \end{bmatrix}$$

(Replacing R_1 by $R_1 - 2R_2$)

$$\sim \begin{bmatrix} 1 & 0 & -21 \\ 0 & 1 & 11 \\ 0 & 0 & -9 \end{bmatrix} \sim \begin{bmatrix} 7 & -2 & -12 \\ -3 & 1 & 6 \\ 3 & -1 & -5 \end{bmatrix}$$

(Replacing R_3 by $R_3 - R_2$)

$$\sim \begin{bmatrix} 1 & 0 & -21 \\ 0 & 1 & 11 \\ 0 & 0 & 1 \end{bmatrix} \sim \begin{bmatrix} 7 & -2 & -12 \\ -3 & 1 & 6 \\ -\frac{1}{3} & \frac{1}{9} & \frac{5}{9} \end{bmatrix}$$

(Replacing R_3 by $-\frac{1}{9}\,R_3$).

$$\sim \begin{bmatrix} 1 & 0 & 0 \\ c & 1 & 11 \\ 0 & 0 & 1 \end{bmatrix} \sim \begin{bmatrix} 0 & \frac{1}{2} & -\frac{1}{3} \\ -3 & 1 & 6 \\ -\frac{1}{3} & \frac{1}{9} & \frac{5}{9} \end{bmatrix}$$

(Replacing R_1 by $R_1 + 21R_3$)

$$\sim \begin{bmatrix} 1 & 0 & 0 \\ 0 & 1 & 0 \\ 0 & 0 & 1 \end{bmatrix} \sim \begin{bmatrix} 0 & \frac{1}{2} & -\frac{1}{3} \\ \frac{2}{3} & -\frac{2}{9} & -\frac{1}{9} \\ -\frac{1}{3} & \frac{1}{9} & \frac{5}{9} \end{bmatrix}$$

(Replacing R_2 by $R_2 - 11R_3$)

$= I \qquad | \qquad = A^{-1}$

Therefore $A^{-1} = \begin{bmatrix} 0 & \frac{1}{2} & -\frac{1}{3} \\ \frac{2}{3} & -\frac{2}{9} & -\frac{1}{9} \\ -\frac{1}{3} & \frac{1}{9} & \frac{5}{9} \end{bmatrix}$ **Ans.**

TRANSPOSED CONJUGATE OF A MATRIX

Definition: *The transpose of conjugate of a matrix A i.e.,* $(\overline{A})'$ *is defined as transposed conjugate or tranjugate of A and is denoted by* A^{Θ} *i.e.,,* $A^{\Theta} = (\overline{A})$.

For example: If $A = \begin{bmatrix} 1+i & 2+3i \\ 2 & 3i \end{bmatrix}$

then $\overline{A} = \begin{bmatrix} 1-i & 2-3i \\ 2 & -3i \end{bmatrix}$

$\therefore A^{\Theta}$ = transpose of $\overline{A} = (\overline{A})'$

$$= \begin{bmatrix} 1-i & 2 \\ 2-3i & -3i \end{bmatrix}$$

THEOREMS ON TRANSPOSED CONJUGATE OF A MATRIX

Theorem 1:

For an matrix **A**, $(A^{\Theta})^{\Theta} = A$.

Proof:

Let $A^{\Theta} = \mathbf{B}$ *i.e.,* $\mathbf{B} = (\overline{\mathbf{A}})'$

Then $\mathbf{B}'$ = tanspose of $\mathbf{B}$

= transpose of $(\overline{\mathbf{A}})'$

= $\overline{\mathbf{A}}$, since we know $(\mathbf{A}')' = A$

$\therefore$ $(\overline{\mathbf{B}})$ = complex conjugate of $\mathbf{B}'$

= complex conjugate of $\overline{\mathbf{A}}$

= $\mathbf{A}$, since we know $\overline{\overline{\mathbf{A}}} = \mathbf{A}$

i.e., $B^{\Theta} = \mathbf{A}$, since $B^{\Theta} = (\overline{\mathbf{B}}) = (\overline{\mathbf{B}})'$

i.e., $(A^{\Theta})^{\Theta} = \mathbf{A}$, since $A^{\Theta} = \mathbf{B}$. **Hence proved.**

ELEMENTARY COLUMN OPERATION AND COLUMN EQUIVALENT MATRICES

Here the letter R in the symbols are to be replaced by C e.g. C_{ij}, C_i (c), C_{ik} (c), where C_{ij} stands for the interchange of ith and jth columns etc. or $C_i \leftrightarrow C_j$; $C_i \to c\, C_i$, $C_i \to C_i + c\, C_k$.

Theorem 1:

If there the two m × n matrices A and B, then $B \overset{col}{\sim} A$ *if B = AT, where T is an n × n non-singular matrix.*

Proof:

If $B \overset{col}{\sim} A$, then $B' \overset{row}{\sim} A'$, where $\mathbf{B}'$ and $\mathbf{A}'$ are the transposed matrices of $\mathbf{B}$ and $\mathbf{A}$ respectively.

Therefore $\mathbf{B}' = \mathbf{SA}'$, where $\mathbf{S}$ is an n × n non-singular matrix.

Consequently $\mathbf{B} = \mathbf{AS}'$, where $\mathbf{S}'$ is the transposed matrix of $\mathbf{S}$.

= $\mathbf{AT}$, where $\mathbf{T} = \mathbf{S}'$, an n × n-singular matrix.

Theorem 2:

Each elementary column operation on an m × n matrix A can be effected by post multiplying A by then n × n matrix obtained from the n × n identity matrix I_n by the same elementary column operation.

Proof:

If $B \overset{col}{\sim} A$,

then $B' \overset{row}{\sim} A'$ where **B'** and **A'** are the transposed matrices of **B** and **A**.

Since if **B** is obtained from **A** by elementary column operation, then **B'** can be obtained from **A'** by an elementary row operation.

Hence **B' = EA'**, where **E** is the elementary matrix obtained from $\mathbf{I_n}$ by an elementary row operation.

Therefore **B = A E'**, where **E'**, the transposed matrix of E, can be obtained from I_n by the same elementary column operation.

Hence the theorem

SOLVED EXAMPLES

Example 1:

Write down the transpose of the matrix.

$$A = \begin{bmatrix} 1 & 2 & 4 \\ 6 & 8 & 1 \end{bmatrix}$$

Solution:

Let A' be the required transpose of the matrix A. then A' = matrix obtained by interchanging the rows and columns of the matrix A

$$= \begin{bmatrix} 1 & 6 \\ 2 & 8 \\ 4 & 1 \end{bmatrix}$$

Ans.

Example 2:

Find the inverse of the matrix $A = \begin{bmatrix} 1 & 2 & -1 \\ -1 & 1 & 2 \\ 2 & -1 & 1 \end{bmatrix}$

Solution:

$$\overset{A}{\begin{bmatrix} 1 & 2 & -1 \\ -1 & 1 & 2 \\ 2 & -1 & 1 \end{bmatrix}} \Bigg\| \sim \overset{I}{\begin{bmatrix} 1 & 0 & 0 \\ 0 & 1 & 0 \\ 0 & 0 & 1 \end{bmatrix}}$$

$$\sim \begin{bmatrix} 1 & 2 & -1 \\ 0 & 3 & 1 \\ 0 & -5 & 3 \end{bmatrix} \sim \begin{bmatrix} 1 & 0 & 0 \\ 1 & 1 & 0 \\ -2 & 0 & 1 \end{bmatrix}$$

(Replacing R_2 by $R_2 + R_1$ and R_3 by $R_3 - 2R_1$)

$$\sim \begin{bmatrix} 1 & 5 & 0 \\ 0 & 3 & 1 \\ 0 & -11 & 1 \end{bmatrix} \sim \begin{bmatrix} 2 & 1 & 0 \\ 1 & 1 & 0 \\ -4 & -2 & 1 \end{bmatrix}$$

(Replacing R_1 by $R_1 + R_2$ and R_3 by $R_3 - 2R_2$)

$$\sim \begin{bmatrix} 1 & 5 & 0 \\ 0 & 14 & 0 \\ 0 & -11 & 1 \end{bmatrix} \sim \begin{bmatrix} 2 & 1 & 0 \\ 5 & 3 & -1 \\ -4 & -2 & 1 \end{bmatrix}$$

(Replacing R_2 by $R_2 - R_3$),

$$\sim \begin{bmatrix} 1 & 5 & 0 \\ 0 & 1 & 0 \\ 0 & -11 & 1 \end{bmatrix} \sim \begin{bmatrix} 2 & 1 & 0 \\ \frac{5}{14} & \frac{3}{14} & -\frac{1}{14} \\ -4 & -2 & 1 \end{bmatrix}$$

(Replacing R_2 by $\frac{1}{14}$ R_2)

$$\sim \begin{bmatrix} 1 & 0 & 0 \\ 0 & 1 & 0 \\ 0 & 0 & 1 \end{bmatrix} = I \begin{bmatrix} \frac{3}{14} & -\frac{1}{14} & \frac{5}{14} \\ \frac{5}{14} & \frac{3}{14} & -\frac{1}{14} \\ -\frac{1}{14} & \frac{5}{14} & \frac{3}{14} \end{bmatrix} = A^{-1}$$

(Replacing R_1 by $R_1 - 5R_2$ and R_3 by $R_3 + 11R_2$)

Therefore $A^{-1} = \begin{bmatrix} \frac{3}{14} & -\frac{1}{14} & \frac{5}{14} \\ \frac{5}{14} & \frac{3}{14} & -\frac{1}{14} \\ -\frac{1}{14} & \frac{5}{14} & \frac{3}{14} \end{bmatrix} \Rightarrow \frac{1}{14} \begin{bmatrix} 3 & -1 & 5 \\ 5 & 3 & -1 \\ -1 & 5 & 3 \end{bmatrix}$ **Ans.**

Example 3(a):

Evaluate the inverses of the following elementary matrices of order four: $E_3(-2)$, $E_{23}(4)$.

Solution:

The identity matrix of order four is given by

$$I_4 = \begin{bmatrix} 1 & 0 & 0 & 0 \\ 0 & 1 & 0 & 0 \\ 0 & 0 & 1 & 0 \\ 0 & 0 & 0 & 1 \end{bmatrix}$$

1. Then $E_3(-2) = \begin{bmatrix} 1 & 0 & 0 & 0 \\ 0 & 1 & 0 & 0 \\ 0 & 0 & -2 & 0 \\ 0 & 0 & 0 & 1 \end{bmatrix}$, replacing R_3 by $-2R_3$

$\therefore$ Inverse of $E_3(-2)$ *i.e.*, $\{E_3(-2)\}^{-1} = \begin{bmatrix} 1 & 0 & 0 & 0 \\ 0 & 1 & 0 & 0 \\ 0 & 0 & -\frac{1}{2} & 0 \\ 0 & 0 & 0 & 1 \end{bmatrix}$ **Ans.**

(obtained by replacing R_3 of I_4 by $-\frac{1}{2} R_3$)

2. $E_{23}(4) = \begin{bmatrix} 1 & 0 & 0 & 0 \\ 0 & 1 & 4 & 0 \\ 0 & 0 & 1 & 0 \\ 0 & 0 & 0 & 1 \end{bmatrix}$, replacing R_2 of I_4 by $R_2 + 4R_3$

Then the inverse of $E_{23}(4)$ *i.e.*, $\{E_{23}(4)\}^{-1}$ is given by

$\begin{bmatrix} 1 & 0 & 0 & 0 \\ 0 & 1 & -4 & 0 \\ 0 & 0 & 1 & 0 \\ 0 & 0 & 0 & 1 \end{bmatrix}$, replacing R_2 of I_4 by $R_2 - 4R_3$

Example 3(b):

Apply successively the row transformations (or operation R_{23}, $R_3(-2)$ and

R_{13} *(4) to the matrix* $\begin{bmatrix} 3 & 1 & 2 & 1 \\ 2 & 0 & 3 & 2 \\ 1 & 2 & 3 & 4 \\ 3 & 1 & 4 & 1 \end{bmatrix}$

Solution:

1. Applying R_{23} operation to the given matrix we have

$$\begin{bmatrix} 3 & 1 & 2 & 1 \\ 1 & 2 & 3 & 4 \\ 2 & 0 & 3 & 2 \\ 3 & 1 & 4 & 1 \end{bmatrix}$$ [Here we have interchanged second and third rows]

2. Applying $R_3(-2)$ operation to the given matrix we have

$$\begin{bmatrix} 3 & 1 & 2 & 1 \\ 2 & 0 & 3 & 2 \\ -2 & -4 & -6 & -8 \\ 2 & 1 & 4 & 1 \end{bmatrix}$$ [Here we have replaced third row R_3 by $-2R_3$]

3. Applying R_{13} (4) operation to the given matrix we have

$$\begin{bmatrix} 7 & 9 & 14 & 17 \\ 2 & 0 & 3 & 2 \\ 1 & 2 & 3 & 4 \\ 3 & 1 & 4 & 1 \end{bmatrix}$$ [Here we have replaced the first row R_1 by $R_1 + 4R_3$]

Example 3(c):

Find the inverse of the matrix $A = \begin{bmatrix} i & -1 & 2i \\ 2 & 0 & 2 \\ -1 & 0 & 1 \end{bmatrix}$

Solution:

$$\overset{A}{\begin{bmatrix} i & -1 & 2i \\ 2 & 0 & 2 \\ -1 & 0 & 1 \end{bmatrix}} \sim \overset{I}{\begin{bmatrix} 1 & 0 & 0 \\ 0 & 1 & 0 \\ 0 & 0 & 1 \end{bmatrix}}$$

$$\sim\begin{bmatrix} i & -1 & 2i \\ 1 & 0 & 1 \\ -1 & 0 & 1 \end{bmatrix} \Bigg| \sim \begin{bmatrix} 1 & 0 & 0 \\ 0 & 1/2 & 0 \\ 0 & 0 & 1 \end{bmatrix}$$

(Replacing R_2 by $\frac{1}{2}$ R_2)

$$\sim\begin{bmatrix} i & -1 & 2i \\ 1 & 0 & 1 \\ 0 & 0 & 2 \end{bmatrix} \Bigg| \sim \begin{bmatrix} 1 & 0 & 0 \\ 0 & 1/2 & 0 \\ 0 & 1/2 & 1 \end{bmatrix}$$

(Replacing R_3 by $R_3 + R_2$)

$$\sim\begin{bmatrix} i & -1 & 2i \\ 1 & 0 & 1 \\ 0 & 0 & 1 \end{bmatrix} \Bigg| \sim \begin{bmatrix} 1 & 0 & 0 \\ 0 & 1/4 & 0 \\ 0 & 1/4 & 1/2 \end{bmatrix}$$

(Replacing R_3 by 1/2 R_3)

$$\sim\begin{bmatrix} i & -1 & 2i \\ 1 & 0 & 0 \\ 0 & 0 & 1 \end{bmatrix} \Bigg| \sim \begin{bmatrix} 1 & 0 & 0 \\ 0 & 1/2 & -1/2 \\ 0 & 1/4 & 1/2 \end{bmatrix}$$

(Replacing R_2 by $R_2 - R_3$)

$$\sim\begin{bmatrix} i & -1 & 0 \\ 1 & 0 & 0 \\ 0 & 0 & 1 \end{bmatrix} \Bigg| \sim \begin{bmatrix} 1 & -\frac{3}{4}i & -\frac{1}{2}i \\ 0 & 1/4 & -1/2 \\ 0 & 1/4 & 1/2 \end{bmatrix}$$

(Replacing R_1 by $R_1 - i\,R_2 - 2i\,R_2$)

$$\sim\begin{bmatrix} 0 & 1 & 0 \\ 1 & 0 & 0 \\ 0 & 0 & 1 \end{bmatrix} \Bigg| \sim \begin{bmatrix} -1 & \frac{3}{4}i & \frac{1}{2}i \\ 0 & 1/4 & -1/2 \\ 0 & 1/4 & 1/2 \end{bmatrix}$$

(Replacing R_1 by $- R_1$)

$$\sim\begin{bmatrix} 1 & 0 & 0 \\ 0 & 1 & 0 \\ 0 & 0 & 1 \end{bmatrix} = I \Bigg| \sim \begin{bmatrix} 0 & 1/4 & -1/2 \\ -1 & \frac{3}{4} & \frac{1}{2}i \\ 0 & 1/4 & 1/2 \end{bmatrix} = A^{-1}$$

(Interchanging R_1 and R_2)

$$\text{Therefore } A^{-1} = \begin{bmatrix} 0 & \frac{1}{4} & -\frac{1}{2} \\ -1 & \frac{3}{4} & \frac{1}{2} \\ 0 & \frac{1}{4} & \frac{1}{2} \end{bmatrix}$$

Ans.

Example 4:

Verify that (B)ᵗ (A)ᵗ = (AB)ᵗ, when

$$A = \begin{bmatrix} 1 & 1 & 2 \\ 2 & 1 & 0 \end{bmatrix} \text{ and } B = \begin{bmatrix} 1 & 2 \\ 2 & 0 \\ -1 & 1 \end{bmatrix}$$

Solution:

$$AB = \begin{bmatrix} 1 & 1 & 2 \\ 2 & 1 & 0 \end{bmatrix} \times \begin{bmatrix} 1 & 2 \\ 2 & 0 \\ -1 & 1 \end{bmatrix}$$

$$= \begin{bmatrix} 1.1 + 1.2 + 2.(-1) & 1.2 + 1.0 + 2.1 \\ 2.1 + 1.2 + 0.(-1) & 2.2 + 1.0 + 0.1 \end{bmatrix} = \begin{bmatrix} 1 & 4 \\ 4 & 4 \end{bmatrix}$$

$\therefore$ $(\mathbf{AB})^t$ = transposed matrix of **AB**

$$= \begin{bmatrix} 1 & 4 \\ 4 & 4 \end{bmatrix}, \text{ by definition} \qquad ...(1)$$

Again $(\mathbf{B})^t = \begin{bmatrix} 1 & 2 & -1 \\ 2 & 0 & 1 \end{bmatrix}$ and $(\mathbf{A})^t = \begin{bmatrix} 1 & 2 \\ 1 & 1 \\ 2 & 0 \end{bmatrix}$

$$\therefore \mathbf{B}^t\,\mathbf{A}^t = \begin{bmatrix} 1 & 2 & -1 \\ 2 & 0 & 1 \end{bmatrix} \times \begin{bmatrix} 1 & 2 \\ 1 & 1 \\ 2 & 0 \end{bmatrix}$$

$$= \begin{bmatrix} 1.1 + 2.1 - 1.2 & 1.2 + 2.1 - 1.0 \\ 2.1 + 0.1 + 1.2 & 2.2 + 0.1 + 1.0 \end{bmatrix}$$

$$= \begin{bmatrix} 1 & 4 \\ 4 & 4 \end{bmatrix} = (\mathbf{AB})^t, \text{ from (1).}$$

Hence proved.

Example 5:

Verify that $(AB)^t = B^t A^t$, *where*

$$A = \begin{bmatrix} 1 & 1 & 1 \\ 2 & 2 & 3 \\ 2 & 4 & 9 \end{bmatrix} \text{ and } B = \begin{bmatrix} 0 & 1 & -1 \\ -3 & 2 & 4 \\ 1 & 1 & 0 \end{bmatrix}$$

Solution:

$$AB = \begin{bmatrix} 1 & 1 & 1 \\ 2 & 2 & 3 \\ 2 & 4 & 9 \end{bmatrix} \times \begin{bmatrix} 0 & 1 & -1 \\ -3 & 2 & 4 \\ 1 & 1 & 0 \end{bmatrix}$$

$$= \begin{bmatrix} 1.0 + 1.(-3) + 1.1 & 1.1 + 1.2 + 1.1 & 1.(-1) + 1.4 + 1.0 \\ 2.0 + 2.(-3) + 3.1 & 2.1 + 2.2 + 3.1 & 2.(-1) + 2.4 + 3.0 \\ 2.0 + 4.(-3) + 9.1 & 2.1 + 4.2 + 9.1 & 2.(-1) + 4.4 + 9.0 \end{bmatrix}$$

$$= \begin{bmatrix} -2 & 4 & 3 \\ -3 & 9 & 6 \\ -3 & 19 & 14 \end{bmatrix}$$

$\therefore$ **$(AB)^t$** = Transposed matrix of **(AB)**

$\Rightarrow$ **$(AB)^t$** $= \begin{bmatrix} -2 & -3 & -3 \\ 3 & 9 & 19 \\ 3 & 6 & 14 \end{bmatrix}$, by defintion

Again $A^t = \begin{bmatrix} 1 & 2 & 2 \\ 1 & 2 & 4 \\ 1 & 3 & 9 \end{bmatrix}$, $B^t = \begin{bmatrix} 0 & -3 & 1 \\ 1 & 2 & 1 \\ -1 & 4 & 0 \end{bmatrix}$

$$\therefore \quad \mathbf{B^tA^t} = \begin{bmatrix} 0 & -3 & 1 \\ 1 & 2 & 1 \\ -1 & 4 & 0 \end{bmatrix} \times \begin{bmatrix} 1 & 2 & 2 \\ 1 & 2 & 4 \\ 1 & 3 & 9 \end{bmatrix}$$

$$= \begin{bmatrix} 0.1 - 3.1 + 1.1 & 0.2 - 3.2 + 1.3 & 0.2 - 3.4 + 1.9 \\ 1.1 + 2.1 + 1.1 & 1.2 + 2.2 + 1.3 & 1.2 + 2.4 + 1.9 \\ -1.1 + 4.1 + 0.1 & -1.2 + 4.2 + 0.3 & -1.2 + 4.4 + 0.9 \end{bmatrix}$$

$$= \begin{bmatrix} -2 & -3 & -3 \\ 4 & 9 & 19 \\ 3 & 6 & 14 \end{bmatrix} = (\mathbf{AB})^t, \text{ from (1)}$$

Hence proved.

Example 6:

Compute the following elementary matrices of order 4 E_{23}, E_2 (4), E_{34} (-2), E'_{34} (–2).

Solution:

$$I_4 = \begin{bmatrix} 1 & 0 & 0 & 0 \\ 0 & 1 & 0 & 0 \\ 0 & 0 & 1 & 0 \\ 0 & 0 & 0 & 1 \end{bmatrix}$$

1. $E_{23} = \begin{bmatrix} 1 & 0 & 0 & 0 \\ 0 & 0 & 1 & 0 \\ 0 & 1 & 0 & 0 \\ 0 & 0 & 0 & 1 \end{bmatrix}$ [Interchanging R_2 and R_3 or C_2 and C_3. (Note]

2. $E_2\ (4) = \begin{bmatrix} 1 & 0 & 0 & 0 \\ 0 & 4 & 0 & 0 \\ 0 & 0 & 1 & 0 \\ 0 & 0 & 0 & 1 \end{bmatrix}$ [Replacing R_3 by $4R_2$ or C_2 by $4C_2$]

3. $E_{34}\ (-2) = \begin{bmatrix} 1 & 0 & 0 & 0 \\ 0 & 1 & 0 & 0 \\ 0 & 0 & 1 & -2 \\ 0 & 0 & 0 & 1 \end{bmatrix}$ [Replacing R_3 by $R_3 - 2R_4$]

4. $E'_{34}\ (-2) = \begin{bmatrix} 1 & 0 & 0 & 0 \\ 0 & 1 & 0 & 0 \\ 0 & 0 & 1 & 0 \\ 0 & 0 & -2 & 1 \end{bmatrix}$ [Replacing C_3 by $C_3 - 2C_4$]

Here students should note that E'_{34} (–2) is nothing but the transpose of E_{34} (–2).

Example 7(a):

Apple successively the column operations C_{13} : C_2 (-4) and C_{23} (–2) to the matrix

$$\begin{bmatrix} 1 & -1 & 2 & 3 & 4 \\ 2 & 1 & -2 & 1 & 3 \\ 3 & 2 & 1 & -2 & 5 \\ 4 & 5 & 6 & 7 & 8 \end{bmatrix}$$

Solution:

1. Applying C_{13} operation to the given matrix, we have

$$\begin{bmatrix} 2 & -1 & 1 & 3 & 4 \\ -2 & 1 & 2 & 1 & 3 \\ 1 & 2 & 3 & -2 & 5 \\ 6 & 5 & 4 & 7 & 8 \end{bmatrix}$$ [Here we have int erchanged C_1 and C_3 i.e. first and third columns].

2. Applying C_2 (–4) operation to the given matrix, we have

$$\begin{bmatrix} 1 & 4 & 2 & 3 & 4 \\ 2 & -4 & -2 & 1 & 3 \\ 3 & -8 & 1 & -2 & 5 \\ 4 & -20 & 6 & 7 & 8 \end{bmatrix}$$ [Here we have replaced sec ond columns C_2 by $-4C_2$]

3. Applying $C_{23}(-2)$ operation to the given matrix, we have

$$\begin{bmatrix} 1 & -5 & 2 & 3 & 4 \\ 2 & 5 & -2 & 1 & 3 \\ 3 & 0 & 1 & -2 & 5 \\ 4 & -7 & 6 & 7 & 8 \end{bmatrix}$$ [Here we have replaced sec ond columns C_2 by $C_2 - 2C_3$]

EXERCISES

1. *Apply the column operation $C_3(4)$ and $C_{12}(-3)$ to the matrix*

$$\begin{bmatrix} 0 & 1 & 2 & 3 & 4 \\ 1 & 2 & 3 & 4 & 0 \\ 3 & 4 & 0 & 1 & 2 \\ 2 & 0 & 1 & 3 & 4 \end{bmatrix}$$

2. *Compute E_{23}, $E_2(-2)$ and $E_{34}(-1)$ for the identity matrix of order 4.*

Ans. $\begin{bmatrix} 1 & 0 & 0 & 0 \\ 0 & 0 & 1 & 0 \\ 0 & 1 & 0 & 0 \\ 0 & 0 & 0 & 1 \end{bmatrix}$; $\begin{bmatrix} 1 & 0 & 0 & 0 \\ 0 & -2 & 0 & 0 \\ 0 & 0 & 1 & 0 \\ 0 & 1 & 0 & 1 \end{bmatrix}$; $\begin{bmatrix} 1 & 0 & 0 & 0 \\ 0 & 1 & 0 & 0 \\ 0 & 0 & 1 & -1 \\ 0 & 0 & 0 & 1 \end{bmatrix}$

3. *Evaluate the inverses of the following elementary matrices of order 4:* E_{14}, $E_4(3)$, $E_{23}(2)$.

4. *Find the inverse of the matrix*

$$A = \begin{bmatrix} 1 & 1 & 1 & 1 \\ 1 & 2 & 3 & -4 \\ 2 & 3 & 5 & -5 \\ 3 & -4 & -5 & 8 \end{bmatrix}$$

Ans. $\frac{1}{18}\begin{bmatrix} 2 & 16 & 6 & 4 \\ 22 & 41 & -30 & -1 \\ -10 & -44 & 30 & -2 \\ 4 & -13 & 6 & -1 \end{bmatrix}$

5. *Find the inverse of the matrix* $\begin{bmatrix} 1 & 3 & 3 & 2 & 1 \\ 1 & 4 & 3 & 3 & -1 \\ 1 & 3 & 4 & 1 & 1 \\ 1 & 1 & 1 & 1 & -1 \\ 1 & -2 & -1 & 2 & 2 \end{bmatrix}$

6. *Has the following matrix an inverse?*

$$\begin{bmatrix} 2 & 1 & 3 & 1 \\ 1 & 2 & -1 & 4 \\ 3 & 3 & 2 & 5 \\ 1 & -1 & 4 & -3 \end{bmatrix}$$

7. *Find A^{-1} if* $A = \begin{bmatrix} 2 & 4 & 3 \\ 0 & 1 & 1 \\ 2 & 2 & -1 \end{bmatrix}$

8. *Find the reciprocal matrix of* $\begin{bmatrix} 1 & 1 & 1 \\ 2 & 2 & 3 \\ 1 & 4 & 9 \end{bmatrix}$

9. *Reduce the matrix* $\begin{bmatrix} 1 & -1 & 1 \\ 2 & 3 & 4 \\ 3 & -1 & 4 \end{bmatrix}$ *to the triangular form.*

10. *Apply the row operations $R_4(-3)$ and $R_{21}(4)$ to the matrix*

$$\begin{bmatrix} 4 & -1 & 2 & 3 \\ -1 & 8 & -3 & -4 \\ 2 & 3 & 4 & -1 \\ -3 & -4 & -1 & 8 \end{bmatrix}$$

6

Integration of Irrational Algebraic Fractions

INTRODUCTION TO INTEGRATION

The problems rationalisation is brought about by multiplying a similar quantity both in numerator and denominator. Sometimes this quantity may differ in sign.

INTEGRATION OF $\dfrac{1}{\left\{\left(Ax^2+B\right)\sqrt{\left(Cx^2+D\right)}\right\}}$

Here we put $x = \dfrac{1}{t}$;

so that $dx = -\left(\dfrac{1}{t^2}\right)dt$.

Then $\displaystyle\int \frac{dx}{\left(Ax^2+B\right)\sqrt{\left(Cx^2+D\right)}} = \int \frac{-\left(\frac{1}{t^2}\right)dt}{\left(\frac{A}{t^2}+B\right)\sqrt{\left(\frac{C}{t^2}+D\right)}}$

$$= -\int \frac{t\,dt}{\left(A+Bt^2\right)\sqrt{\left(C+Dt^2\right)}}.$$

Now substituting $C + Dt^2 = u^2$

or $2D\,t\,dt = 2u\,du$,

the given integral $\displaystyle = -\int \frac{\left(\frac{1}{D}\right)u\,du}{\left[A+\left\{\frac{B\left(u^2-C\right)}{D}\right\}\right]u}$

$$= -\int \frac{du}{\left(Bu^2+AD-BC\right)}.$$

This integral is of the standard form and so it can be easily evaluated.

TO EVALUATE $\int \frac{dx}{(px+q)\sqrt{(ax^2+bx+c)}}$

Example 1:

Integrate $(3x-2)\sqrt{(x^2+x+1)}$.

Solution:

Let (3x – 2) = A. diff. coeff. of $(x^2 + x + 1)$ + B, where A and B are constants,

i.e., 3x – 2 = A . (2x + 1) + B.

On comparing the coefficients, we get 3 + 2A; $\therefore\ A = \frac{3}{2}$.

Also – 2 = A + B; $\therefore\ B = -2 - \frac{3}{2} = -\frac{7}{2}$.

Thus the given integral

$$= \int \left[\frac{3}{2}(2x+1) - \frac{7}{2}\right]\sqrt{(x^2+x+1)}\,dx$$

$$= \int \frac{3}{2}(2x+1)\sqrt{(x^2+x+1)}\,dx - \frac{7}{2}\int\sqrt{(x^2+x+1)}\,dx$$

$$= \frac{3}{2}\int (x^2+x+1)^{1/2}(2x+1)dx - \frac{7}{2}\int\sqrt{\left\{\left(x+\frac{1}{2}\right)^2+\frac{3}{4}\right\}}\,dx,$$

[Applying power formula], $\left[\text{Form} \int\sqrt{(x^2+a^2)}\,dx\right]$

$$= \frac{3}{2}\cdot\frac{2}{3}(x^2+x+1)^{3/2} - \frac{7}{2}\left[\frac{1}{2}\left(x+\frac{1}{2}\right)\sqrt{\left\{\left(x+\frac{1}{2}\right)^2+\frac{3}{4}\right\}} + \frac{1}{2}\cdot\frac{3}{4}\sinh^{-1}\left\{\left(x+\frac{1}{2}\right)\Big/\sqrt{\left(\frac{3}{4}\right)}\right\}\right]$$

$$= (x^2+x+1)^{3/2} - \frac{7}{4}\left(x+\frac{1}{2}\right)\sqrt{(x^2+x+1)} - \frac{21}{16}\sinh^{-1}\left(\frac{2x+1}{\sqrt{3}}\right).$$

Example 2:

Integrate $(x+1)\sqrt{(x^2-x+1)}$.

Solution:

We have $x+1 = \frac{1}{2}(2x-1)+1+\frac{1}{2} = \frac{1}{2}(2x-1)+\frac{3}{2}$.

$$\therefore \int (x+1)\sqrt{(x^2-x+1)}\,dx$$

$$= \int \left[\frac{1}{2}(2x-1)+\frac{3}{2}\right]\sqrt{(x^2-x+1)}\,dx$$

$$= \int \frac{1}{2}(2x-1)\sqrt{(x^2-x+1)}\,dx + \frac{3}{2}\int \sqrt{(x^2-x+1)}\,dx$$

$$= \frac{1}{2}\cdot\frac{2}{3}\cdot\left(x^2-x+1\right)^{3/2} + \frac{3}{2}\int \sqrt{\left\{\left(x+\frac{1}{2}\right)^2+\frac{3}{4}\right\}}\,dx$$

$$= \frac{1}{3}\left(x^2-x+1\right)^{3/2} + \frac{3}{2}\cdot\frac{1}{2}\left(x-\frac{1}{2}\right)\sqrt{\left\{\left(x-\frac{1}{2}\right)^2+\frac{3}{4}\right\}}$$

$$+\frac{3}{2}\cdot\frac{3}{4}\cdot\frac{1}{2}\sinh^{-1}\left\{\frac{x-\frac{1}{2}}{\left(\sqrt{\frac{3}{4}}\right)}\right\}$$

$$= \left(\frac{1}{24}\right)\left(8x^2+10x-1\right)\sqrt{(x^2-x+1)} + \left(\frac{9}{16}\right)\sinh^{-1}\left\{\frac{(2x-1)}{\sqrt{3}}\right\}.$$

Example 3:

Evaluate $\int \frac{dx}{x+\sqrt{(x-1)}}$.

Solution:

Put $x - 1 = t^2$

so that $dx = 2t\,dt$.

Then the given integral

$$I = \int \frac{2t\,dt}{t^2+1+t} = \int \frac{(2t+1)-1}{t^2+t+1}dt$$

$$= \int \frac{2t+1}{t^2+t+1}dt - \int \frac{1}{t^2+t+1}dt$$

$$= \log\left(t^2+t+1\right) - \int \frac{1}{\left(t+\frac{1}{2}\right)^2+\left(\frac{\sqrt{3}}{2}\right)^2}dt$$

$$= \log\left(t^2+t+1\right) - \frac{2}{\sqrt{3}}\tan^{-1}\frac{t+\frac{1}{2}}{\frac{\sqrt{3}}{2}}$$

$$= \log\left(t^2 + t + 1\right) - \frac{2}{\sqrt{3}}\tan^{-1}\frac{2t+1}{\sqrt{3}}$$

$$= \log\left[x + \sqrt{(x-1)}\right] - \frac{2}{\sqrt{3}}\tan^{-1}\frac{2\sqrt{(x-1)+1}}{\sqrt{3}}.$$

INTEGRATION OF $\dfrac{(px+q)}{\sqrt{\left(ax^2+bx+c\right)}}$

This integral is evaluated by first breaking the numerator into two parts such that one part is a constant multiple of the differential coefficient of $ax^2 + bx + c$ *i.e.*, of $2ax + b$ and the other part is a constant. Thus we put

$px + q = L \times$ diff. coeff. of $(ax^2 + bx + c) + M$

$= L\,(2ax + b) + M$, L and M are suitable constants.

On comparing coefficients on both sides, we have

$$L = \frac{p}{(2a)} \text{ and } M = q - \left\{\frac{(pb)}{2a}\right\}.$$

$$\therefore \int \frac{(px+q)\,dx}{\sqrt{\left(ax^2+bx+c\right)}}$$

$$= \frac{p}{2a}\int \frac{(2ax+b)\,dx}{\sqrt{\left(ax^2+bx+c\right)}} + \int \frac{\{q-(pb)/(2a)\}}{\sqrt{\left(ax^2+bx+c\right)}}dx$$

$$= \frac{p}{a}\sqrt{\left(ax^2+bx+c\right)} + \left(q - \frac{pb}{2a}\right)\int \frac{dx}{\sqrt{\left(ax^2+bx+c\right)}},$$

[Applying power formula to evaluate the first integral]

The integral on the right can be evaluated by methods discussed earlier.

Example 1:

Evaluate $\dfrac{1}{2}\displaystyle\int \frac{dx}{\sqrt{\{x(1-x)\}}}.$

Solution:

We have $\dfrac{1}{2}\displaystyle\int \frac{dx}{\sqrt{\{x(1-x)\}}} = \frac{1}{2}\int \frac{dx}{\sqrt{\left(x-x^2\right)}}$

$$=\frac{1}{2}\int\frac{dx}{\sqrt{\{0-(x^2-x)\}}}$$

$$=\frac{1}{2}\int\frac{dx}{\sqrt{\left\{0-\left(x-\frac{1}{2}\right)^2+\frac{1}{4}\right\}}}=\frac{1}{2}\int\frac{dx}{\sqrt{\left\{\frac{1}{4}-\left(x-\frac{1}{2}\right)^2\right\}}}$$

$$=\frac{1}{2}\sin^{-1}\frac{x-\frac{1}{2}}{\frac{1}{2}}=\frac{1}{2}\sin^{-1}(2x-1).$$

INTEGRATION OF $\sqrt{(ax^2+bx+c)}$

$\sqrt{(ax^2+bx+c)}$ can be integrated by reducing $ax^2 + bx + c$ to the form $a\{(x+\alpha)^2 \pm \beta^2\}$. The given integral can then be easily evaluated by applying one of the following standard results.

$$\int\sqrt{(x^2+a^2)}\,dx=\frac{1}{2}x\sqrt{(x^2+a^2)}+\frac{1}{2}a^2\sinh^{-1}\left(\frac{x}{a}\right)$$

$$=\frac{1}{2}x\sqrt{(x^2+a^2)}+\frac{1}{2}a^2\log\left\{x+\sqrt{(x^2+a^2)}\right\};$$

$$\int\sqrt{(x^2-a^2)}\,dx=\frac{1}{2}x\sqrt{(x^2-a^2)}-\frac{1}{2}a^2\cosh^{-1}\left(\frac{x}{a}\right)$$

$$=\frac{1}{2}x\sqrt{(x^2-a^2)}-\frac{1}{2}a^2\log\left[x+\sqrt{(x^2-a^2)}\right];$$

and $\int\sqrt{(a^2-x^2)}\,dx=\frac{1}{2}x\sqrt{(a^2-x^2)}+\frac{1}{2}a^2\sin^{-1}\left(\frac{x}{a}\right).$

Note that in each result the sign before a^2 in the scond term is the same as the sign before a^2 in the expression under the radical sign.

Example:

Integrate $\dfrac{1}{\sqrt{(1-x-x^2)}}$.

Solution:

We have $\displaystyle\int\frac{dx}{\sqrt{(1-x-x^2)}}=\int\frac{dx}{\sqrt{\{1-(x^2+x)\}}}$

$$=\int \frac{dx}{\sqrt{\left\{1-\left(x+\frac{1}{2}\right)^2+\frac{1}{4}\right\}}}$$

$$=\int \frac{dx}{\sqrt{\left\{\frac{5}{4}-\left(x+\frac{1}{2}\right)^2\right\}}}=\sin^{-1}\left\{\frac{x+\frac{1}{2}}{\frac{1}{2}\sqrt{5}}\right\}=\sin^{-1}\left(\frac{2x+1}{\sqrt{5}}\right).$$

INTEGRATION OF $\frac{1}{(ax+b)\sqrt{(cx+d)}}$

In such problems, put $cx + d = t^2$, so that $c\,dx = 2t\,dt$; then the fraction reduces to a form which can be easily integrated.

INTEGRATION OF $(px+q)\sqrt{(ax^2+bx+c)}$

We written $px+q=\frac{p}{2a}(2ax+b)+\left(q-\frac{pb}{2a}\right)$

$$\therefore \int (px+q)\sqrt{(ax^2+bx+c)}\,dx$$

$$=\int\left(\frac{p}{2a}\right)(2ax+b)\sqrt{(ax^2+bx+c)}\,dx$$

$$+\int\left\{q-\left(\frac{pb}{2a}\right)\right\}\sqrt{(ax^2+bx+c)}\,dx$$

$$=\frac{p}{2a}\cdot\frac{2}{3}\left(ax^2+bx+c\right)^{3/2}+\left(q-\frac{pb}{2a}\right)\int\sqrt{(ax^2+bx+c)}\,dx,$$

the first integral being evaluated by applying the power formula.

The integral left on the right hand side can now be evaluated by methods discussed earlier.

Example:

Integrate $(x+1)\sqrt{(2x^2+3)}$.

Solution:

We have $\int (x+1)\sqrt{(2x^2+3)}\,dx$

$$= \int x\sqrt{(2x^2+3)}\,dx + \int \sqrt{(2x^2+3)}\,dx$$

$$= \frac{1}{4}\int (4x)\sqrt{(2x^2+3)}\,dx + \int \sqrt{2}.\sqrt{\left(x^2+\frac{3}{2}\right)}\,dx$$

$$= \frac{1}{4}\cdot\frac{(2x^2+3)^{3/2}}{\frac{3}{2}} + \sqrt{2}\left\{\frac{1}{2}x.\sqrt{\left(x^2+\frac{3}{2}\right)} + \frac{1}{2}\cdot\frac{3}{2}\sinh^{-1}\left(\frac{x\sqrt{2}}{\sqrt{3}}\right)\right\}$$

$$= \frac{1}{6}(2x^2+3)^{3/2} + \frac{1}{2}x\sqrt{(2x^2+3)} + \frac{3}{4}\sqrt{2}.\sinh^{-1}\left\{\frac{(x\sqrt{2})}{\sqrt{3}}\right\}.$$

INTEGRATION $\frac{1}{\{(ax^2+bx+c)\sqrt{(Ax+B)}\}}$

Such fractions are integrated by putting $Ax + B = t^2$.

Example 1:

Evaluate $\int \frac{dx}{x\sqrt{(1+x^n)}}$.

Solution:

The given integral

$I = \int \frac{x^{n-1}dx}{x^n\sqrt{(1+x^n)}}$, multiplying above and below by x^{n-1}.

Put $1 + x^n = t^2$

so that $nx^{n-1}\,dx = 2t\,dt$.

$$\therefore\ I = \int \frac{2t\,dt}{n(t^2-1)t} = \frac{2}{n}\int \frac{dt}{t^2-1}$$

$$= \frac{2}{n}\cdot\frac{1}{2}\log\frac{t-1}{t+1} = \frac{1}{n}\log\frac{\sqrt{(1+x^n)}-1}{\sqrt{(1+x^n)}+1}$$

Example 2:

Integrate $\frac{1}{(x-1)\sqrt{(x^2+x+1)}}$, *(x > 1).*

Solution:

Put $(x-1) = \frac{1}{t}$,

so that $dx = -\left(\frac{1}{t^2}\right)dt$.

Also $x = 1 + \left(\frac{1}{t}\right) = \frac{(1+t)}{t}$. $\quad \therefore \int \frac{dx}{(x-1)\sqrt{(x^2+x+1)}}$

$$= \int \frac{-\left(\frac{1}{t^2}\right)dt}{\frac{1}{t}\sqrt{\left[\left(\frac{1+t}{t}\right)^2 + \left(\frac{1+t}{t}\right) + 1\right]}}$$

$$= -\int \frac{dt}{\sqrt{(3t^2+3t+1)}} = -\frac{1}{\sqrt{3}}\int \frac{dt}{\sqrt{\left\{\left(t+\frac{1}{2}\right)^2 + \frac{1}{12}\right\}}}$$

$$= -\frac{1}{\sqrt{3}}\sinh^{-1}\left[\frac{t+\frac{1}{2}}{\frac{1}{(2\sqrt{3})}}\right]$$

$$= -\frac{1}{\sqrt{3}}\sinh^{-1}\left\{2\sqrt{3}\left(t+\frac{1}{2}\right)\right\} = -\frac{1}{\sqrt{3}}\sinh^{-1}\left[\frac{x+1}{x-1}\sqrt{3}\right].$$

3.4 INTEGRATION OF $\frac{1}{\sqrt{(ax^2+bx+c)}}$

We can express $ax^2 + bx + c$ as $a\left\{x^2 + \left(\frac{b}{a}\right)x + \left(\frac{c}{a}\right)\right\}$.

or $a\left\{\left(x+\frac{b}{2a}\right)^2 + \frac{c}{a} - \frac{b^2}{4a^2}\right\}$

or $a\left[\left(x+\frac{b}{2a}\right)^2 + \frac{4ac-b^2}{4a^2}\right]$.

This is of the form $a\{(x+\alpha)^2 \pm \beta^2\}$.

Thus the given integral can be reduced to one of the standard forms

$$\int \frac{dx}{\sqrt{(x^2+a^2)}}, \int \frac{dx}{\sqrt{(x^2-a^2)}} \text{ or } \int \frac{dx}{\sqrt{(a^2-x^2)}}.$$

So it can be easily evaluated.

SOLVED EXAMPLES

Example 1:

Integrate $\dfrac{1}{\sqrt{(4x+3x-2x^2)}}$.

Solution:

We have $\displaystyle\int\frac{dx}{\sqrt{(4+3x-2x^2)}}=\frac{1}{\sqrt{2}}\int\frac{dx}{\sqrt{\left\{2+\left(\frac{3}{2}\right)x-x^2\right\}}}$

$$=\frac{1}{\sqrt{2}}\int\frac{dx}{\sqrt{\left\{2-\left(x^2-\frac{3}{2}x\right)\right\}}}=\frac{1}{\sqrt{2}}\int\frac{dx}{\sqrt{\left(2+\frac{9}{16}-\left(x^2-\frac{3}{2}x+\frac{9}{16}\right)\right)}}$$

$$=\frac{1}{\sqrt{2}}\int\frac{dx}{\left\{\frac{41}{16}-\left(x-\frac{3}{4}\right)^2\right\}}=\frac{1}{\sqrt{2}}\int\frac{dx}{\sqrt{\left\{\left(\frac{\sqrt{41}}{4}\right)^2-\left(x-\frac{3}{4}\right)^2\right\}}}$$

$$=\frac{1}{\sqrt{2}}\sin^{-1}\left\{\frac{x-\frac{3}{4}}{\left(\frac{\sqrt{41}}{4}\right)}\right\}=\frac{1}{\sqrt{2}}\sin^{-1}\left(\frac{4x-3}{\sqrt{41}}\right).$$

Example 2:

Integrate $\dfrac{(x^2+2x+3)}{\sqrt{(x^2+x+1)}}$.

Solution:

We have $\displaystyle\int\frac{(x^2+2x+3)dx}{\sqrt{(x^2+x+1)}}=\int\frac{(x^2+x+1)+(x+2)}{\sqrt{(x^2+x+1)}}dx$

$$=\int\sqrt{(x^2+x+1)}\,dx+\int\frac{\frac{1}{2}(2x+1)+2-\frac{1}{2}}{\sqrt{(x^2+x+1)}}dx$$

$$=\int\sqrt{\left\{\left(x+\frac{1}{2}\right)^2+\frac{3}{4}\right\}}dx+\frac{1}{2}\int\frac{2x+1}{\sqrt{(x^2+x+1)}}dx$$

$$+\frac{3}{2}\int\frac{dx}{\sqrt{\left\{\left(x+\frac{1}{2}\right)^2+\frac{3}{4}\right\}}}$$

$$=\frac{1}{2}\left(x+\frac{1}{2}\right)\sqrt{\left\{\left(x+\frac{1}{2}\right)^2+\frac{3}{4}\right\}}+\frac{1}{2}\cdot\frac{3}{4}\sinh^{-1}\left\{\frac{\left(x+\frac{1}{2}\right)}{\sqrt{\left(\frac{3}{4}\right)}}\right\}$$

$$+\sqrt{\left(x^2+x+1\right)}+\frac{3}{2}\sinh^{-1}\left\{\frac{\left(x+\frac{1}{2}\right)}{\sqrt{\left(\frac{3}{4}\right)}}\right\}$$

$$=\frac{1}{4}(2x+1)\sqrt{\left(x^2+x+1\right)}+\frac{15}{8}\sinh^{-1}\left\{\frac{(2x+1)}{\sqrt{3}}\right\}+\sqrt{(x^2+x+1}$$

$$=\frac{1}{4}(2x+5)\sqrt{\left(x^2+x+1\right)}+\frac{15}{8}\sinh^{-1}\left\{\frac{(2x+1)}{\sqrt{3}}\right\}$$

Example 3(a):

Evaluate $\int\frac{1}{\left(3+4x^2\right)\sqrt{\left(4-3x^2\right)}}dx$.

Solution:

Put $x=\frac{1}{t}$,

so that $dx=-\left(\frac{1}{t^2}\right)dt$.

$$\therefore\quad I=\int\frac{dx}{\left(3+4x^2\right)\sqrt{\left(4-3x^2\right)}}=\int\frac{-\left(\frac{1}{t^2}\right)dt}{\left\{3+\left(\frac{4}{t^2}\right)\right\}\sqrt{\left\{4-\left(\frac{3}{t^2}\right)\right\}}}$$

$$=-\int\frac{t\,dt}{\left(3t^2+4\right)\sqrt{\left(4t^2-3\right)}}.$$

Now putting $4t^2-3=z^2$;

so that $8t\,dt=2z\,dz$, we get

$$I = -\frac{1}{4}\int \frac{z\,dz}{\left[3.\left\{\frac{(z^2+3)}{4}\right\}+4\right]z} = -\int \frac{dz}{3z^2+25}$$

$$= -\frac{1}{2}\int \frac{dz}{z^2+\left(\frac{5}{\sqrt{3}}\right)^2} = -\frac{1}{3}\cdot\frac{\sqrt{3}}{5}\tan^{-1}\left(\frac{z}{5/\sqrt{3}}\right)$$

$$= -\frac{1}{5\sqrt{3}}\tan^{-1}\left(\frac{z\sqrt{3}}{5}\right)$$

$$= -\frac{1}{5\sqrt{3}}\tan^{-1}\left\{\frac{\sqrt{3}}{5}\sqrt{(4t^2-3)}\right\}, \qquad [\because z^2 = 4t^2 - 3]$$

$$= -\frac{1}{5\sqrt{3}}\tan^{-1}\left[\frac{\sqrt{3}}{5}\sqrt{\left\{\left(\frac{4}{x^2}\right)-3\right\}}\right]$$

$$= -\frac{1}{5\sqrt{3}}\tan^{-1}\left\{\frac{\sqrt{3}\sqrt{(4-3x^2)}}{5x}\right\}$$

$$= -\frac{1}{5\sqrt{3}}\tan^{-1}\left\{\frac{\sqrt{(12-9x^2)}}{5x}\right\}.$$

Example 3(b):

Evaluate $\int \frac{x\sqrt{(a^2-x^2)}}{a^2+x^2}dx.$

Solution:

Put $a^2 - x^2 = t^2$,

so that $-x\,dx = t\,dt$.

$$\therefore \int \frac{x\sqrt{(a^2-x^2)}}{a^2+x^2}dx = \int \frac{t(-t\,dt)}{a^2+(a^2-t^2)} = \int \frac{-t^2dt}{2a^2-t^2}$$

$$= \int \frac{(2a^2-t^2)-2a^2}{2a^2-t^2}dt = \int dt - 2a^2\int \frac{dt}{2a^2-t^2}$$

$$= t - 2a^2\cdot\frac{1}{2.\sqrt{(2a^2)}}\log\frac{a\sqrt{2}+t}{a\sqrt{2}-t}$$

$$= \sqrt{\left(a^2 - x^2\right)} - \frac{1}{\sqrt{2}} a \log \frac{a\sqrt{2} + \sqrt{\left(a^2 - x^2\right)}}{a\sqrt{2} - \sqrt{\left(a^2 - x^2\right)}}.$$

Example 3(c):

Evaluate $\int_0^{1/2} \frac{dx}{\left(1 - 2x^2\right)\sqrt{\left(1 - x^2\right)}}$.

Solution:

Put $x = \frac{1}{t}$;

so that $dx = -\left(\frac{1}{t^2}\right) dt$.

Also when x = 0,

t = ∞ and

when $x = \frac{1}{2}$, t = 2.

$$\therefore \; I = \int_0^{1/2} \frac{dx}{\left(1 - 2x^2\right)\sqrt{\left(1 - x^2\right)}}$$

$$= \int_\infty^2 \frac{\left(-\frac{1}{t^2}\right) dt}{\left\{1 - \left(\frac{2}{t^2}\right)\right\} \sqrt{\left\{1 - \left(\frac{1}{t^2}\right)\right\}}}$$

$$= \int_\infty^2 \frac{-\frac{1}{t^2}}{(t^2 - 2)\sqrt{(t^2 - 1)}}$$

Now put $t^2 - 1 = z^2$

so that 2t dt = 2z dz.

The limits for z are ∞ to $\sqrt{3}$.

$$\therefore \; I = -\int_\infty^{\sqrt{3}} \frac{z\,dz}{\left(z^2 - 1\right) z} = -\int_\infty^{\sqrt{3}} \frac{dz}{\left(z^2 - 1\right)} = -\frac{1}{3}\left[\log \frac{z - 1}{z + 1}\right]_\infty^{\sqrt{3}}$$

$$= -\frac{1}{2}\left[\log \frac{\sqrt{3} - 1}{\sqrt{3} + 1} - \lim_{z \to \infty} \log \frac{z - 1}{z + 1}\right]$$

$$= -\frac{1}{2}\log\frac{\sqrt{3}-1}{\sqrt{3}+1} + \lim_{z\to\infty}\log\left\{\frac{1-\left(\frac{1}{z}\right)}{1+\left(\frac{1}{z}\right)}\right\}$$

$$= \frac{1}{2}\log\left(\frac{\sqrt{3}-1}{\sqrt{3}+1}\right)^{-1} + \log 1 = \frac{1}{2}\log\frac{\sqrt{3}+1}{\sqrt{3}-1}, \qquad [\because \log 1 = 0]$$

$$= \frac{1}{2}\log\left\{\frac{(\sqrt{3}+1)(\sqrt{3}+1)}{(\sqrt{3}-1)(\sqrt{3}+1)}\right\} = \frac{1}{2}\log\left(\frac{4+2\sqrt{3}}{2}\right) = \frac{1}{2}\log(2+\sqrt{3}).$$

Example 4:

Integrate $\frac{1}{\left[(x+2)\sqrt{(x+1)}\right]}$.

Solution:

Put $(x + 1) = t^2$,

so that $dx = 2t\ dt$.

$$\therefore \int\frac{dx}{(x+2)\sqrt{(x+1)}} = \int\frac{2t\,dt}{(t^2+1).t} = 2\int\frac{dt}{t^2+1}$$

$$= 2\tan^{-1}t = 2\tan^{-1}\left[\sqrt{(x+1)}\right].$$

Example 5:

Evaluate $\int\frac{dx}{(x+2)\sqrt{(x+3)}}$.

Solution:

Do your self.

Ans. $\log\frac{\sqrt{(x+3))}-1}{\sqrt{(x+3)}+1}$

Example 6(a):

Integrate $\frac{1}{\sqrt{\left(2+x-3x^2\right)}}$.

Solution:

We have $\int\frac{dx}{\sqrt{\left(2+x-3x^2\right)}} = \frac{1}{\sqrt{3}}\int\frac{dx}{\sqrt{\left\{\frac{2}{3}-\left(x^2-\frac{1}{3}x\right)\right\}}}$

$$= \frac{1}{\sqrt{3}} \int \frac{dx}{\sqrt{\left\{\frac{2}{3} + \frac{1}{36} - \left(x^2 - \frac{1}{3}x + \frac{1}{36}\right)\right\}}} = \frac{1}{\sqrt{3}} \int \frac{dx}{\sqrt{\left\{\frac{25}{36} - \left(x - \frac{1}{6}\right)^2\right\}}}$$

$$= \frac{1}{\sqrt{3}} \int \frac{dx}{\sqrt{\left\{\left(\frac{5}{6}\right)^2 - \left(x - \frac{1}{6}\right)^2\right\}}}$$

$$= \frac{1}{\sqrt{3}} \sin^{-1} \left\{\frac{\left(x - \frac{1}{6}\right)}{\frac{5}{6}}\right\} = \frac{1}{\sqrt{3}} \sin^{-1}\left(\frac{6x-1}{5}\right).$$

Note: Remember that $\int \frac{dx}{\sqrt{\{a^2 - (x-b)^2\}}} = \sin^{-1}\left(\frac{x-b}{a}\right).$

Example 6(b):

Integrate $\frac{1}{\sqrt{(2x^2 - x + 2)}}.$

Solution:

We have $\int \frac{dx}{\sqrt{(2x^2 - x + 2)}} = \frac{1}{\sqrt{2}} \int \frac{dx}{\sqrt{\left(x^2 - \frac{1}{2}x + 1\right)}}$

$$= \frac{1}{\sqrt{2}} \int \frac{dx}{\sqrt{\left(x^2 + \frac{1}{2}x + \frac{1}{16} + \frac{15}{16}\right)}} = \frac{1}{\sqrt{2}} \int \frac{dx}{\sqrt{\left\{\left(x - \frac{1}{4}\right)^2 + \left(\frac{\sqrt{15}}{4}\right)^2\right\}}}$$

$$= \frac{1}{\sqrt{2}} \sinh^{-1} \frac{x - \frac{1}{4}}{\frac{\sqrt{(15)}}{4}} = \frac{1}{\sqrt{2}} \sin^{-1}\left[\frac{4x-1}{\sqrt{15}}\right].$$

Example 7:

Evaluate $\int \sqrt{\left(\frac{1+x}{1-x}\right)}\, dx.$

Solution:

Multiplying the numerator and the denominator by $\sqrt{(1+x)}$, we have the given integral

$$= \int \frac{\sqrt{(1+x)}}{\sqrt{(1-x)}} \cdot \frac{\sqrt{(1+x)}}{\sqrt{(1+x)}} dx = \int \frac{(1+x)}{\sqrt{(1-x^2)}} dx$$

$$= \int \frac{1.dx}{\sqrt{\left(1-x^2\right)}} + \int \frac{x.dx}{\sqrt{\left(1-x^2\right)}} = \sin^{-1} x + \int \frac{x\,dx}{\sqrt{\left(1-x^2\right)}} s$$

$$= \sin^{-1} x - \frac{1}{2}\int \left(1-x^2\right)^{-1/2} (-2x)dx$$

$$= \left(\sin^{-1} x\right) - \frac{1}{2}\frac{\left\{\left(1-x^2\right)^{1/2}\right\}}{\left(\frac{1}{2}\right)}$$ [By power formula]

$$= \sin^{-1} x - \sqrt{\left(1-x^2\right)}.$$

Example 8:

Integrate $\frac{\sqrt{x}}{(1+x)}$.

Solution:

Rationalising the numerator, we have

$$I = \int \frac{\sqrt{x}\,dx}{(1+x)} = \int \frac{x\,dx}{(1+x)\sqrt{x}} = \int \frac{\{(1+x)-1\}}{(1+x)\sqrt{x}} dx$$

$$= \int \frac{(1+x)}{(1+x)\sqrt{x}} dx - \int \frac{1}{(1+x)\sqrt{x}} dx$$

$$= \int \frac{1}{\sqrt{x}} dx - \int \frac{1}{(1+x)\sqrt{x}} dx$$

$$= \int x^{-1/2} dx - \int \frac{1}{(1+x)\sqrt{x}} dx = \frac{x^{1/2}}{(1/2)} - \int \frac{1}{(1+x)\sqrt{x}} dx.$$

Now let $I_1 = \int \frac{1}{(1+x)\sqrt{x}} dx.$

Putting $x = t^2$

so that $dx = 2t\,dt$,

we get $I_1 = \int \frac{2t\,dt}{\left(1+t^2\right)t} = 2\int \frac{dt}{1+t^2}$

$$= 2\tan^{-1} t = 2\tan^{-1}\sqrt{x}.$$

$\therefore\ I = 2\sqrt{x} - I_1 = 2\sqrt{x} - 2\tan^{-1}\sqrt{x}$.

Example 9:

Integrate $\sqrt{\left[\dfrac{(x-1)}{(x+1)}\right]}$.

Solution:

Rationalising the numerator by multiplying the numerator and the denomerator by $\sqrt{(x-1)}$, we get

$$\int\sqrt{\left(\frac{x-1}{x+1}\right)}dx = \int\frac{(x-1)\,dx}{\sqrt{(x^2-1)}} = \int\frac{x\,dx}{\sqrt{(x^2-1)}} - \int\frac{dx}{\sqrt{(x^2-1)}}$$

$$= \frac{1}{2}\int(x^2-1)^{-1/2}(2x)\,dx - \cosh^{-1}x$$

$$= \frac{1}{2}\frac{\left\{(x^2-1)^{1/2}\right\}}{\left(\frac{1}{2}\right)} - \cosh^{-1}x = \sqrt{(x^2-1)} - \cosh^{-1}x.$$

Example 10:

Evaluate $\int\sqrt{\left(\dfrac{x+1}{x-1}\right)}dx$.

Solution:

Do your self.

Ans. $\sqrt{(x^2-1)} + \cosh^{-1}x$.

Example 11:

Evaluate $\int\dfrac{dx}{\sqrt{(1+x)}+\sqrt{x}}$.

Solution:

Rationalising the denominator, we have

$$\int\frac{dx}{\sqrt{(1+x)}+\sqrt{x}} = \int\frac{\left\{\sqrt{(1+x)}-\sqrt{x}\right\}dx}{\left\{\sqrt{(1+x)}+\sqrt{x}\right\}\left\{\sqrt{(1+x)}-\sqrt{x}\right\}}$$

$$= \int\frac{\left\{\sqrt{(1+x)}-\sqrt{x}\right\}}{(1+x)-x}dx = \int\left[\sqrt{(1+x)} - \sqrt{x}\right]dx$$

$$= \frac{2}{3} \cdot (1+x)^{3/2} - \frac{2}{3} x^{3/2}.$$

Example 12:

Integrate $\dfrac{1}{\left[x + \sqrt{(x^2 - 1)}\right]}$.

Solution:

Rationalising the denominator, we have

$$\int \frac{dx}{\left[x + \sqrt{(x^2 - 1)}\right]} = \int \frac{\left\{x - \sqrt{x^2 - 1}\right\}}{x^2 - (x^2 - 1)} dx$$

$$= \int \left[x - \sqrt{(x^2 - 1)}\right] dx = \int x\,dx - \int \sqrt{(x^2 - 1)}\,dx$$

$$= \frac{1}{2} x^2 - \frac{1}{2} x \sqrt{(x^2 - 1)} + \frac{1}{2} \log\left\{x + \sqrt{(x^2 - 1)}\right\}.$$

Example 13:

Evaluate $\displaystyle\int \frac{dx}{\sqrt{(x+a)} + \sqrt{(x+b)}}$.

Solution:

Rationalising the denominator, we have

$$\int \frac{dx}{\sqrt{(x+a)} + \sqrt{(x+b)}} = \int \frac{\sqrt{(x+b)} - \sqrt{(x+a)}}{(x+b) - (x+a)} dx$$

$$= \int \frac{(x+b)^{1/2} - (x+a)^{1/2}}{b-a} dx = \frac{1}{b-a}\left[\frac{2}{3}(x+b)^{3/2} - \frac{2}{3}(x+a)^{3/2}\right]$$

$$= \frac{2}{3} \cdot \frac{1}{(b-a)}\left[(x+b)^{3/2} - (x+a)^{3/2}\right].$$

Example 14:

Evaluate $\displaystyle\int \frac{1}{x} \sqrt{\left(\frac{x-1}{x+1}\right)}\, dx$.

Solution:

$$\int \frac{1}{x} \sqrt{\left(\frac{x-1}{x+1}\right)} dx = \int \frac{1}{x} \cdot \frac{(x-1)}{\sqrt{(x^2 - 1)}} dx$$

$$=\int \frac{x}{x\sqrt{(x^2-1)}}dx-\int \frac{1}{x\sqrt{(x^2-1)}}dx$$

$$=\int \frac{dx}{\sqrt{(x^2-1)}}-\int \frac{dx}{x\sqrt{(x^2-1)}}=\cosh^{-1}x-\sec^{-1}x.$$

Example 15:

Evaluate $\int \sqrt{(\sec x-1)}\,dx$.

Solution:

We have $\int \sqrt{(\sec x-1)}\,dx$

$$=\int \frac{\sqrt{(1-\cos x)}}{\sqrt{(\cos x)}}dx=\int \frac{\sqrt{2}.\sin\frac{1}{2}x.dx}{\sqrt{\left(2\cos^2\frac{1}{2}x-1\right)}}.$$

Now putting $\sqrt{2}\cos\frac{1}{2}x=t$

so that $-\frac{1}{2}\sqrt{2}\sin\frac{1}{2}x\,dx=dt$,

we get the required integral $-2\int \frac{dt}{\sqrt{(t^2-1)}}=-2\cosh^{-1}t$

$$=-2\cosh^{-1}\left(\sqrt{2}\cos\frac{1}{2}x\right).$$

Example 16:

Integrate $\sqrt{(x^2-x+1)}$.

Solution:

We have

$$\int \sqrt{(x^2-x+1)}\,dx=\int \sqrt{\left\{\left(x-\frac{1}{2}\right)^2+\frac{3}{4}\right\}}dx,\ \left[\text{form}\int \sqrt{(x^2+a^2)}\,dx\right]$$

$$=\frac{1}{2}\left(x-\frac{1}{2}\right)\sqrt{\left\{\left(x-\frac{1}{2}\right)^2+\frac{3}{4}\right\}}+\frac{1}{2}\cdot\left(\frac{3}{4}\right)\sinh^{-1}\left\{\frac{\left(x-\frac{1}{2}\right)}{\left(\frac{1}{2}\sqrt{3}\right)}\right\}$$

$$=\frac{1}{2}\left(x-\frac{1}{2}\right)\sqrt{\left(x^2-x+1\right)}+\frac{3}{8}\sinh^{-1}\left\{\frac{(2x-1)}{\sqrt{3}}\right\}.$$

Example 17:

Integrate $\sqrt{\left(x^2+x-1\right)}\,dx$.

Solution:

We have $\int\sqrt{\left(x^2+x-1\right)}dx=\int\sqrt{\left\{\left(x+\frac{1}{2}\right)^2\right\}-\left(\frac{5}{4}\right)},$

$$\left[\text{form}\int\sqrt{\left(x^2-a^2\right)}\,dx\right]$$

$$=\frac{1}{2}\left(x+\frac{1}{2}\right)\sqrt{\left\{\left(x+\frac{1}{2}\right)^2-\left(\frac{5}{4}\right)\right\}}-\frac{1}{2}\cdot\frac{5}{4}\cosh^{-1}\sqrt{\left\{\frac{\left(x+\frac{1}{2}\right)}{\left(\frac{1}{2}\sqrt{5}\right)}\right\}}$$

$$=\frac{1}{4}(2x+1)\sqrt{\left(x^2+x-1\right)}-\frac{5}{8}\cosh^{-1}\left\{\frac{(2x+1)}{\sqrt{5}}\right\}.$$

Example 18(a):

Integrate $\sqrt{\left(2x^2+3x+4\right)}$.

Solution:

We have $\int\sqrt{\left(2x^2+3x+4\right)}\,dx=\sqrt{2}\int\sqrt{\left(x^2+\frac{3}{2}x+2\right)}\,dx$

$$=\sqrt{2}\int\sqrt{\left\{\left(x+\frac{3}{4}\right)^2+\left(\frac{23}{16}\right)\right\}}\,dx,$$

$$\left[\text{form}\int\sqrt{\left(x^2+a^2\right)}\,dx\right]$$

$$=\sqrt{2}\left[\frac{1}{2}\left(x+\frac{3}{4}\right)\sqrt{\left\{\left(x+\frac{3}{4}\right)^2+\frac{23}{16}\right\}}+\frac{1}{2}\cdot\frac{23}{16}\sinh^{-1}\left\{\frac{\left(x+\frac{3}{4}\right)}{\left(\frac{1}{4}\sqrt{23}\right)}\right\}\right],$$

$$=\frac{1}{8}\,(4x+3)\sqrt{(2x^2+3x+4)}+(23\sqrt{2}/23)\sinh^{-1}\left\{\frac{(4x+3)}{\sqrt{23}}\right\}$$

Example 18(b):

Integrate $\sqrt{(4-3x-2x^2)}$.

Solution:

We have $\int\sqrt{(4-3x-2x^2)}\,dx=\sqrt{2}\int\sqrt{\left(2-\frac{3}{2}x-x^2\right)}dx$

$$=\sqrt{2}\int\sqrt{\left\{2-\left(x^2+\frac{3}{2}x\right)\right\}}\,dx=\sqrt{2}\int\sqrt{\left\{2+\frac{9}{16}-\left(x^2+\frac{3}{2}x+\frac{9}{16}\right)\right\}}\,dx$$

$$=\sqrt{2}\int\sqrt{\left\{\frac{41}{16}-\left(x+\frac{3}{4}\right)^2\right\}}\,dx, \qquad \left[\text{form} \int\sqrt{(a^2-x^2)}\,dx\right]$$

$$=\sqrt{2}\int\sqrt{\left(\frac{41}{16}-t^2\right)}\,dt, \text{ putting } x+\frac{3}{4}=t \text{ so that } dx = dt$$

$$=\sqrt{2}\cdot\frac{t}{2}\sqrt{\left(\frac{41}{16}-t^2\right)}+\sqrt{2}\cdot\frac{41}{32}\sin^{-1}\left\{\frac{t}{(\sqrt{41/4})}\right\}$$

$$=\sqrt{2}\cdot\frac{x+\frac{3}{4}}{2}\sqrt{\left\{\frac{41}{16}-\left(x+\frac{3}{4}\right)^2\right\}}+\sqrt{2}\cdot\frac{41}{32}\sin^{-1}\left\{\frac{x+\frac{3}{4}}{\sqrt{\frac{41}{4}}}\right\}$$

$$=\frac{4x+3}{8}\sqrt{(4-3x-2x^2)}+\frac{41\sqrt{2}}{32}\sin^{-1}\left(\frac{4x+3}{\sqrt{41}}\right).$$

Example 18(c):

Evaluate $\int\sqrt{\{(x-1)(2-x)\}}\,dx$.

Solution:

We have $\int\sqrt{\{(x-1)(2-x)\}}\,dx=\int\sqrt{(-x^2+3x-2)}\,dx$

$$=\int\sqrt{\{-2-(x^2-3x)\}}\,dx$$

$$=\int\sqrt{\left\{-2-\left(x-\frac{3}{2}\right)^2+\frac{9}{4}\right\}}dx=\int\sqrt{\left\{\frac{1}{4}-\left(x-\frac{3}{2}\right)^2\right\}}\,dx,$$

$$\left[\text{form} \int\sqrt{(a^2-x^2)}\,dx\right]$$

$$=\frac{1}{2}\left(x-\frac{3}{2}\right)\sqrt{\left\{\frac{1}{4}-\left(x-\frac{3}{2}\right)^2\right\}}+\frac{1}{2}\cdot\frac{1}{4}\sin^{-1}\left\{\frac{\left(x-\frac{3}{2}\right)}{\left(\frac{1}{2}\right)}\right\}$$

$$=\frac{1}{4}(2x-3)\sqrt{(3x-x^2-2)}+\frac{1}{8}\sin^{-1}(2x-3).$$

Example 19:

Integrate $\dfrac{(2x+3)}{\sqrt{(x^2+x+1)}}$

Solution:

We have $\left(\frac{d}{dx}\right)(x^2+x+1)=2x+1.$

$$\therefore \int\frac{(2x+3)dx}{\sqrt{(x^2+x+1)}}=\int\frac{(2x+1)dx}{\sqrt{(x^2+x+1)}}+\int\frac{2\,dx}{\sqrt{(x^2+x+1)}}$$

$$=\int(x^2+x+1)^{-1/2}(2x+1)dx+\int\frac{2\,dx}{\sqrt{(x^2+x+1)}}$$

$$=2\sqrt{(x^2+x+1)}+2\int\frac{dx}{\sqrt{\left\{\left(x+\frac{1}{2}\right)^2+1-\frac{1}{4}\right\}}}$$

$$=2\sqrt{(x^2+x+1)}+2\int\frac{dx}{\sqrt{\left\{\left(x+\frac{1}{2}\right)^2+\left(\frac{\sqrt{3}}{2}\right)^2\right\}}}$$

$$=2\sqrt{(x^2+x+1)}+2\sinh^{-1}\left(\frac{x+\frac{1}{2}}{\frac{\sqrt{3}}{2}}\right)$$

$$=2\sqrt{(x^2+x+1)}+2\sinh^{-1}\left\{\frac{(2x+1)}{\sqrt{3}}\right\}.$$

Example 20:

Evaluate $\displaystyle\int\frac{x+2}{\sqrt{(x^2+3x+1)}}dx.$

Solution:

Here the given integral $I = \int \frac{\frac{1}{2}(2x+3)+\frac{1}{2}}{\sqrt{(x^2+3x+1)}}dx$

$$= \frac{1}{2}\int (x^2+3x+1)^{-1/2}(2x+3)dx + \frac{1}{2}\int \frac{dx}{\sqrt{\left[\{x+(3/2)\}^2-(\sqrt{5}/2)^2\right]}}$$

$$= (x^2+3x+1)^{1/2} + \frac{1}{2}\cosh^{-1}\left[\frac{x+\left(\frac{3}{2}\right)}{\frac{\sqrt{5}}{2}}\right]$$

$$= \sqrt{(x^2+3x+1)} + \frac{1}{2}\cosh^{-1}\left[\frac{(2x+3)}{\sqrt{5}}\right].$$

Example 21:

Integrate $\frac{(2x+5)}{\sqrt{(x^2+3x+1)}}$.

Solution:

We have $\left(\frac{d}{dx}\right)(x^2+3x+1) = 2x+3$.

Now $2x + 5 = (2x + 3) + 2$.

$$\therefore \int \frac{(2x+5)dx}{\sqrt{(x^2+3x+1)}} = \int \frac{(2x+3)+2}{\sqrt{(x^2+3x+1)}}dx$$

$$= \int \frac{(2x+3)dx}{\sqrt{(x^2+3x+1)}} + 2\int \frac{dx}{\sqrt{(x^2+3x+1)}}$$

$$= \int (x^2+3x+1)^{-1/2}(2x+3)dx + 2\int \frac{dx}{\sqrt{(x^2+3x+1)}}$$

$$= 2\sqrt{(x^2+3x+1)} + 2\int \frac{dx}{\sqrt{\left\{\left(x+\frac{3}{2}\right)^2-\left(\frac{\sqrt{5}}{2}\right)^2\right\}}},$$

[evaluating the first integral by power formula]

$$= 2\sqrt{\left(x^2+3x+1\right)} + 2\cosh^{-1}\left\{\frac{\left(x+\frac{3}{2}\right)}{\sqrt{\left(\frac{5}{4}\right)}}\right\}$$

$$= 2\sqrt{\left(x^2+3x+1\right)} + 2\cosh^{-1}\left\{\frac{(2x+3)}{\sqrt{5}}\right\}.$$

Example 22:

Evaluate $\int \frac{(x+2)\,dx}{\sqrt{\left(5-12x-9x^2\right)}}$.

Solution:

We have $\int \frac{(x+2)dx}{\sqrt{\left(5-12x-9x^2\right)}}$

$$= \frac{1}{3}\int \frac{(x+2)dx}{\sqrt{\left(\frac{5}{9}-\frac{4}{3}x-x^2\right)}}$$, taking 9 common from the denominator

$$= \frac{1}{3}\int \frac{-\frac{1}{2}\left(-2x-\frac{4}{3}\right)+2-\frac{2}{3}}{\sqrt{\left(\frac{5}{9}-\frac{4}{3}x-x^2\right)}}dx$$

$$= -\frac{1}{6}\int\left(\frac{5}{9}-\frac{4}{3}x-x^2\right)^{-1/2}\left(-2x-\frac{4}{3}\right)dx + \frac{4}{9}\int \frac{dx}{\sqrt{\left\{\frac{5}{9}-\left(x^2+\frac{4}{3}x\right)\right\}}}$$

$$= -\frac{1}{6}\frac{\left(\frac{5}{9}-\frac{4}{3}x-x^2\right)^{1/2}}{\frac{1}{2}} + \frac{4}{9}\int \frac{dx}{\sqrt{\left\{\frac{5}{9}-\left(x+\frac{2}{3}\right)^2+\frac{4}{9}\right\}}}$$

$$= -\frac{1}{3}\sqrt{\left(\frac{5}{9}-\frac{4}{3}x-x^2\right)} + \frac{4}{9}\int \frac{dx}{\sqrt{\left\{1-\left(x+\frac{2}{3}\right)^2\right\}}}$$

$$=-\frac{1}{9}\sqrt{\left(5-12x-9x^2\right)}+\frac{4}{9}\sin^{-1}\left(\frac{x+\frac{2}{3}}{1}\right)$$

$$=-\frac{1}{9}\sqrt{\left(5-12x-9x^2\right)}+\frac{4}{9}\sin^{-1}\left(\frac{3x+2}{3}\right).$$

Example 23:

Evaluate $\int \frac{dx}{(x+2)\sqrt{(x-1)}}$

Solution:

Do your self.

Ans. $\frac{2}{\sqrt{3}}\tan^{-1}\left[\frac{\sqrt{(x-1)}}{\sqrt{3}}\right].$

Example 24:

Integrate $\frac{1}{\left[(x-3)\sqrt{(x+2)}\right]}.$

Solution:

Put $x + 2 = t^2$

so that $dx = 2t\,dt$.

$$\therefore \int \frac{dx}{(x-3)\sqrt{(x+2)}}=\int \frac{2t\,dt}{\left[\left(t^2-2\right)-3\right].t}$$

$$=2\int \frac{dt}{t^2-5}=2\int \frac{dt}{t^2-\left(\sqrt{5}\right)^2}=2\cdot\frac{1}{2\sqrt{5}}\log\left\{\frac{t-\sqrt{5}}{t+\sqrt{5}}\right\}$$

$$=\frac{1}{\sqrt{5}}\log\left\{\frac{\sqrt{(x+2)}-\sqrt{5}}{\sqrt{(x+2)}+\sqrt{5}}\right\}.$$

Example 25:

Integrate $\frac{1}{\left[(2x+1)\sqrt{(4x+3)}\right]}.$

Solution:

Put $4x + 3 = t^2$,

so that 4 dx = 2t dt and

$$(2x+1)=\frac{2\left(t^2-3\right)}{4}+1=\frac{t^2-3}{2}+1=\frac{t^2-1}{2}.$$

$$\therefore \int\frac{dx}{\left[(2x+1)\sqrt{(4x+3)}\right]}=\int\frac{\frac{1}{2}t\,dt}{\frac{1}{2}\left(t^2-1\right)t}=\int\frac{dt}{\left(t^2-1\right)}$$

$$=\frac{1}{2}\log\left\{\frac{t-1}{t+1}\right\}=\frac{1}{2}\log\frac{\sqrt{(4x+3)}-1}{\sqrt{(4x+3)}+1}.$$

Example 26:

Integrate $\dfrac{1}{\sqrt{\left(2x^2+3x+4\right)}}$.

Solution:

We have

$$\int\frac{dx}{\sqrt{\left(2x^2+3x+4\right)}}=\int\frac{dx}{\sqrt{2}.\sqrt{\left\{x^2+\left(\frac{3}{2}\right)x+2\right\}}}$$

$$\int\frac{dx}{\sqrt{2}.\sqrt{\left\{\left(x+\frac{3}{4}\right)^2+\left(2-\frac{9}{16}\right)\right\}}}=\int\frac{dx}{\sqrt{2}.\sqrt{\left\{\left(x+\frac{3}{4}\right)^2+\frac{23}{16}\right\}}}$$

$$=\int\frac{dx}{\sqrt{2}.\sqrt{\left\{\left(x+\frac{3}{4}\right)^2+\left(\frac{\sqrt{23}}{4}\right)^2\right\}}}$$

$$=\frac{1}{\sqrt{2}}\int\frac{dt}{\sqrt{\left\{t^2+\left(\frac{\sqrt{23}}{4}\right)^2\right\}}},$$

putting $x+\frac{3}{4}=t$

so that dx = dt

$$=\frac{1}{\sqrt{2}}\sinh^{-1}\left\{\frac{t}{\left(\sqrt{23}/4\right)}\right\}=\frac{1}{\sqrt{2}}\sin^{-1}\left\{\frac{4t}{\sqrt{(23)}}\right\}$$

$$=\frac{1}{\sqrt{2}}\sinh^{-1}\left\{\frac{4x+3}{\sqrt{23}}\right\}.$$

Example 27:

Integrate $\dfrac{1}{\sqrt{(x^2+x+1)}}$.

Solution:

We have $\dfrac{1}{\sqrt{(x^2+x+1)}}$

$$=\int \frac{dx}{\sqrt{\left\{\left(x+\frac{1}{2}\right)^2+\frac{3}{4}\right\}}}=\int \frac{dx}{\sqrt{\left\{\left(x+\frac{1}{2}\right)^2+\left(\frac{1}{2}\sqrt{3}\right)^2\right\}}}$$

$$=\sinh^{-1}\left(\frac{x+\frac{1}{2}}{\frac{1}{2}\sqrt{3}}\right)=\sinh^{-1}\left(\frac{2x+1}{\sqrt{3}}\right).$$

Example 28:

Integrate $\dfrac{(x^2+x+1)}{\sqrt{(x^2+2x+3)}}$.

Solution:

We have $\displaystyle\int \frac{(x^2+x+1)dx}{\sqrt{(x^2+2x+3)}}=\int \frac{(x^2+2x+3)-x-2}{\sqrt{(x^2+2x+3)}}dx$

$$=\int \frac{x^2+2x+3}{\sqrt{(x^2+2x+3)}}dx-\int \frac{x+2}{\sqrt{(x^2+2x+3)}}dx$$

$$=\int \frac{(x^2+2x+3)dx}{\sqrt{(x^2+2x+3)}}-\int \frac{\frac{1}{2}(2x+2)+2-1}{\sqrt{(x^2+2x+3)}}dx$$

$$=\int \sqrt{(x^2+2x+3)}\,dx-\frac{1}{2}\int \frac{(2x+2)dx}{\sqrt{(x^2+2x+3)}}-\int \frac{dx}{\sqrt{(x^2+2x+3)}}$$

$$=\int \sqrt{\{(x+1)^2+2\}}\,dx-\frac{1}{2}\int \frac{(2x+2)dx}{\sqrt{(x^2+2x+3)}}-\int \frac{dx}{\sqrt{\{(x+1)^2+2\}}}$$

$$=I_1-\frac{1}{2}I_2-I_3.$$

For the 1st integral, put (x + 1) = t

so that dx = dt.

$$\therefore\ I_1 = \int\sqrt{(t^2+2)}\,dt = \frac{1}{2}t.\sqrt{(t^2+2)} + \frac{1}{2}\cdot 2\sinh^{-1}\left(\frac{t}{\sqrt{2}}\right)$$

$$= \frac{1}{2}(x+1)\sqrt{\{(x+1)^2+2\}} + \sinh^{-1}\left\{\frac{(x+1)}{\sqrt{2}}\right\}.$$

For the 2nd integral,

put (x^2 + 2x + 3) = t so that (2x + 2) dx = dt.

$$\therefore\ I_2 = \int\left(\frac{1}{\sqrt{t}}\right)dt = \int t^{-1/2}dt = 2\sqrt{t} = 2\sqrt{(x^2+2x+3)}.$$

For the 3rd integral, put (x + 1) = t so that dx = dt.

$$\therefore\ I_3 = \int\frac{dt}{\sqrt{(t^2+2)}} = \sinh^{-1}\left(\frac{t}{\sqrt{2}}\right) = \sinh^{-1}\left\{\frac{x+1}{\sqrt{2}}\right\}.$$

Hence the required integral

$$= \frac{1}{2}(x+1)\sqrt{\{(x+1)^2+2\}} + \sinh^{-1}\left\{\frac{(x+1)}{\sqrt{2}}\right\}$$

$$-\frac{1}{2}\cdot 2\sqrt{(x^2+2x+3)} - \sinh^{-1}\left\{\frac{(x+1)}{\sqrt{2}}\right\}$$

$$= \frac{1}{2}(x-1)\sqrt{(x^2+2x+3)}.$$

Note: The values of the integrals I_1, I_2 and I_3 can also be written directly without making the substitutions.

Example 29:

Integrate $\dfrac{(x^3+3)}{\sqrt{(x^2+1)}}$.

Solution:

We have $\displaystyle\int\frac{(x^3+3)dx}{\sqrt{(x^2+1)}} = \int\frac{x(x^2+1)-x+3}{\sqrt{(x^2+1)}}dx$

$$= \int\frac{x(x^2+1)}{\sqrt{(x^2+1)}}dx - \int\frac{x\,dx}{\sqrt{(x^2+1)}} + 3\int\frac{dx}{\sqrt{(x^2+1)}}$$

$$=\frac{1}{2}\int(2x)\sqrt{(x^2+1)}\,dx-\int\frac{2x\,dx}{\sqrt{(x^2+1)}}+3\int\frac{dx}{\sqrt{(x^2+1)}}$$

$$=\frac{1}{2}\left[\frac{2}{3}\,x^2+1^{3/2}\right]-\frac{1}{2}\left[2\sqrt{(x^2+1}\right]+3\sinh^{-1}x$$

$$=\frac{1}{3}\left(x^2+1\right)^{3/2}-\sqrt{\left(x^2+1\right)}+3\sinh^{-1}x.$$

Example 30(a):

Integrate $(2x-5)\sqrt{\left(2+3x-x^2\right)}$.

Solution:

We have (2x – 5) = – (3 – 2x) – 2.

$$\therefore\ \int(2x-5)\sqrt{\left(2+3x-x^2\right)}\,dx$$

$$=-\int(3-2x)\sqrt{\left(2+3x-x^2\right)}\,dx-2\int\sqrt{\left(2+3x-x^2\right)}\,dx$$

$$=-\frac{\left(2+3x-x^2\right)^{3/2}}{\frac{3}{2}}-2\int\left\{\frac{17}{4}-\left(x-\frac{3}{2}\right)^2\right\}dx$$

$$=-\frac{2}{3}\left(2+3x-x^2\right)^{3/2}-2\left[\frac{1}{2}\left(x-\frac{3}{2}\right)\sqrt{\left\{\frac{17}{4}-\left(x-\frac{3}{2}\right)\right\}^2}\right.$$

$$\left.+\frac{1}{2}\cdot\frac{17}{4}\sinh^{-1}\left\{\frac{x-\frac{3}{2}}{\sqrt{\left(\frac{17}{4}\right)}}\right\}\right]$$

$$=-\frac{2}{3}\left(2+3x-x^2\right)^{3/2}-\frac{1}{2}(2x-3)\sqrt{\left(2+3x-x^2\right)}$$

$$-\frac{17}{4}\sin^{-1}\left(\frac{2x-3}{\sqrt{17}}\right)$$

The integral of this form can be easily evaluated by putting $px+q=\frac{1}{t}$ so that $p\,dx=-\frac{1}{t^2}dt$. Also $x=\frac{1}{p}\left(\frac{1}{t}-q\right)$.

Now expressing (ax^2+bx+c) in terms of t, we have

$$ax^2 + bx + c = \frac{a}{p^2}\left(\frac{1}{t} - q\right)^2 + \frac{b}{p}\left(\frac{1}{t} - q\right) + c$$

$= \frac{\left(At^2 + Bt + C\right)}{t^2}$, say where A, B, and C are some constants.

Thus $\int \left[\frac{1}{\left\{(px+q)\sqrt{\left(ax^2+bx+c\right)}\right\}}\right] dx$

$$= -\frac{1}{p}\int \frac{\left(\frac{1}{t^2}\right)dt}{\left(\frac{1}{t}\right).\sqrt{\left\{\frac{\left(At^2+Bt+C\right)}{t^2}\right\}}} = -\frac{1}{p}\int \frac{dt}{\sqrt{\left(At^2+Bt+C\right)}}.$$

Example 30(b):

Integrate $\frac{1}{\left\{(x+1)\sqrt{\left(x^2+1\right)}\right\}}$.

Solution:

Put $(x + 1) = \frac{1}{t}$

so that $dx = \left(-\frac{1}{t^2}\right)dt$.

Also $x = \left(\frac{1}{t}\right) - 1$.

Hence the required integral

$$= \int \frac{-\left(\frac{1}{t^2}\right)dt}{\left(\frac{1}{t}\right).\sqrt{\left[\left\{\left(\frac{1}{t}\right)-1\right\}^2+1\right]}} = \int \frac{-\left(\frac{1}{t^2}\right)dt}{\left(\frac{1}{t}\right).\sqrt{\left[\frac{\left\{(1-t)^2+t^2\right\}}{t^2}\right]}}$$

$$= -\int \frac{dt}{\sqrt{\left(1-2t+t^2+t^2\right)}} = -\int \frac{dt}{\sqrt{\left(1-2t+2t^2\right)}}$$

$$= -\int \frac{dt}{\sqrt{2}.\sqrt{\left(t^2+t+\frac{1}{2}\right)}} = -\frac{1}{\sqrt{2}}\int \frac{dt}{\sqrt{\left\{\left(t-\frac{1}{2}\right)^2+\frac{1}{2}-\frac{1}{4}\right\}}}$$

$$= -\frac{1}{\sqrt{2}}\int \frac{dt}{\sqrt{\left\{\left(t-\frac{1}{2}\right)^2+\left(\frac{1}{2}\right)^2\right\}}} = -\frac{1}{\sqrt{2}}\sinh^{-1}\left(\frac{t-\frac{1}{2}}{\frac{1}{2}}\right)$$

$$= -\frac{1}{\sqrt{2}}\sinh^{-1}(2t-1) = -\frac{1}{\sqrt{2}}\sinh^{-1}\left\{\frac{2}{x+1}-1\right\}$$

$$= -\frac{1}{\sqrt{2}}\sinh^{-1}\left(\frac{1-x}{1+x}\right).$$

Example 31:

Integrate $\frac{1}{\left\{(x-1)\sqrt{(x^2+1)}\right\}}$.

Solution:

Put $(x-1) = \frac{1}{t}$,

so that $dx = -\left(\frac{1}{t^2}\right)dt$.

Also $x = 1+\left(\frac{1}{t}\right)$.

$$\therefore \int \frac{dx}{(x-1)\sqrt{(x^2+1)}} = \int \frac{-\left(\frac{1}{t^2}\right)dt}{\left(\frac{1}{t}\right)\cdot\sqrt{\left[\left\{1+\left(\frac{1}{t}\right)\right\}^2+1\right]}}$$

$$= -\int \frac{dt}{\sqrt{\left\{(1+t)^2+t^2\right\}}} = -\int \frac{dt}{\sqrt{(1+2t+2t^2)}}$$

$$= -\frac{1}{\sqrt{2}}\int \frac{dt}{\sqrt{\left(t^2+t+\frac{1}{2}\right)}} = -\frac{1}{\sqrt{2}}\int \frac{dt}{\sqrt{\left\{\left(t+\frac{1}{2}\right)^2+\left(\frac{1}{2}\right)^2\right\}}}$$

$$= -\frac{1}{\sqrt{2}}\sinh^{-1}\left(\frac{t+\frac{1}{2}}{\frac{1}{2}}\right) = -\frac{1}{\sqrt{2}}\sinh^{-1}(2t+1)$$

$$= -\frac{1}{\sqrt{2}}\sinh^{-1}\left\{\frac{2}{x-1}+1\right\} = -\frac{1}{\sqrt{2}}\sinh^{-1}\left(\frac{x+1}{x-1}\right).$$

Example 32(a):

Integrate $\frac{1}{\left\{(x+1)\sqrt{(x^2-1)}\right\}}$.

Solution:

Put $(x + 1) = \frac{1}{t}$,

so that $dx = -\left(\frac{1}{t^2}\right)dt$.

$$\therefore \int\frac{dx}{(x+1)\sqrt{(x^2-1)}} = \int\frac{-\left(\frac{1}{t^2}\right)dt}{\left(\frac{1}{t}\right)\sqrt{\left[\left\{\left(\frac{1}{t}\right)-1\right\}^2-1\right]}}$$

$$= -\int\frac{dt}{\sqrt{\left\{(1-t)^2-t^2\right\}}} = -\int\frac{dt}{\sqrt{(1-2t)}}$$

$$= \frac{1}{2}\int(1-2t)^{-1/2}(-2)dt = \frac{1}{2}\cdot 2(1-2t)^{1/2}$$

$$= (1-2t)^{1/2} = \int\left\{1-\frac{2}{(x+1)}\right\} = \int\left(\frac{x-1}{x+1}\right).$$

Example 32(b):

Integrate $\frac{1}{\left[(1+x)\sqrt{(1-x^2)}\right]}$.

Solution:

Put $(1 + x) = \frac{1}{t}$,

so that $dx = -\left(\frac{1}{t^2}\right)dt$.

Also $x = \left(\frac{1}{t}\right)-1$.

$$\therefore \int\frac{dx}{(1+x)\sqrt{(1-x^2)}} = \int\frac{-\left(\frac{1}{t^2}\right)dt}{\left(\frac{1}{t}\right)\sqrt{\left[1-\left\{\left(\frac{1}{t}\right)-1\right\}^2\right]}}$$

$$= -\int \frac{dt}{\sqrt{\left[t^2 - (1-t)^2\right]}} = -\int \frac{dt}{\sqrt{(2t-1)}}$$

$$= -\frac{1}{2}\int (2t-1)^{-1/2}.2\,dt = -\sqrt{(2t-1)}$$

$$= -\sqrt{\left[\frac{2}{1+x} - 1\right]} = -\sqrt{\left(\frac{1-x}{1+x}\right)}.$$

Example 33:

Integrate $\dfrac{1}{\left\{(1-x)\sqrt{(1-x^2)}\right\}}$.

Solution:

Put $(1 - x) = \left(\frac{1}{t}\right)$,

so that $-\,dx = -\left(\frac{1}{t^2}\right)dt$.

Also $x = 1 - \left(\frac{1}{t}\right)$.

$$\therefore \int \frac{dx}{(1-x)\sqrt{(1-x^2)}} = \int \frac{\left(\frac{1}{t^2}\right)dt}{\left(\frac{1}{t}\right)\sqrt{\left[1 - \left\{1 - \left(\frac{1}{t}\right)\right\}^2\right]}}$$

$$= \int \frac{dt}{\sqrt{(2r-1)}} = \frac{1}{2}\int (2t-1)^{-1/2}.2\,dt$$

$$= \sqrt{(2t-1)} = \sqrt{\left[\frac{2}{(1-x)} - 1\right]} = \sqrt{\left(\frac{1+x}{1-x}\right)}.$$

Example 34:

Integrate $\dfrac{1}{\left\{(x-1)\sqrt{(x^2-1)}\right\}}$.

Solution:

Put $(x - 1) = \left(\frac{1}{t}\right)$,

so that $dx = -\left(\frac{1}{t^2}\right)dt$.

Also $x = 1 + \left(\frac{1}{t}\right)$.

$$\therefore \int \frac{dx}{(x-1)\sqrt{(x^2-1)}} = \int \frac{-\left(\frac{1}{t^2}\right)dt}{\left(\frac{1}{t}\right)\sqrt{\left[\left\{\left(\frac{1}{t}\right)+1\right\}^2 - 1\right]}}$$

$$= \int \frac{-dt}{\sqrt{(1-2t)}}$$

$$= -\frac{1}{2}\int (2t+1)^{-1/2}.2\,dt = -\frac{1}{2}(2t+1)^{1/2}.2 = -\sqrt{(2t+1)}$$

$$= -\sqrt{\left[\frac{2}{x-1}+1\right]} = -\sqrt{\left(\frac{x+1}{x-1}\right)}.$$

Example 35:

Integrate $\frac{1}{\left\{(x-a)\sqrt{(x^2-a^2)}\right\}}$.

Solution:

$(x - a) = \left(\frac{1}{t}\right)$,

so that $dx = -\left(\frac{1}{t^2}\right)dt$.

Also $x = a + \left(\frac{1}{t}\right)$.

$$\therefore \int \frac{dx}{(x-a)\sqrt{(x^2-a^2)}} = \int \frac{-\left(\frac{1}{t^2}\right)dt}{\left(\frac{1}{t}\right)\sqrt{\left[\left\{\left(\frac{1}{t}\right)+a\right\}^2 - a^2\right]}}$$

$$= -\int \frac{dt}{\sqrt{(1+2at)}} = -\int (1+2at)^{-1/2}\,dt = -\frac{(1+2at)^{1/2}}{\frac{1}{2}}\cdot\frac{1}{2a}$$

$$= -\frac{1}{a}\sqrt{(1+2at)} = -\frac{1}{a}\sqrt{\left\{1+\frac{2a}{x-a}\right\}} = -\frac{1}{a}\sqrt{\left(\frac{x+a}{x-a}\right)}.$$

Example 36(a):

Integrate $\dfrac{1}{\left\{(1+x)\sqrt{(1+2x-x^2)}\right\}}$.

Solution:

Put $1 + x = \left(\dfrac{1}{t}\right)$

i.e., $x = \left(\dfrac{1}{t}\right) - 1$

and $dx = -\left(\dfrac{1}{t^2}\right)dt$.

Hence $\left(1+2x-x^2\right) = 1 + 2\left(\dfrac{1}{t}-1\right) - \left(\dfrac{1}{t}-1\right)^2 = \dfrac{4t-2t^2-1}{t^2}$.

∴ the given integral

$$\int \frac{1}{(1+x)\sqrt{\left(1+2x-x^2\right)}}dx$$

$$= \int \frac{1.\left(-\dfrac{1}{t^2}\right)dt}{\left(\dfrac{1}{t}\right).\left\{\sqrt{\left(4t-2t^2-1\right)}\right\}.\left(\dfrac{1}{t}\right)}$$

$$= -\frac{1}{\sqrt{2}}\int \frac{dt}{\sqrt{\left(2t-t^2-\dfrac{1}{2}\right)}} = -\frac{1}{\sqrt{2}}\int \frac{dt}{\sqrt{\left\{-\dfrac{1}{2}-\left(t^2-2t\right)\right\}}}$$

$$= -\frac{1}{\sqrt{2}}\int \frac{dt}{\sqrt{\left\{-\dfrac{1}{2}-(t-1)^2+1\right\}}} = \frac{1}{\sqrt{2}}\int \frac{dt}{\sqrt{\left[\dfrac{1}{2}-(t-1)^2\right]}}$$

$$= -\frac{1}{\sqrt{2}}\sin^{-1}\left\{\frac{t-1}{1/\sqrt{2}}\right\} = -\frac{1}{\sqrt{2}}\sin^{-1}\left[\left\{\frac{1}{(1+x)}-1\right\}\cdot\sqrt{2}\right]$$

$$= -\frac{1}{\sqrt{2}}\sin^{-1}\left(\frac{-x\sqrt{2}}{1+x}\right) = \frac{1}{\sqrt{2}}\sin^{-1}\left(\frac{x\sqrt{2}}{1+x}\right),$$

$\left[\because \sin^{-1}(-x) = -\sin^{-1}x\right]$.

Example 36(b):

Integrate $\dfrac{1}{\left\{x\sqrt{\left(x^2+x+1\right)}\right\}}$.

Solution:

Put $x = \frac{1}{t}$,

so that $dx = -\left(\frac{1}{t^2}\right)dt$.

$$\therefore \int \frac{dx}{x\sqrt{(x^2+x+1)}} = -\int \frac{\left(\frac{1}{t^2}\right)dt}{\left(\frac{1}{t}\right)\sqrt{\left[\left(\frac{1}{t}\right)^2+\left(\frac{1}{t}\right)+1\right]}}$$

$$= -\int \frac{dt}{\sqrt{(t^2+t+1)}} = -\int \frac{dt}{\sqrt{\left[\left(t+\frac{1}{2}\right)^2+\left(\frac{3}{4}\right)\right]}}$$

$$= -\sinh^{-1}\left\{\frac{t+\frac{1}{2}}{\frac{1}{2}\sqrt{3}}\right\} = -\sinh^{-1}\left(\frac{2t+1}{\sqrt{3}}\right)$$

$$= -\sinh^{-1}\left[\frac{2\left(\frac{1}{x}\right)+1}{\sqrt{3}}\right] = -\sinh^{-1}\left[\frac{2+x}{x\sqrt{3}}\right].$$

Example 37:

Evaluate $\int \frac{dx}{(x+2)\sqrt{(x^2+6x+7)}}$.

Solution:

Let $I = \int \frac{dx}{(x+2)\sqrt{(x^2+6x+7)}}$.

Put $x+2 = \frac{1}{t}$,

so that $dx = -\left(\frac{1}{t^2}\right)dt$.

Also $x = \frac{1}{t} - 2 = \frac{1-2t}{t}$.

$$\therefore \quad I = \int \frac{-\left(\frac{1}{t^2}\right)dt}{\frac{1}{t}\sqrt{\left[\left(\frac{1-2t}{t}\right)^2 + 6\left(\frac{1-2t}{t}\right) + 7\right]}}$$

$$= -\int \frac{dt}{\sqrt{\left\{(1-2t)^2 + 6t(1-2t) + 7t^2\right\}}}$$

$$= -\int \frac{dt}{\sqrt{\left(1+2t-t^2\right)}} = -\int \frac{dt}{\sqrt{\left\{1-\left(t^2-2t\right)\right\}}}$$

$$= -\int \frac{dt}{\sqrt{\left\{1-(t-1)^2+1\right\}}} = -\int \frac{dt}{\sqrt{\left\{2-(t-1)^2\right\}}}$$

$$= -\sin^{-1}\left(\frac{t-1}{\sqrt{2}}\right) = -\sin^{-1}\left[\frac{\{1/(x+2)\}-1}{\sqrt{2}}\right]$$

$$= -\sin^{-1}\left[\frac{-(x+1)}{\sqrt{2}(x+2)}\right] = \sin^{-1}\left[\frac{x+1}{\sqrt{2}(x+2)}\right].$$

Example 38:

Integrate $\dfrac{1}{\left\{(2-x)\sqrt{\left(1-2x+3x^2\right)}\right\}}$.

Solution:

Put $(2 - x) = \dfrac{1}{t}$

so that $dx = \left(\dfrac{1}{t^2}\right)dt$.

Also $x = 2 - \left(\dfrac{1}{t}\right) = \dfrac{(2t-1)}{t}$.

$$\therefore \left(1-2x+3x^2\right) = 1 - 2\left(\frac{2t-1}{t}\right) + 3\left(\frac{2t-1}{t}\right)^2$$

$$= \frac{1}{t^2}\left\{t^2 - 2t(2t-1) + 3(2t-1)^2\right\} = \frac{1}{t^2}\left\{9t^2 - 10t + 3\right\}$$

$$\therefore \int \frac{dx}{(2-x)\sqrt{(1-2x+3x^2)}} = \int \frac{\left(\frac{1}{t^2}\right)dt}{\left(\frac{1}{t}\right)\left(\frac{1}{t}\right)\sqrt{(9t^2+10t+3)}}$$

$$= \int \frac{dt}{\sqrt{(9t^2-10t+3)}} = \frac{1}{3}\int \frac{dt}{\sqrt{\left[t^2-\left(\frac{10}{9}\right)t+\frac{1}{3}\right]}}$$

$$= \frac{1}{3}\int \frac{dt}{\sqrt{\left\{\left(t-\frac{5}{9}\right)^2+\frac{2}{81}\right\}}} = \frac{1}{3}\sinh^{-1}\left\{\frac{t-\frac{5}{9}}{\sqrt{\left(\frac{2}{81}\right)}}\right\}$$

$$= \frac{1}{3}\sinh^{-1}\left\{\frac{9\left(t-\frac{5}{9}\right)}{\sqrt{2}}\right\} = \frac{1}{3}\sinh^{-1}\left(\frac{9t-5}{\sqrt{2}}\right)$$

$$= \frac{1}{3}\sinh^{-1}\left\{\frac{1}{\sqrt{2}}\left(\frac{9}{2-x}-5\right)\right\} = \frac{1}{3}\sinh^{-1}\left\{\frac{1}{\sqrt{2}}\left(\frac{5x-1}{2-x}\right)\right\}.$$

Example 39(a):

Evaluate $\int \frac{\sqrt{(1+x^2)}}{(1-x^2)}dx$.

Solution:

Multiplying the numerator and the denominator by $\sqrt{(1+x^2)}$, we get

$$\int \frac{\sqrt{(1+x^2)}}{1-x^2}dx = \int \frac{(1+x^2)dx}{(1-x^2)\sqrt{(1+x^2)}}$$

$$= \int \frac{2-(1-x^2)}{(1-x^2)\sqrt{(1+x^2)}}dx = 2\int \frac{dx}{(1-x^2)\sqrt{(1+x^2)}} - \int \frac{dx}{\sqrt{(1+x^2)}}$$

$$= -2\int \frac{dx}{(x^2-1)\sqrt{(x^2+1)}} - \sinh^{-1} x$$

$$= -2 \cdot \frac{1}{2\sqrt{2}} \log \frac{\sqrt{(1+x^2)} - x\sqrt{2}}{\sqrt{(1+x^2)} + x\sqrt{2}} - \sinh^{-1} x.$$

Example 39(b):

Evaluate $\int \frac{dx}{(2x^2+3)\sqrt{(x^2-4)}}$.

Solution:

Put $x = \frac{1}{t}$,

so that $dx = -\left(\frac{1}{t^2}\right)dt$.

$$\therefore \text{ Given integral } I = \int \frac{\left(-\frac{1}{t^2}\right)dt}{\left[\left(\frac{2}{t^2}\right)+3\right]\sqrt{\left[\left(\frac{1}{t^2}\right)-4\right]}}$$

$$= -\int \frac{t\,dt}{(2+3t^2)\sqrt{(1-4t^2)}}.$$

Now put $1 - 4t^2 = z^2$,

so that $-8t\ dt = 2z\ dz$

and $2 + 3t^2 = 2 + \frac{3}{4}(1-z^2) = \frac{1}{4}(11-3z^2)$.

Then the integral $I = \frac{1}{4}\int \frac{z\,dz}{\frac{1}{4}(11-3z^2)z} = \int \frac{dz}{3\left(\frac{11}{3} - z^2\right)}$

$$= \frac{1}{3} \cdot \frac{1}{2}\sqrt{\left(\frac{3}{11}\right)} \cdot \log \frac{\sqrt{\left(\frac{11}{3}\right)} + z}{\sqrt{\left(\frac{11}{3}\right)} - z}$$

$$= \frac{1}{2\sqrt{(33)}} \log \frac{\sqrt{(11)} + \sqrt{[3(1-4t^2)]}}{\sqrt{(11)} - \sqrt{[3(1-4t^2)]}}$$

$$=\frac{1}{2\sqrt{(33)}}\log\frac{\sqrt{11}+\sqrt{\left\{3-\left(\frac{12}{x^2}\right)\right\}}}{\sqrt{11}-\sqrt{\left\{3-\left(\frac{12}{x^2}\right)\right\}}}, \qquad \left[\because t=\frac{1}{x}\right]$$

$$=\frac{1}{2\sqrt{(33)}}\log\frac{x\sqrt{(11)}+\sqrt{(3x^2-12)}}{x\sqrt{(11)}-\sqrt{(3x^2-12)}}.$$

Example 39(c):

Evaluate $\int\frac{dx}{\sqrt{(1+x)}+\sqrt{x}}$.

Solution:

Rationalising the denominator, we have

$$\int\frac{dx}{\sqrt{(1+x)}+\sqrt{x}}=\int\frac{\left\{\sqrt{(1+x)}-\sqrt{x}\right\}dx}{\left\{\sqrt{(1+x)}+\sqrt{x}\right\}\left\{\sqrt{(1+x)}-\sqrt{x}\right\}}$$

$$=\int\frac{\left\{\sqrt{(1+x)}-\sqrt{x}\right\}}{(1+x)-x}dx=\int\left[\sqrt{(1+x)}-\sqrt{x}\right]dx$$

$$=\frac{2}{3}\cdot(1+x)^{3/2}-\frac{2}{3}x^{3/2}.$$

Example 39(d):

Evaluate $\int\frac{dx}{(x^2+1)\sqrt{(x^2-1)}}$.

Solution:

Put $x=\frac{1}{t}$,

so that $dx=-\left(\frac{1}{t^2}\right)dt$.

$\therefore$ the given integral $I=\int\frac{-\left(\frac{1}{t^2}\right)dt}{\left(\frac{1}{t^2}+1\right)\sqrt{\left(\frac{1}{t^2}-1\right)}}$

$$= -\int \frac{t\,dt}{(1+t^2)\sqrt{(1-t^2)}}.$$

Now put $1 - t^2 = z^2$,

so that $-2t\,dt = 2z\,dz$

and $1 + t^2 = 1 + (1 - z^2) = 2 - z^2$.

Then $I = \int \frac{z\,dz}{(2-z^2)z} = \int \frac{dz}{2-z^2}$

$$= \frac{1}{2\sqrt{2}} \log \frac{\sqrt{2}+z}{\sqrt{2}-z} = \frac{1}{2\sqrt{2}} \log \frac{\sqrt{2}+\sqrt{(1-t^2)}}{\sqrt{2}-\sqrt{(1-t^2)}}$$

$$= \frac{1}{2\sqrt{2}} \log \frac{\sqrt{2}+\sqrt{\left(1-\frac{1}{x^2}\right)}}{\sqrt{2}-\sqrt{\left(1-\frac{1}{x^2}\right)}}, \qquad \left[\because t = \frac{1}{x}\right]$$

$$= \frac{1}{2\sqrt{2}} \log \frac{\sqrt{2}x+\sqrt{(x^2-1)}}{\sqrt{2}x-\sqrt{(x^2-1)}}.$$

Example 40:

Integrate $\dfrac{1}{\sqrt{(3x-x^2-2)}}$.

Solution:

We have $\int \frac{dx}{\sqrt{(3x-x^2-2)}} = \int \frac{dx}{\sqrt{\{-2-(x^2-3x)\}}}$

$$= \int \frac{dx}{\sqrt{\left\{-2-\left(x-\frac{3}{2}\right)^2+\frac{9}{4}\right\}}}$$

$$= \int \frac{dx}{\sqrt{\left\{\frac{1}{4}-\left(x-\frac{3}{2}\right)^2\right\}}} = \sin^{-1}\left\{\frac{x-\frac{3}{2}}{\frac{1}{2}}\right\} = \sin^{-1}(2x-3)$$

Example 41:

Evaluate $\int \sqrt{\left(\frac{x+1}{x+2}\right)} \frac{dx}{x+3}$.

Solution:

Put $\frac{x+1}{x+2} = t^2$; $\therefore\ x = \frac{2t^2-1}{1-t^2} = -2 + \frac{1}{1-t^2}$.

$$\therefore \quad x+3 = \frac{2-t^2}{1-t^2}$$

and $dx = \frac{2t\,dt}{\left(1-t^2\right)^2}$.

$\therefore$ the given integral

$$I = \int t \cdot \frac{1-t^2}{\left(2-t^2\right)} \cdot \frac{2t\,dt}{\left(1-t^2\right)^2} = \int \frac{2t^2}{\left(2-t^2\right)\left(1-t^2\right)} dt$$

$$= \int \left[\frac{2}{1-t^2} - \frac{4}{2-t^2}\right] dt, \text{ resolving into partial fractions}$$

$$= \log\frac{1+t}{1-t} - \sqrt{2}\log\frac{\sqrt{2}+t}{\sqrt{2}-t},$$

where $t = \sqrt{\left(\frac{x+1}{x+2}\right)}$.

7

Integration of Rational Fractions

INTRODUCTION

A fraction whose numerator and denominator are both rational and algebraic functions is defined as a rational fraction.

Thus $\dfrac{f(x)}{\phi(x)} = \dfrac{a_0x^m + a_1 x^{m-1}x \ldots + a_{m-1}x + a_m}{b_0x^n + b_1x^{n-1} + \ldots + b_{n-1}x + b_n}$

in which $a_0, a_1, \ldots, a_m, b_0, b_1, \ldots, b_n$ are constants and m and n are positive integers, is a rational algebraic fraction. Such fractions can always be integrated by splitting the given fraction into *partial fractions.*

Case I : Integration of fractions with non-repeated factors only in the denominator.

Working Rule

(i) The degree of the Nr. f(x) must be less than the degree of the Dr. $\phi(x)$ and if not so, then divide f(x) by $\phi(x)$ till the remainder is of a lower degree than $\phi(x)$.

(ii) Now break the denominator $\phi(x)$ into linear and quadratic factors.

(iii) Express the fraction as the sum of partial fractions. For example the partial fractions of the fraction

$$\frac{x^3 - 4x^2 + 5x + 7}{(x-a)(x-b)\left(lx^2 + mx + n\right)}$$ will be written as

$$\frac{A}{x-a} + \frac{B}{x-b} + \frac{Cx + D}{lx^2 + mx + n}.$$

(iv) To obtain the partial fraction corresponding to the factor (x – a) in the denominator we put (x – a) = 0 and we get the value of x *i.e.*, x = a. Now we put x = a everywhere in the given fraction except in the factor (x – a) itself. Thus, the value of A is obtained.

Case II: Repeated Linear Factors: If the denominator contains a repeated linear factor, the application of long division method is simpler to evaluate the partial fractions.

Working Rule

Put the repeated linear factor equal to y, (say). Find x in terms of y and then transform x everywhere in terms of y. The repeated factor ($1/y^r$) should be taken out. Arrange the numerator and the denominator in *ascending powers* of y and then divide the numerator by the denominator till y^r comes as a factor of the remainder.

Thus the fraction should be written as

$$\frac{1}{y^r}\left[\text{Quotient}+\frac{\text{Remainder}}{\text{divisor}}\right].$$

After removing the brackets express y in terms of x.

Note: When repeated factor is of degree two only, we can directly find out the partial fractions and need not apply the long division method.

TO INTEGRATE $(px + q)/(ax^2 + bx + c)^n$

The above integral can be evaluated by breaking it into a sum of two integrals such that in the first integral the numerator is the differential coefficient of $(ax^2 + bx + c)$ and in the second integral there is no term of x in the numerator. For this we have to find numbers L and M such that $(px + q) = L(2ax + b) + M$.

Thus we write

$$px+q=\frac{p}{2a}(2ax+b)+q-\frac{pb}{2a}.$$

Example:

Integrate $\dfrac{(x^2+1)}{(x^4+1)}$.

Solution:

Let $\quad I=\int\frac{x^2+1}{x^4+1}dx.$

Here both the numerator and the denominator do not contain odd powers of x. Also the numerator is of degree 2 and the denominator is of degree 4. So dividing the numerator and the denominator by x^2, we get

$$I = \int \frac{1 + (1/x^2)}{x^2 + (1/x^2)} dx$$

$$= \int \frac{1 + (1/x^2)}{[x - (1/x)]^2 + 2} dx, \qquad \left[\text{Note that } \frac{d}{dx}\left\{x - \frac{1}{x}\right\} = 1 + \frac{1}{x^2}\right].$$

Now put $x - \left(\frac{1}{x}\right) = t$, so that $\left\{1 + \left(\frac{1}{x^2}\right)\right\} dx = dt$.

$$\therefore\ I = \int \frac{dt}{t^2 + 2} = \frac{1}{\sqrt{2}} \tan^{-1}\left(\frac{t}{\sqrt{2}}\right) = \frac{1}{\sqrt{2}} \tan^{-1}\left\{\frac{x - (1/x)}{\sqrt{2}}\right\}$$

$$= \frac{1}{\sqrt{2}} \tan^{-1}\left(\frac{x^2 - 1}{x\sqrt{2}}\right).$$

INTEGRATION OF $1/(x^2 + k)^n$

This function is integrated by the method of successive reduction.

To obtain a reduction formula, we integrate $\frac{1}{(x^2 + k)^{n-1}}$ by parts, taking unity as the second function.

$$\text{Thus } \int \frac{1}{(x^2 + k)^{n-1}} \cdot 1\, dx = \frac{x}{(x^2 + k)^{n-1}} - \int x \cdot \frac{-(n-1)}{(x^2 + k)^n} \cdot 2x\, dx$$

$$\text{or } I_{n-1} = \frac{x}{(x^2 + k)^{n-1}} + 2(n-1) \int \frac{(x^2 + k) - k}{(x^2 + k)^n} \qquad [\because x^2 = (x^2 + k) - k]$$

$$\text{or } I_{n-1} = \frac{x}{(x^2 + k)^{n-1}} + 2(n-1)\left[\int \frac{dx}{(x^2 + k)^{n-1}} - k \int \frac{dx}{(x^2 + k)^n}\right]$$

$$\text{or } I_{n-1} = \frac{x}{(x^2 + k)^{n-1}} + 2(n-1) I_{n-1} - 2k(n-1) I_n.$$

$$\therefore\ 2k(n-1)\, I_n = \frac{x}{(x^2 + k)^{n-1}} + \{2(n-1) - 1\} I_{n-1}$$

$$\text{or } 2k(n-1) I_n = \frac{x}{(x^2 + k)^{n-1}} + (2n-3) I_{n-1}. \text{ Hence}$$

$$\int \frac{dx}{\left(x^2+k\right)^n} = \frac{x}{2k(n-1)\left(x^2+k\right)^{n-1}} + \frac{(2n-3)}{2k(n-1)} \int \frac{dx}{\left(x^2+k\right)^{n-1}}.$$

Above is the reduction formula for $\int \left[\frac{1}{\left(x^2+k\right)^n}\right] dx$. By repeated application of this formula the integral shall reduce to that of $\frac{1}{\left(x^2+k\right)}$ which is $\frac{1}{\sqrt{k}} \tan^{-1}\left(\frac{x}{\sqrt{3}}\right)$.

INTEGRATION OF $\frac{(px+q)}{\left(ax^2+bx+c\right)}$

To integrate such integrals break the given fraction into two fractions such that in one the numerator is the differential coefficient of the denominator, and in the other the numerator is merely a constant. Thus

$$\int \frac{(px+q)dx}{ax^2+bx+c} = \int \frac{(p/2a)\,(2ax+b)+q-\{(pb)/2a\}}{ax^2+bx+c} dx$$

$$= \frac{p}{2a} \int \frac{2ax+b}{ax^2+bx+c} dx + \int \frac{q=\{(pb)/(2a)\}}{ax^2+bx+c} dx$$

$$= \frac{p}{2a} \log\left(ax^2+bx+c\right) + \int \frac{q-\{(pb)/(2a)\}}{ax^2+bx+c} dx.$$

The 2nd integral can now be easily evaluated.

Example 1:

Integrate $\frac{(5x-2)}{\left(1+2x+3x^2\right)}$.

Solution:

Here $\left(\frac{d}{dx}\right)$ $(1 + 2x + 3x^2) = 6x + 2$.

$$\therefore\ I = \int \frac{5x-2}{1+2x+3x^2} dx = \int \frac{\frac{5}{6}(6x+2)-2-\frac{5}{3}}{3x^2+2x+1} dx$$

$$= \int \frac{\frac{5}{6}(6x+2)-\frac{11}{3}}{3x^2+2x+1} dx$$

$$=\frac{5}{6}\int\frac{6x+2}{3x^2+2x+1}dx-\frac{11}{3}\int\frac{1}{3x^2+2x+1}dx$$

$$=\frac{5}{6}\log\left(3x^2+2x+1\right)-\frac{11}{9}\int\frac{dx}{x^2+\frac{2}{3}x+\frac{1}{3}}.$$

Now $\int\frac{dx}{x^2+\frac{2}{3}x+\frac{1}{3}}=\int\frac{dx}{\left(x+\frac{1}{3}\right)^2+\frac{1}{3}-\frac{1}{9}}=\int\frac{dx}{\left(x+\frac{1}{3}\right)^2+\frac{2}{9}}$

$$=\int\frac{9\,dx}{(3x+1)^2+\left(\sqrt{2}\right)^2}=\frac{9}{3\sqrt{2}}\tan^{-1}\left(\frac{3x+1}{\sqrt{2}}\right).$$

Hence the required integral

$$I=\frac{5}{6}\log\left(3x^2+2x+1\right)-\frac{11}{9}\cdot\frac{9}{3\sqrt{2}}\tan^{-1}\left(\frac{3x+1}{\sqrt{2}}\right)$$

$$=\frac{5}{6}\log\left(3x^2+2x+1\right)-\frac{11}{6}\sqrt{2}\tan^{-1}\left\{\frac{(3x+1)}{\sqrt{2}}\right\}.$$

Example 2:

Integrate $\frac{x^2}{\left(x^4+x^2+1\right)}.$

Solution:

Here we shall give a very interesting method for evaluating the integral of a fraction in which the denominator is of degree 4 and numerator is either of degree 2 or is constant. Moreover the odd powers of x occur neither in the numerator nor in the denominator.

Let $I=\int\frac{x^2}{x^4+x^2+1}dx,$ [Note the form of the integrand]

$=\int\frac{1}{x^2+1+\left(\frac{1}{x^2}\right)}dx$, dividing the numerator and the denominator both by x^2.

Now the denominator $x^2+1+\frac{1}{x^2}$ can be written either as $\left(x-\frac{1}{x}\right)^2+3$ or as $\left(x+\frac{1}{x}\right)^2-1$. The diff. coeff of $x-\frac{1}{x}$ is $1+\frac{1}{x^2}$ and that of $x+\frac{1}{x}$ is $1-\frac{1}{x^2}$. So we write

$$I=\frac{1}{2}\int\frac{\left(1+\frac{1}{x^2}\right)+\left(1-\frac{1}{x^2}\right)dx}{x^2+1+\left(\frac{1}{x^2}\right)}dx$$

$$=\frac{1}{2}\int\frac{\left(1+\frac{1}{x^2}\right)dx}{\left(x-\frac{1}{x}\right)^2+3}+\frac{1}{2}\int\frac{\left(1-\frac{1}{x^2}\right)dx}{\left(x+\frac{1}{x}\right)^2-1}.$$

In the first integral put

$$x-\frac{1}{x}=t$$

so that $\left(1+\frac{1}{x^2}\right)dx=dt$, and in the second integral put $x+\frac{1}{x}=z$

so that $\left(1-\frac{1}{x^2}\right)dx=dz$.

$$\therefore\quad I=\frac{1}{2}\int\frac{dt}{t^2+\left(\sqrt{3}\right)^2}+\frac{1}{2}\int\frac{dz}{z^2-1}$$

$$=\frac{1}{2\sqrt{3}}\tan^{-1}\frac{t}{\sqrt{3}}+\frac{1}{2}\frac{1}{2\times 1}\log\frac{z-1}{z+1}$$

$$=\frac{1}{2\sqrt{3}}\tan^{-1}\left\{\frac{\left(x-\frac{1}{x}\right)}{\sqrt{3}}\right\}+\frac{1}{4}\log\frac{\left(x+\frac{1}{x}\right)-1}{\left(x+\frac{1}{x}\right)+1}$$

$$=\frac{1}{2\sqrt{3}}\tan^{-1}\left\{\frac{x^2-1}{\left(\sqrt{3}\right)x}\right\}+\frac{i}{4}\log\frac{x^2-x+1}{x^2+x+1}.$$

INTEGRATION OF $\frac{1}{\left(ax^2+bx+c\right)}$

To evaluate such integrals put the denominator in the form $a\{(x+\alpha)^2 \pm \beta^2\}$ and then integrate.

SOLVED EXAMPLES

Example 1:

Integrate $\frac{x^2}{\{(x-1)(3x-1)(3x-2)\}}$.

Solution:

Let $\frac{x^2}{(x-1)(3x-1)(3x-2)} \equiv \frac{A}{(x-1)} + \frac{B}{(3x-1)} + \frac{C}{(3x-2)}$.

Here $A = \frac{1^2}{(3\times 1-1)(3\times 1-2)} = \frac{1}{2}$,

$$B = \frac{\left(\frac{1}{3}\right)^2}{\left(\frac{1}{3}-1\right)\left\{3.\left(\frac{1}{3}\right)-2\right\}} = \frac{\frac{1}{9}}{\frac{2}{3}} = \frac{1}{6}$$

and $C = \frac{\left(\frac{2}{3}\right)^2}{\left(\frac{2}{3-1}\right)\left\{3.\left(\frac{2}{3}\right)-1\right\}} = -\frac{4}{3}$.

$\therefore$ required integral $= \int\left[\frac{1}{2(x-1)} + \frac{1}{6(3x-1)} - \frac{4}{3(3x-2)}\right]dx$

$= \frac{1}{2}\log(x-1) + \frac{1}{18}\log(3x-1) - \frac{4}{9}\log(3x-2)$.

Example 2:

Evaluate $\int_0^1 \frac{x^3 dx}{\left(x^2+1\right)\left(x^2+7x+12\right)}$.

Solution:

We have

$$\int_0^1 \frac{x^3 dx}{\left(x^2+1\right)\left(x^2+7x+12\right)} = \int_0^1 \frac{x^3 dx}{\left(x^2+1\right)(x+4)(x+3)}.$$

Let $\frac{x^3}{\left(x^2+1\right)(x+4)(x+3)} \equiv \frac{A}{(x+4)} + \frac{B}{(x+3)} + \frac{Cx+D}{\left(x^2+1\right)}$. ...(1)

We have $A = \frac{(-4)^3}{\left\{(-4)^2-1\right\}(-4+3)} = \frac{-64}{-17} = \frac{64}{17}$.

and $$B=\frac{(-3)^3}{\{(-3)^2+1\}\{-3+4\}}=\frac{-27}{10}.$$

Now putting x = 0 on both sides of (1), we get $0=\frac{A}{4}+\frac{B}{3}+D$.

Therefore $D=-\frac{A}{4}-\frac{B}{3}=-\frac{16}{17}+\frac{9}{10}=-\frac{7}{170}$.

Again multiplying both sides of (1) by x and taking limit when $x \to \infty$, we get

$$1 = A + B + C \text{ so that } C = 1 - A - B = 1 - \frac{64}{17}+\frac{27}{10}=\frac{-11}{170}.$$

The students must note carefully the method we have adopted to find the values of C and D. This method is very helpful to find the values of two constants when the values of all other constants have been found by some other methods.

∴ the required integral

$$I=\int_0^1\left\{\frac{64}{17(x+4)}-\frac{27}{10(x+3)}-\frac{1}{170}\left(\frac{11x+7}{x^2+1}\right)\right\}dx$$

$$=\frac{64}{17}\{\log(x+4)\}_0^1-\frac{27}{10}\{\log(x+3)\}_0^1-\frac{1}{170}\int_0^1\frac{11x+7}{\left(x^2+1\right)}dx$$

$$=\frac{64}{17}\log\frac{5}{4}-\frac{27}{10}\log\frac{4}{3}-\frac{1}{170}\int_0^1\frac{11x+7}{\left(x^2+1\right)}dx.$$

Now $\int_0^1\frac{11x+7}{x^2+1}dx=11\int_0^1\frac{x}{x^2+1}dx+7\int_0^1\frac{dx}{x^2+1}$

$$=\frac{11}{2}\int_0^1\frac{2x}{x^2+1}dx+7\int_0^1\frac{dx}{x^2+1}$$

$$=\frac{11}{2}\left[\log\left(x^2+1\right)\right]_0^1+7\left[\tan^{-1}x\right]_0^1=\frac{11}{2}\log 2+\frac{7}{4}\pi.$$

∴ required integral $=\frac{64}{17}\log\frac{5}{4}-\frac{27}{10}\log\frac{4}{3}-\frac{11}{340}\log 2-\frac{7\pi}{680}$.

Example 3:

Evaluate $\int_0^\infty\frac{\left(3+3x+x^2\right)}{\left(2+2x+x^2\right)^2}dx$.

Solution:

Here $\left(\frac{d}{dx}\right)$ (x + 2x + x²) = 2 + 2x.

Now $3 + 3x + x^2 = (2 + 2x + x^2) + (x + 1)$

$= (2 + 2x + x^2) + \frac{1}{2}(2x + 2).$

Thus $\int_0^{\infty} \frac{(3+3x+x^2)dx}{(2+2x+x^2)^2}$

$$= \int_0^{\infty} \frac{(2+2x+x^2)dx}{(2+2x+x^2)^2} + \frac{1}{2}\int_0^{\infty} \frac{(2+2x)dx}{(2+2x+x^2)^2}$$

$$= \int_0^{\infty} \frac{dx}{(x^2+2x+2)} + \frac{1}{2}\int_0^{\infty} \frac{(2+2x)dx}{(x^2+2x+2)^2}$$

$$= \int_0^{\infty} \frac{dx}{\{(x+1)^2+1\}} + \frac{1}{2}\left[-\frac{1}{(x^2+2x+2)}\right]_0^{\infty}$$

$$= \left[\tan^{-1}(x+1)\right]_0^{\infty} + \frac{1}{2}\left[0-\left(-\frac{1}{2}\right)\right]$$

$$= \left(\tan^{-1}\infty - \tan^{-1}1\right) + \frac{1}{4} = \frac{1}{2}\pi - \frac{1}{4}\pi + \frac{1}{4} = \frac{1}{4}(\pi+1).$$

Example 4:

Integrate $\frac{1}{(1+3e^x+2e^{2x})}$.

Solution:

Multiplying numerator and denominator by e^{-x}, we have

$$I = \int \frac{dx}{1+3e^x+2e^{2x}} = \int \frac{e^{-x}dx}{e^{-x}+3+2e^x}.$$

Now put $y = e^{-x}$

so that $dy = -e^{-x}\,dx$.

$$\text{Thus } I = \int \frac{-dy}{y+3+2\left(\frac{1}{y}\right)} = \int \frac{-y\,dy}{y^2+3y+2} = \int \frac{-\frac{1}{2}(2y+3)+\frac{3}{2}}{y^2+3y+2}dy$$

$$= -\frac{1}{2}\int \frac{(2y+3)dy}{y^2+3y+2} + \frac{3}{2}\int \frac{dy}{\left(y+\frac{3}{2}\right)^2 - \left(\frac{1}{2}\right)^2}$$

$$= -\frac{1}{2}\log\left(y^2+3y+2\right) + \frac{3}{2}\frac{1}{2\cdot\left(\frac{1}{2}\right)}\log\left\{\frac{\left(y+\frac{3}{2}\right)-\frac{1}{2}}{\left(y+\frac{3}{2}\right)+\frac{1}{2}}\right\}$$

$$= -\frac{1}{2}\log\left(e^{-2x}+3e^{-x}+2\right) + \frac{3}{2}\log\left\{\frac{y+1}{y+2}\right\}$$

$$= -\frac{1}{2}\log\left\{\frac{1+3e^x+2e^{2x}}{e^{2x}}\right\} + \frac{3}{2}\log\left\{\frac{e^{-x}+1}{e^{-x}+2}\right\}$$

$$= -\frac{1}{2}\log\left(1+3e^x+2e^{2x}\right) + \frac{1}{2}\log\left(e^{2x}\right) + \frac{3}{2}\log\left(\frac{1+e^x}{1+2e^x}\right)$$

$$= -\frac{1}{2}\log\left\{\left(1+e^x\right)\left(1+2e^x\right)\right\} + \frac{1}{2}(2x)$$

$$+\frac{3}{2}\log\left(1+e^x\right) - \frac{3}{2}\log\left(1+2e^x\right)$$

$= \log(1+e^x) - 2\log(1+2e^x) + x.$

Example 5(a):

Integrate $\frac{x}{\{(x-a)(x-b)(x-c)\}}$.

Solution:

Let $\frac{x}{(x-a)(x-b)(x-c)} \equiv \frac{A}{(x-a)} + \frac{B}{(x-b)} + \frac{C}{(x-c)}$.

Now to find the partial fraction corresponding to the factor (x – a) in the denominator, put x = a everywhere in the given fraction except in the factor (x – a) itself.

Thus we get $A = \frac{a}{\{(a-b)(a-c)\}}$.

Similarly we get $B = \frac{b}{\{(b-c)(b-a)\}}$

and $C = \frac{c}{\{(c-a)(c-b)\}}$.

Now $\int \frac{x}{(x-a)(x-b)(x-c)}dx = \int\left[\frac{A}{x-a} + \frac{B}{x-b} + \frac{C}{x-c}\right]dx$

$$= A \log(x-a) + B \log(x-b) + C \log(x-c)$$

$$= \frac{a}{(a-b)(a-c)}\log(x-a) + \frac{b}{(b-c)(b-a)}\log(x-b)$$

$$+\frac{c}{(c-a)(c-b)}\log(x-c) = \Sigma\left[\frac{a\log(x-a)}{(a-b)(a-c)}\right].$$

Example 5(b):

Integrate $\dfrac{\{(x-a)(x-b)(x-c)\}}{\{(x-\alpha)(x-\beta)(x-\gamma)\}}$.

Solution:

Since the numerator is not of a lower degree than the denominator, we first divide Nr. by Deno. Dividing the numerator by the denominator orally we see that the quotient is 1. So let

$$\frac{(x-a)(x-b)(x-c)}{(x-\alpha)(x-\beta)(x-\gamma)} \equiv 1 + \frac{A}{(x-\alpha)} + \frac{B}{(x-\beta)} + \frac{C}{(x-\gamma)}.$$

Then $A = \dfrac{(\alpha-a)(\alpha-b)(\alpha-c)}{(\alpha-\beta)(\alpha-\gamma)}$,

$$B = \frac{(\beta-a)(\beta-b)(\beta-c)}{(\beta-\alpha)(\beta-\gamma)},$$

and $C = \dfrac{(\gamma-a)(\gamma-b)(\gamma-c)}{(\gamma-\alpha)(\gamma-\beta)}$.

Now the given integral

$$I = \int\left[1 + \frac{A}{x-\alpha} + \frac{B}{x-\beta} + \frac{C}{x-\gamma}\right]dx$$

$$= x + A\log(x-\alpha) + B\log(x-\beta) + C\log(x-\gamma)$$

$$= x + \Sigma\left[\frac{(\alpha-a)(\alpha-b)(\alpha-c)}{(\alpha-\beta)(\alpha-\gamma)}\log(x-\alpha)\right],$$

[On putting the values of A, B and C].

Example 5(c):

Integrating $\dfrac{(x^2+x+2)}{\{(x-2)(x-1)\}}$.

Solution:

Here since numerator is not of a lower degree than the denominator, we first divide the numerator by the denominator. As we see orally the quotient is 1. So let

$$\frac{x^2+x+2}{(x-2)(x-1)} \equiv 1+\frac{A}{x-2}+\frac{B}{x-1}.$$

We have $A=\frac{2^2+2+2}{(2-1)}=8,\ B=\frac{1^2+1+2}{(1-2)}=-4.$

Hence $\int \frac{x^2+x+2}{(x-2)(x-1)}dx = \int\left\{1+4\left(\frac{2}{x-2}-\frac{1}{x-1}\right)\right\}dx$

$= x + 4\ \{2 \log (x - 2) - \log (x - 1)\}$

$= x + 4 \log \{(x - 2)^2/(x - 1)\}.$

Example 6:

Integrate $\frac{x^3}{\{(x-1)(x-2)(x-3)\}}.$

Solution:

Here since the numerator is not of a lower degree than the denominator, we divide the numerator by the denominator till the remainder is of lesser degree than the denominator.

We need not find out the actual value of the remainder because ultimately we have to break the fraction into partial fractions. Note that the denominators of the partial fractions depend only upon the denominator of the given fraction. So let

$$\frac{x^3}{(x-1)(x-2)(x-3)} \equiv 1+\frac{A}{x-1}+\frac{B}{x-1}+\frac{C}{x-3}.$$

We have $A=\frac{1^3}{(1-2)(1-3)}=\frac{1}{2},\quad B=\frac{8}{(2-1)(2-3)}=-8,$

and $\quad C=\frac{3^3}{(3-1)(3-2)}=\frac{27}{2}.$

$$\therefore \frac{x^3}{(x-1)(x-2)(x-3)}=1+\frac{1}{2(x-1)}-\frac{8}{(x-2)}+\frac{27}{2(x-3)}.$$

Hence $\int \frac{x^3dx}{(x-1)(x-2)(x-3)}$

$$= \int 1.dx + \int \frac{dx}{2(x-1)} - \int \frac{8dx}{(x-2)} + \int \frac{27dx}{2(x-3)}$$

$$= x + \frac{1}{2}\log(x-1) - 8\log(x-2) + \left(\frac{27}{2}\right)\log(x-3).$$

Example 7:

Evaluate $\int \frac{(x^2+1)(x^2+2)}{(x^2+3)(x^2+4)}dx.$

Solution:

We have $\frac{(x^2+1)(x^2+2)}{(x^2+3)(x^2+4)} = \frac{(y+1)(y+2)}{(y+3)(y+4)}$, where $y = x^2$.

Now let $\frac{(y+1)(y+2)}{(y+3)(y+4)} = 1 + \frac{A}{y+3} + \frac{B}{y+4}$,

resolving into partial fractions.

We have $A = \frac{(-3+1)(-3+2)}{(-3+4)} = 2,$

$B = \frac{(-4+1)(-4+2)}{(-4+3)} = -6.$

$\therefore \frac{(y+1)(y+2)}{(y+3)(y+4)} = 1 + \frac{2}{y+3} - \frac{6}{y+4}.$

$\therefore$ the given integral $I = \int \left[1 + \frac{2}{x^2+3} - \frac{6}{x^2+4}\right]dx$

$$= \int dx + 2\int \frac{dx}{x^2+3} - 6\int \frac{dx}{x^2+4}$$

$$= x + 2.\frac{1}{\sqrt{3}}\tan^{-1}\frac{x}{\sqrt{3}} - 6.\frac{1}{2}\tan^{-1}\frac{x}{2}$$

$$= x + \frac{2}{\sqrt{3}}\tan^{-1}\left(\frac{x}{\sqrt{3}}\right) - 3\tan^{-1}\left(\frac{x}{2}\right).$$

Example 8(a):

Evaluate $\int \frac{(x^2+a^2)(x^2+b^2)}{(x^2+c^2)(x^2+d^2)}dx.$

Solution:

Do your self

Ans. The given integral

$$I = x + \frac{(a^2-c^2)(b^2-c^2)}{(d^2-c^2)} \cdot \frac{1}{c}\tan^{-1}\left(\frac{x}{c}\right) + \frac{(a^2-d^2)(b^2-d^2)}{(c^2-d^2)} \cdot \frac{1}{d}\tan^{-1}\left(\frac{x}{d}\right).$$

Example 8(b):

Integrate $\frac{x}{\{(x-1)^3(x-2)\}}$.

Solution:

Putting (x – 1) = y

or x = y + 1, we get

$$\frac{x}{(x-1)^3(x-2)} = \frac{y+1}{y^3(y+1-2)} = \frac{y+1}{y^3(y-1)} = \frac{1+y}{y^3(-1+y)},$$

[Note that we have arranged the Nr. and the Dr. in ascending powers of y]

$$= \frac{1}{y^3}\left[-1-2y-2y^2+\frac{2y^3}{-1+y}\right], \text{ by actual division}$$

$$= -\frac{1}{y^3} - \frac{2}{y^2} - \frac{2}{y} + \frac{2}{(y-1)}$$

$$= -\frac{1}{(x-1)^3} - \frac{2}{(x-1)^2} - \frac{2}{(x-1)} + \frac{2}{(x-2)}, \qquad [\because y = x-1].$$

Hence the required integral of the given fraction

$$= -\int \frac{dx}{(x-1)^3} - \int \frac{2\,dx}{(x-1)^2} - \int \frac{2\,dx}{(x-1)} + \int \frac{2\,dx}{(x-2)}$$

$$= \frac{1}{x(x-1)^2} + \frac{2}{(x-1)} - 2\log(x-1) + 2\log(x-2).$$

Example 9:

Evaluate $\int \frac{dx}{(x-1)^2(x^2+4)}$.

Solution:

Let $\frac{1}{(x-1)^2(x^2+4)} = \frac{A}{(x-1)} + \frac{B}{(x-1)^2} + \frac{Cx+D}{x^2+4}$.

$\therefore\ 1 \equiv A\ (x - 1)\ (x^2 + 4) + B\ (x^2 + 4) + (Cx + D)\ (x - 1)^2$ ...(1)

Putting $x = 1$ on both sides of (1), we get $B = \frac{1}{5}$.

Now to get the values of C and D we put $x^2 = -4$ on both sides of (1) and we get

$1 \equiv (Cx + D)\ (x^2 - 2x + 1)$

$\Rightarrow 1 \equiv (Cx + D)\ (-4 - 2x + 1)$. $\qquad [\because x^2 = -4]$

$\Rightarrow 1 \equiv (Cx + D)\ (-2x - 3)$

$\Rightarrow 1 \equiv -2Cx^2 - 3Cx - 2Dx - 3D$

$\Rightarrow 1 \equiv 8C - 3Cx - 2Dx - 3D$, $\qquad [\because x^2 = -4]$.

Now equating the coefficients of x and constant terms on both sides, we get

$$-3C - 2D = 0,\ 8C - 3D = 1.$$

Solving these we get $C = \frac{2}{25}, D = -\frac{3}{25}$.

Putting the values of B, C and D in (1), we get

$1 \equiv A\ (x - 1)\ (x^2 + 4) + \frac{1}{5}\left(x^2 + 4\right) + \left(\frac{2}{25} - \frac{3}{25}x\right)\ (x - 1)^2$ ...(2)

Equating the coefficients of x^3 on both sides of (2), we get

$0 = A + \frac{2}{25}$ or $A = -\frac{2}{25}$.

$$\therefore \int \frac{dx}{(x-1)^2\left(x^2+4\right)} = \frac{-2}{25}\int \frac{dx}{x-1} + \frac{1}{5}\int \frac{dx}{(x+1)^2} + \frac{1}{25}\int \frac{2x-3}{x^2+4}dx$$

$$= -\frac{2}{25}\log(x-1) - \frac{1}{5(x-1)} + \frac{1}{25}\int \frac{2x\,dx}{x^2+4} - \frac{3}{25}\int \frac{dx}{x^2+4}$$

$$= -\frac{2}{25}\log(x-1) - \frac{1}{5(x-1)} + \frac{1}{25}\log\left(x^2+4\right) - \frac{3}{50}\tan^{-1}\frac{x}{2}.$$

Example 10:

Evaluate $\int \frac{\left(x^2+x+1\right)dx}{(x+1)^2\,(x+2)}$.

Solution:

Let $\frac{x^2+x+1}{(x+1)^2(x+2)} = \frac{A}{x+1} + \frac{B}{(x+1)^2} + \frac{C}{x+2}$.

$\therefore\ x^2 + x + 1 \equiv A\ (x + 1)\ (x + 2) + B\ (x + 2) + C\ (x + 1)^2$ ...(1)

Putting x = – 1 on both sides of (1), we get C = 3.

Putting the values of B and C in (1), we get

$x^2 + x + 1 \equiv A(x + 1)(x + 2) + (x + 2) + 3(x + 1)^2$ ...(2)

Equating the coefficients of x^2 on both sides of (2), we get

1 + A + 3, or A = – 2.

$$\therefore \int \frac{(x^2+x+1)dx}{(x+1)^2(x+2)}$$

$$= \int \left[\frac{-2}{x+1} + \frac{1}{(x+1)^2} + \frac{3}{x+2}\right]dx$$

$$= -2\log(x+1) - \frac{1}{x+1} + 3\log(x+2)$$

$$= \log\frac{(x+2)^3}{(x+1)^2} - \frac{1}{x+1}.$$

Example 11:

Evaluate $\int \frac{dx}{x^3(x-1)^2(x+1)}$.

Solution:

Let $\frac{1}{x^3(x-1)^2(x+1)}$

$$\equiv \frac{A}{x} + \frac{B}{x^2} + \frac{C}{x^3} + \frac{D}{(x-1)} + \frac{E}{(x-1)^2} + \frac{F}{(x+1)}.$$

$\therefore 1 \equiv Ax^2(x-1)^2(x+1) + Bx(x-1)^2(x+1) + C(x-2)^2(x+1)$
$+ Dx^3(x-1)(x+1) + Ex^3(x+1) + Fx^3(x-1)^2$ ...(1)

Putting x = 0, 1 and – 1 successively on both sides of (1),

we get $C = 1,\ E = \frac{1}{2}$

and $F = -\frac{1}{4}.$

With these values of C, E and F, (1) becomes

$1 \equiv Ax^2(x-1)^2(x+1) + Bx(x-1)^2(x+1) + (x-1)^2(x+1)$

$+ Dx^3(x-1)(x+1) + \frac{1}{2}x^3(x+1) - \frac{1}{4}x^3(x-1)^2.$ ...(2)

To obtain A, B, D, equating the coefficients of x^3, x^4, x^5 on both sides of (2), we get

$$-A - B + 1 - D + \frac{1}{2} - \frac{1}{4} = 0,$$

$$-A + B + \frac{1}{2} + \frac{1}{2} = 0$$

and $\quad A + D - \frac{1}{4} = 0.$

Solving these, we get A = 2, B = 1

and $\quad D = -\frac{7}{4}.$

$$\therefore \int \frac{1}{x^3(x-1)^2(x+1)} dx$$

$$= \int \left[\frac{2}{x} + \frac{1}{x^2} + \frac{1}{x^3} - \frac{7}{4(x-1)} + \frac{1}{2(x-1)^2} - \frac{1}{4(x+1)}\right] dx$$

$$= 2\log x - \frac{1}{x} - \frac{1}{2x^2} - \frac{7}{4}\log(x-1) - \frac{1}{2(x-1)} - \frac{1}{4}\log(x+1).$$

Example 12:

We have $\int \frac{(x^2+1)}{(x^2-1)} dx = \int \frac{(x^2-1+2)}{(x^2-1)} dx$

$$= \int \frac{(x^2-1)}{(x^2-1)} dx + \int \frac{2}{(x^2-1)} dx = \int dx + 2\int \frac{dx}{x^2-1}$$

$$= x + 2 \cdot \frac{1}{2 \times 1} \log \frac{x-1}{x+1} = x + \log \frac{x-1}{x+1}.$$

Example 13:

Integrate $\frac{(x+1)}{(x^3+x^2-6x)}.$

Solution:

Here $\frac{x+1}{x^3+x^2-6x} = \frac{x+1}{x(x-2)(x+3)} \equiv \frac{A}{x} + \frac{B}{x-2} + \frac{C}{x+3}$, (say).

To find A suppress x in the given fraction and put x = 0 in the remaining fraction.

Thus $A = \frac{0+1}{(0-2)(0+3)} = -\frac{1}{6}.$

To find B suppress (x – 2) in the given fraction and put x = 2 in the remaining fraction. Thus $B = \frac{2+1}{2(2+3)} = \frac{3}{10}$.

Similarly $C = \frac{-3+1}{-3(-3-2)} = -\frac{2}{15}$.

Thus $\frac{x+1}{x(x-2)(x+3)} = -\frac{1}{6x} + \frac{3}{10(x-2)} - \frac{2}{15(x+3)}$.

Obviously $\int \frac{(x+1)}{x(x-2)(x+3)} = -\int \frac{1.dx}{6x} + \int \frac{3dx}{10(x-2)} - \int \frac{2dx}{15(x+3)}$

$= -\frac{1}{6}\log x + \frac{3}{10}\log(x-2) - \frac{2}{15}\log(x+3).$

Example 14:

Integrate $\frac{(3x+1)}{\{(x-1)^3(x+1)\}}$.

Solution:

Putting x – 1 = y

so that x = 1 + y, we get

$$\frac{3x+1}{(x-1)^3(x+1)} = \frac{3(1+y)+1}{y^3(2+y)} = \frac{4+3y}{y^3(2+y)},$$

arranging the Nr. and the Dr. in ascending powers of y

$$= \frac{1}{y^3}\left[2 + \frac{1}{2}y - \frac{1}{4}y^2 + \frac{1}{4}\frac{y^3}{2+y}\right], \text{ by actual division}$$

$$= \frac{2}{y^3} + \frac{1}{2y^2} - \frac{1}{4y} + \frac{1}{4}.\frac{1}{(2+y)}$$

$$= \frac{2}{(x-1)^3} + \frac{2}{2(x-1)^2} - \frac{1}{4(x-1)} + \frac{1}{4(x+1)}.$$

Hence the required integral of the given fraction

$$= \int\left[\frac{2}{(x-1)^3} + \frac{1}{2(x-1)^2} - \frac{1}{4(x-1)} + \frac{1}{4(x+1)}\right]dx$$

$$= \frac{-1}{(x-1)^2} - \frac{1}{2(x-1)} - \frac{1}{4}\log(x-1) + \frac{1}{4}\log(x+1)$$

$$= \frac{-1}{(x-1)^2} - \frac{1}{2(x-1)} + \frac{1}{4}\log\frac{x+1}{x-1}.$$

Example 15:

Evaluate $\int \frac{x}{x^4 + x^2 + 1} dx.$

Solution:

Put $x^2 = t$, so that $2x\, dx = dt$.

Then the given integral $I = \frac{1}{2}\int \frac{dt}{t^2 + t + 1}$

$$= \frac{1}{2}\int \frac{dt}{\left(t+\frac{1}{2}\right)^2 + \left(\frac{3}{4}\right)} = \frac{1}{2}\cdot\frac{2}{\sqrt{3}}\tan^{-1}\left[\frac{t+\frac{1}{2}}{\frac{\sqrt{3}}{2}}\right]$$

$$= \frac{1}{\sqrt{3}}\tan^{-1}\left(\frac{2t+1}{\sqrt{3}}\right) = \frac{1}{\sqrt{3}}\tan^{-1}\left(\frac{2x^2+1}{\sqrt{3}}\right).$$

Example 16(a):

Integrate $\frac{1}{(9x^2 - 12x + 8)}.$

Solution:

We have $\int \frac{dx}{9x^2 - 12x + 8} = \frac{1}{9}\int \frac{dx}{x^2 - \frac{4}{3}x + \frac{8}{9}},$

making the coeff. of x^2 in the denominator as 1

$$= \frac{1}{9}\int \frac{dx}{\left(x^2 - \frac{4}{3}x + \frac{4}{9} + \frac{8}{9} - \frac{4}{9}\right)} = \frac{1}{9}\int \frac{dx}{\left(x - \frac{2}{3}\right)^2 + \frac{4}{9}}$$

$$= \frac{1}{9}\int \frac{dx}{\left(x - \frac{2}{3}\right)^2 + \left(\frac{2}{3}\right)^2}$$

$$= \frac{1}{9}\cdot\frac{3}{2}\cdot\tan^{-1}\frac{\left(x - \frac{2}{3}\right)}{\frac{2}{3}} = \frac{1}{6}\tan^{-1}\frac{3x-2}{2}.$$

Example 16(b):

Evaluate $\int_0^1 \left\{ \frac{1}{\left(1-x+x^2\right)} \right\} dx.$

Solution:

$$\text{Dr.} = 1 - x + x^2 = \left(x - \frac{1}{2}\right)^2 + \left(\frac{\sqrt{3}}{2}\right)^2.$$

$$\therefore \int_0^1 \frac{dx}{1-x+x^2} = \int_0^1 \frac{dx}{\left(x-\frac{1}{2}\right)^2 + \left(\frac{\sqrt{3}}{2}\right)^2}$$

$$= \frac{2}{\sqrt{3}}\left[\tan^{-1}\left(\frac{x-\frac{1}{2}}{\frac{\sqrt{3}}{2}}\right)\right]_0^1$$

$$= \frac{2}{\sqrt{3}}\left[\tan^{-1}\left(\frac{2x-1}{\sqrt{3}}\right)\right]_0^1 = \frac{2}{\sqrt{3}}\left[\tan^{-1}\left(\frac{1}{\sqrt{3}}\right) - \tan^{-1}\left(-\frac{1}{\sqrt{3}}\right)\right]$$

$$= \frac{2}{\sqrt{3}}\left[\tan^{-1}\left(\frac{1}{\sqrt{3}}\right) + \tan^{-1}\left(\frac{1}{\sqrt{3}}\right)\right], \qquad [\because \tan^{-1}(-x) = -\tan^{-1} x]$$

$$= \frac{4}{\sqrt{3}}\tan^{-1}\left(\frac{1}{\sqrt{3}}\right) = \frac{4}{\sqrt{3}}\cdot\frac{\pi}{6} = \frac{2\pi}{3\sqrt{3}}.$$

Example 16(c):

Integrate $\frac{1}{\left(2x^2+x+1\right)}.$

Solution:

We have

$$\int \frac{dx}{2x^2+x+1} = \frac{1}{2}\int \frac{dx}{x^2+\frac{1}{2}x+\frac{1}{2}} = \frac{1}{2}\int \frac{dx}{\left(x+\frac{1}{4}\right)^2 + \frac{1}{2} - \frac{1}{16}}$$

$$= \frac{1}{2}\int \frac{dx}{\left(x+\frac{1}{4}\right)^2 + \frac{7}{16}} = \frac{1}{2}\int \frac{dx}{\left(x+\frac{1}{4}\right)^2 + \left(\frac{\sqrt{7}}{4}\right)^2}$$

$$= \frac{1}{2} \cdot \frac{4}{\sqrt{7}} \tan^{-1} \frac{x + \frac{1}{4}}{\frac{(\sqrt{7})}{4}} = \frac{2}{\sqrt{7}} \tan^{-1}\left(\frac{4x+1}{\sqrt{7}}\right).$$

Example 17:

Evaluate $\int \frac{dx}{2x^2 + 3x + 5}$.

Solution:

$$\frac{2}{\sqrt{31}} \cdot \tan^{-1}\left[\frac{4x+3}{\sqrt{31}}\right].$$

Example 18:

Evaluate $\int_{-\infty}^{\infty} \left\{ \frac{1}{(x^2 + 2x + 2)} \right\} dx$.

Solution:

We have $\int_{-\infty}^{\infty} \frac{dx}{x^2 + 2x + 2} = \int_{-\infty}^{\infty} \frac{dx}{(x+1)^2 + 1}$

$$= \tan^{-1}\left[\frac{(x+1)}{1}\right]_{-\infty}^{\infty} = \tan^{-1}(\infty) - \tan^{-1}(-\infty) = \frac{\pi}{2} - \left(-\frac{\pi}{2}\right)$$

$$= \frac{\pi}{2} + \frac{\pi}{2} = \pi.$$

Example 19:

Integrate $\frac{1}{(2x^2 + x - 1)}$.

Solution:

We have $\int \frac{dx}{(2x^2 + x - 1)} = \frac{1}{2} \int \frac{dx}{\left(x^2 + \frac{1}{2}x - \frac{1}{2}\right)}$

$$= \frac{1}{2} \int \frac{dx}{\left(x + \frac{1}{4}\right)^2 - \frac{1}{2} - \frac{1}{16}} = \frac{1}{2} \int \frac{dx}{\left(x + \frac{1}{4}\right)^2 - \frac{9}{16}} = \frac{1}{2} \cdot \frac{1}{2} \cdot \frac{4}{3} \log \frac{x + \frac{1}{4} - \frac{3}{4}}{x + \frac{1}{4} + \frac{3}{4}}$$

$$= \frac{1}{3} \log \left\{ \frac{2x-1}{2(x+1)} \right\} = \frac{1}{3} \log \frac{2x-1}{x+1} - \frac{1}{3} \log 2$$

$= \frac{1}{3}\log\left\{\frac{(2x-1)}{(x-1)}\right\}$, omitting the constant term $-\frac{1}{3}\log 2$ which may be added to constant of integration.

Example 20:

Integrate $\frac{1}{(x^2-3x+2)}$.

Solution:

We have $\int \frac{dx}{(x^2-3x+2)} = \int \frac{dx}{\left(x-\frac{3}{2}\right)^2+2-\frac{9}{4}} = \int \frac{dx}{\left(x-\frac{3}{2}\right)^2-\frac{1}{4}}$

$$= \int \frac{dx}{\left(x-\frac{3}{2}\right)^2-\left(\frac{1}{2}\right)^2} = \frac{1}{2.\left(\frac{1}{2}\right)}\log\left[\frac{\left(x-\frac{3}{2}\right)-\left(\frac{1}{2}\right)}{\left(x-\frac{3}{2}\right)+\left(\frac{1}{2}\right)}\right] = \log\left[\frac{(x-2)}{(x-1)}\right].$$

Example 21:

Integrate $\int_0^1 \frac{(x-3)dx}{x^2+2x-4}$.

Solution:

We have $\int_0^1 \frac{(x-3)dx}{x^2+2x-4} = \int_0^1 \frac{3-x}{4-2x-x^2}dx$

$$= \int_0^1 \frac{\left\{4-\frac{1}{2}(2x+2)\right\}}{(4-2x-x^2)}dx, \qquad \left[\because \frac{d}{dx}(\text{deno min ator}) = -(2x+2)\right]$$

$$= \int_0^1 \frac{4dx}{(4-2x-x^2)} - \frac{1}{2}\int_0^1 \frac{(2x+2)dx}{(4-2x-x^2)}$$

$$= \int_0^1 \frac{4dx}{4-(x^2+2x)} + \frac{1}{2}\left[\log(4-2x-x^2)\right]_0^1$$

$$= \int_0^1 \frac{4dx}{(\sqrt{5})^2-(x+1)^2} + \frac{1}{2}[\log 1 - \log 4]$$

$$= 4\cdot\frac{1}{2\sqrt{5}}\left[\log\frac{\sqrt{5}+(x+1)}{\sqrt{5}-(x+1)}\right]_0^1 - \frac{1}{2}\log(2^2)$$

$$= \frac{2}{\sqrt{5}}\left[\log\left\{\frac{\sqrt{5}+2}{\sqrt{5}-2}\right\} - \log\left\{\frac{\sqrt{5}+1}{\sqrt{5}-1}\right\}\right] - \log 2$$

$$= \frac{2}{\sqrt{5}}\left[\log\left\{\frac{(\sqrt{5}+2)(\sqrt{5}-1)}{(\sqrt{5}-2)(\sqrt{5}+1)}\right\}\right] - \log 2$$

$$= \frac{2}{\sqrt{5}}\left[\log\left\{\frac{3+\sqrt{5}}{3-\sqrt{5}}\right\}\right] - \log 2$$

$$= \frac{2}{\sqrt{5}}\left[\log\left\{\frac{(3+\sqrt{5})(3+\sqrt{5})}{(3-\sqrt{5})(3+\sqrt{5})}\right\}\right] - \log 2$$

$$= \frac{2}{\sqrt{5}}\left[\log\left(\frac{3+\sqrt{5}}{2}\right)^2\right] - \log 2 = \frac{4\sqrt{5}}{5}\left[\log\left(\frac{3+\sqrt{5}}{2}\right)\right] - \log 2.$$

Example 22:

Integrate $\dfrac{1}{(x^3-1)}$.

Solution:

We have $\dfrac{1}{x^3-1} = \dfrac{1}{(x-1)(x^2+x+1)}$.

Let $\dfrac{1}{(x-1)(x^2+x+1)} \equiv \dfrac{A}{x-1} + \dfrac{Bx+C}{x^2+x+1}$. ...(1)

Then $A = \dfrac{1}{1^2+1+1} = \dfrac{1}{3}$.

Now putting x = 0 on both sides of (1), we get 1 = – A + C

so that C = – 1 + A = – 1 + $\dfrac{1}{3} = -\left(\dfrac{2}{3}\right)$.

Also multiplying both sides of (1) by x and taking limit

when $x \to \infty$, we get 0 = A + B so that B = – A = – $\left(\dfrac{1}{3}\right)$.

$\left[\text{Note that } \lim\limits_{x\to\infty} \dfrac{x}{(x-1)(x^2+x+1)}\right.$

$$= \lim_{x\to\infty} \frac{1}{x^2\left\{1-\left(\frac{1}{x}\right)\right\}\left\{1+\left(\frac{1}{x}\right)+\left(\frac{1}{x^2}\right)\right\}} = 0,$$

$$\lim_{x\to\infty} \frac{Ax}{x-1} = \lim_{x\to\infty} \frac{A}{\left\{1-\left(\frac{1}{x}\right)\right\}} = A, \ \lim_{x\to\infty} \frac{Bx^2+Cx}{x^2+x+1}$$

$$\left. = \lim_{x\to\infty} \frac{x^2\left[B+\left(\frac{C}{x}\right)\right]}{x^2\left[1+\left(\frac{1}{x}\right)+\left(\frac{1}{x^2}\right)\right]} = \lim_{x\to\infty} \frac{B+\left(\frac{C}{x}\right)}{1+\left(\frac{1}{x}\right)+\left(\frac{1}{x^2}\right)} = B \right].$$

$$\therefore \int \frac{1}{x^3-1}dx = \int\left[\frac{1}{3(x-1)} - \frac{1}{3}\frac{x+2}{x^2+x+1}\right]dx$$

$$= \frac{1}{3}\int \frac{1}{x-1}dx - \frac{1}{3}\int \frac{x+2}{x^2+x+1}dx$$

$$= \frac{1}{3}\int \frac{1}{x-1}dx - \frac{1}{3}\int \frac{\frac{1}{2}(2x+1)+\frac{3}{2}}{x^2+x+1}dx$$

$$= \frac{1}{3}\log(x-1) - \frac{1}{6}\int \frac{2x+1}{x^2+x+1}dx - \frac{1}{2}\int \frac{dx}{\left(x+\frac{1}{2}\right)^2+\left(\frac{\sqrt{3}}{2}\right)^2}$$

$$= \frac{1}{3}\log(x-1) - \frac{1}{6}\log\left(x^2+x+1\right) - \frac{1}{2.\left(\frac{\sqrt{3}}{2}\right)}\tan^{-1}\left[\frac{\left(x+\frac{1}{2}\right)}{\frac{\sqrt{3}}{2}}\right]$$

$$= \frac{1}{3}\log(x-1) - \frac{1}{6}\log\left(x^2+x+1\right) - \frac{1}{\sqrt{3}}\tan^{-1}\left(\frac{2x+1}{\sqrt{3}}\right).$$

Example 23(a):

Integrate $\frac{(1-3x)}{\left\{\left(1+x^2\right)(1+x)\right\}}$.

Solution:

Let $\frac{(1-3x)}{\left(1+x^2\right)(1+x)} = \frac{A}{(1+x)} + \frac{Bx+C}{\left(x^2+1\right)}$.

$\therefore\ 1 - 3x \equiv A\,(x^2 + 1) + (Bx + C)\,(1 + x).$

Now putting $x = -1,$

we get $A = 2;$

$x = 0,$

$C = -1$

and $x = 1,$

$B = -2.$

$$\therefore \int \frac{(1-3x)dx}{\left(1+x^2\right)(1+x)} = \int \left[\frac{2}{(1+x)} - \frac{(2x+1)}{\left(x^2+1\right)}\right]dx$$

$$= 2\log(1+x) - \int \frac{2x}{\left(x^2+1\right)}dx - \int \frac{1}{\left(x^2+1\right)}dx$$

$$= 2\log(1+x) - \log(x^2+1) - \tan^{-1}x$$

$$= \log(1+x)^2 - \log(x^2+1) - \tan^{-1}x$$

$$= \log\left\{\frac{\left(1+x^2\right)}{\left(1+x^2\right)}\right\} - \tan^{-1}x.$$

Example 23(b):

Integrate $\dfrac{x}{\left(x^2+x-6\right)}.$

Solution:

Let $I = \int \frac{x}{x^2+x-6}dx.$

Here $\frac{d}{dx}$ (denominator) $= \frac{d}{dx}\,(x^2 + x - 6) = 2x + 1.$

$$\therefore\ I = \int \frac{\frac{1}{2}(2x+1) - \frac{1}{2}}{x^2+x-6}dx = \frac{1}{2}\int \frac{2x+1}{x^2+x-6}dx - \frac{1}{2}\int \frac{dx}{x^2+x-6}$$

$$= \frac{1}{2}\log\left(x^2+x-6\right) - \frac{1}{2}\int \frac{dx}{\left(x+\frac{1}{2}\right)^2 - 6 - \frac{1}{4}}$$

$$= \frac{1}{2}\log\left(x^2+x-6\right) - \frac{1}{2}\int \frac{dx}{\left(x+\frac{1}{2}\right)^2 - \frac{25}{4}}$$

$$=\frac{1}{2}\log\left(x^2+x-6\right)-\frac{1}{2}\int\frac{dx}{\left(x+\frac{1}{2}\right)^2-\left(\frac{5}{2}\right)^2}$$

$$=\frac{1}{2}\log\left(x^2+x-6\right)-\frac{1}{2}\cdot\frac{1}{2.\left(\frac{5}{2}\right)}\log\frac{x+\frac{1}{2}-\frac{5}{2}}{x+\frac{1}{2}+\frac{5}{2}}$$

$$=\frac{1}{2}\log\left(x^2+x-6\right)-\frac{1}{10}\log\frac{x-2}{x+3}.$$

Example 23(c):

Integrate $\frac{3x}{\left(x^2-x-2\right)}$.

Solution:

Here $\left(\frac{d}{dx}\right)(x^2 - x - 2) = 2x - 1.$

$$\therefore \int\frac{3x\,dx}{\left(x^2-x-2\right)}=\int\frac{\frac{3}{2}(2x-1)+\frac{3}{2}}{\left(x^2-x-2\right)}dx$$

$$=\frac{3}{2}\int\frac{(2x-1)dx}{\left(x^2-x-2\right)}+\frac{3}{2}\int\frac{dx}{\left(x^2-x-2\right)}$$

$$=\frac{3}{2}\log\left(x^2-x-2\right)+\frac{3}{2}\int\frac{dx}{\left(x-\frac{1}{2}\right)^2-\left(\frac{3}{2}\right)^2}$$

$$=\frac{3}{2}\log\left(x^2-x-2\right)+\frac{3}{2}\cdot\frac{1}{2\left(\frac{3}{2}\right)}\log\left\{\frac{\left(x-\frac{1}{2}\right)-\left(\frac{3}{2}\right)}{\left(x-\frac{1}{2}\right)+\left(\frac{3}{2}\right)}\right\}$$

$$=\frac{3}{2}\log\left(x^2-x-2\right)+\frac{1}{2}\log\left\{\frac{(x-2)}{(x+1)}\right\}.$$

Example 24:

Integrate $\frac{(3x+1)}{\left(2x^2-2x+3\right)}$.

Solution:

Here $\left(\frac{d}{dx}\right)$ $(2x^2 - 2x + 3) = 4x - 2.$

$$\therefore\ I=\int\frac{3x+1}{2x^2-2x+3}dx=\int\frac{\frac{3}{4}(4x-2)+1+\frac{3}{2}}{\left(2x^2-2x+3\right)}dx$$

$$=\frac{3}{4}\int\frac{4x-2}{2x^2-2x+3}dx+\frac{5}{2}\int\frac{dx}{2x^2-2x+3}dx$$

$$=\frac{3}{4}\log\left(2x^2-2x+3\right)+\frac{5}{2.2}\int\frac{dx}{x^2-x+\left(\frac{3}{2}\right)}$$

$$=\frac{3}{4}\log\left(2x^2-2x+3\right)+\frac{5}{4}\int\frac{dx}{\left(x-\frac{1}{2}\right)^2+\left(\frac{3}{2}\right)-\left(\frac{1}{4}\right)}$$

$$=\frac{3}{4}\log\left(2x^2-2x+3\right)+\frac{5}{4}\int\frac{dx}{\left(x-\frac{1}{2}\right)^2+\left(\frac{\sqrt{5}}{2}\right)^2}$$

$$=\frac{3}{4}\log\left(2x^2-2x+3\right)+\frac{5}{4}\frac{1}{\left(\frac{\sqrt{5}}{2}\right)}\tan^{-1}\left\{\frac{x-\frac{1}{2}}{\left(\frac{\sqrt{5}}{2}\right)}\right\}$$

$$=\frac{3}{4}\log\left(2x^2-2x+3\right)+\frac{\sqrt{5}}{2}\tan^{-1}\left(\frac{2x-1}{\sqrt{5}}\right).$$

Example 25:

Integrate $\dfrac{1}{\left(x^2+3\right)^3}$.

Solution:

By the reduction formula of 2.3, we get

$$\int\frac{dx}{\left(x^2+3\right)^3}=\frac{x}{12\left(x^2+3\right)^2}+\frac{3}{12}\int\frac{dx}{\left(x^2+3\right)^2},$$

[Putting n = 3 and k = 3 in the formula]

$$=\frac{x}{12\left(x^2+3\right)^2}+\frac{1}{4}\left\{\frac{x}{6\left(x^2+3\right)}+\frac{1}{6}\int\frac{dx}{\left(x^2+3\right)}\right\},$$

on applying the same reduction formula by putting n = 2 and k = 3

$$=\frac{x}{12\left(x^2+3\right)^2}+\frac{x}{24\left(x^2+3\right)}+\frac{1}{24\sqrt{3}}\tan^{-1}\frac{x}{\sqrt{3}}.$$

Note: Before applying the reduction formula of 2.3, the students must first derive it.

Example 26:

Evaluate $\int\left[\frac{1}{\left(x^2+4\right)^3}\right]dx.$

Solution:

By the reduction formula of 2.3, (putting n = 3, k = 4) we get

$$I=\int\frac{dx}{\left(x^2+4\right)^3}=\frac{x}{(6-2)\cdot 4\left(x^2+4\right)^2}+\frac{(2.3-3)}{2.(3-1)4}\int\frac{dx}{\left(x^2+4\right)^2}$$

$$=\frac{x}{16\left(x^2+4\right)^2}+\frac{3}{16}\int\frac{dx}{\left(x^2+4\right)^2}.$$

Again applying the same reduction formula (putting n = 2, k = 4), we get

$$I=\frac{x}{16\left(x^2+4\right)^2}+\frac{3}{16}\cdot\frac{x}{2.(2-1).4\left(x^2+4\right)}+\frac{3}{16}\cdot\frac{(2.2-3)}{2.(2-1)}\int\frac{dx}{\left(x^2+4\right)}$$

$$=\frac{x}{16\left(x^2+4\right)^2}+\frac{3x}{128\left(x^2+4\right)}+\frac{3}{32}\int\frac{dx}{x^2+4}$$

$$=\frac{x}{16\left(x^2-4\right)^2}+\frac{3x}{128\left(x^2+4\right)}+\frac{3}{32}\tan^{-1}\left(\frac{x}{2}\right)$$

$$=\frac{x}{16\left(x^2+4\right)^2}+\frac{3x}{128\left(x^2+4\right)}+\frac{3}{64}\tan^{-1}\left(\frac{x}{2}\right).$$

Example 27:

Integrate $\frac{x^2-1}{x^4+1}.$

Solution:

Let $I=\int\frac{x^2-1}{x^4+1}dx$

$$=\int\frac{1-\left(\frac{1}{x^2}\right)}{x^2+\left(\frac{1}{x^2}\right)}dx\text{, dividing the numerator and the denominator by } x^2$$

$$= \int \frac{1-\left(\frac{1}{x^2}\right)}{\left[x+\left(\frac{1}{x}\right)\right]^2 - 2} dx, \qquad \left[\text{Note that } \frac{d}{dx}\left\{x+\frac{1}{x}\right\} = 1-\frac{1}{x^2}\right]$$

Now put $x+\left(\frac{1}{x}\right) = t$

so that $\left\{1-\left(\frac{1}{x^2}\right)\right\}dx = dt.$

$$\therefore\ I = \int \frac{dt}{t^2-2} = \int \frac{dt}{t^2-\left(\sqrt{2}\right)^2}$$

$$= \frac{2}{2\sqrt{2}} \log \frac{t-\sqrt{2}}{t+\sqrt{2}} = \frac{1}{2\sqrt{2}} \log \frac{x+\left(\frac{1}{x}\right)-\sqrt{2}}{x+\left(\frac{1}{x}\right)+\sqrt{2}}$$

$$= \frac{1}{2\sqrt{2}} \log \frac{x^2-\sqrt{2}x+1}{x^2+\sqrt{2}x+1}.$$

Example 28:

Integrate $\dfrac{(x+2)}{\left(2x^2+4x+3\right)^2}$.

Solution:

Here $\left(\frac{d}{dx}\right)$ $(2x^2 + 4x + 3) = 4x + 4.$

$$\therefore \int \frac{(x+2)dx}{\left(2x^2+4x+3\right)^2} = \int \frac{1/4(4x+4)+2-1}{\left(2x^2+4x+3\right)^2} dx$$

$$= \frac{1}{4}\int \frac{(4x+4)dx}{\left(2x^2+4x+3\right)^2} + \frac{1}{4}\int \frac{(2-1)dx}{\left(x^2+2x+\frac{3}{2}\right)^2}$$

$$= \frac{1}{4}\int \left(2x^2+4x+3\right)^{-2}(4x+4)dx + \frac{1}{4}\int \frac{dx}{\left(x^2+2x+\frac{3}{2}\right)^2}$$

$$= -\frac{1}{4\left(2x^2+4x+3\right)} + \frac{1}{4}\int \frac{dx}{\left\{(x+1)^2+\frac{1}{2}\right\}^2}.$$

Now put x + 1 = t and then applying the reduction formula of 2.3, we get

$$I = -\frac{1}{4\left(2x^2+4x+3\right)} + \frac{1}{4}\left[\frac{(x+1)}{(x+1)^2+\frac{1}{2}} + \sqrt{2}\tan^{-1}\left\{\sqrt{2}(x+1)\right\}\right].$$

Example 29:

Integrate $\dfrac{(2x+3)}{\left(x^2+2x+3\right)^2}$.

Solution:

Here $\left(\frac{d}{dx}\right)$ $(x^2 + 2x + 3) = 2x + 2.$

$$\therefore\quad I = \int \frac{(2x+3)dx}{\left(x^2+2x+3\right)^2} = \int \frac{(2x+2+1)dx}{\left(x^2+2x+3\right)^2}$$

$$= \int \frac{(2x+2)dx}{\left(x^2+2x+3\right)^2} + \int \frac{dx}{\left(x^2+2x+3\right)^2}$$

$$= -\frac{1}{\left(x^2+2x+3\right)} + \int \frac{dx}{\left\{(x+1)^2+2\right\}^2}. \qquad ...(1)$$

Now let $I_1 = \int \dfrac{dx}{\left[(x+1)^2+2\right]^2}$.

Put $x + 1 = \sqrt{2}\tan t$,

so that $dx = \sqrt{2}\sec^2 t\ dt$.

$$\therefore\quad I_1 = \int \frac{\sqrt{2}\sec^2 t\,dt}{\left(2\tan^2 t+2\right)^2} = \frac{\sqrt{2}}{4}\int \cos^2 t\,dt = \frac{\sqrt{2}}{4}\int \frac{1}{2}(1+\cos 2t)dt$$

$$= \frac{\sqrt{2}}{8}\left[t + \frac{1}{2}\sin 2t\right] = \frac{\sqrt{2}}{8}[t + \sin t\cos t].$$

Now $\tan t = \dfrac{x+1}{\sqrt{2}}$.

Therefore $\sin t = \dfrac{x+1}{\sqrt{\left\{(x+1)^2+2\right\}}} = \dfrac{x+1}{\sqrt{\left(x^2+2x+3\right)}}$.

and $\cos t = \frac{\sqrt{2}}{\sqrt{(x^2+2x+3)}}$;

also $t = \tan^{-1}\left(\frac{x+1}{\sqrt{2}}\right)$.

Hence $I_1 = \frac{\sqrt{2}}{8}\tan^{-1}\left(\frac{x+1}{\sqrt{2}}\right) + \frac{\sqrt{2}}{8}\cdot\frac{x+1}{\sqrt{(x^2+2x+3)}}\cdot\frac{\sqrt{2}}{\sqrt{(x^2+2x+3)}}$

$= \frac{\sqrt{2}}{8}\tan^{-1}\left(\frac{x+1}{\sqrt{2}}\right) + \frac{1}{4}\frac{x+1}{(x^2+2x+3)}$.

$\therefore\ I = -\frac{1}{x^2+2x+3} + \frac{1}{4}\frac{x+1}{x^2+2x+3} + \frac{\sqrt{2}}{8}\tan^{-1}\left(\frac{x+1}{\sqrt{2}}\right)$, from (1)

$= \frac{x+1-4}{4(x^2+2x+3)} + \frac{\sqrt{2}}{8}\tan^{-1}\left(\frac{x+1}{\sqrt{2}}\right)$

$= \frac{x-3}{4(x^2+2x+3)} + \frac{\sqrt{2}}{8}\tan^{-1}\left(\frac{x+1}{\sqrt{2}}\right)$.

Example 30(a):

Integrate $\frac{x^2}{\{(x+1)(x-2)(x+3)\}}$.

Solution:

Let $\frac{x^2}{(x+1)(x-2)(x+3)} \equiv \frac{A}{x+1} + \frac{B}{x-2} + \frac{C}{x+3}$.

We have $A = \frac{(-1)^2}{(-1-2)(-1+3)} = \frac{1}{(-3)\times(2)} = -\frac{1}{6}$,

$B = \frac{2^2}{(2+1)(2+3)} = \frac{4}{15}$, and

$C = \frac{(-3)^2}{(-3+1)(-3-2)} = \frac{9}{-2\times(-5)} = \frac{9}{10}$.

$\therefore$ required integral

$= \int\left[-\frac{1}{6(x+1)} + \frac{4}{15(x-2)} + \frac{9}{10(x+3)}\right]dx$

$= -\frac{1}{6}\log(x+1) + \frac{4}{15}\log(x-2) + \frac{9}{10}\log(x+3)$

$$=\frac{9}{10}\log(x+3)+\frac{4}{15}\log(x-2)-\frac{1}{6}\log(x+1).$$

Example 30(b):

Integrate $\dfrac{x^3}{\left\{(x+1)^4\,(x+2)\,(x-1)\right\}}$.

Solution:

First we shall break the given fraction into partial fractions. Putting $x + 1 = y$, we get

$$\frac{x^3}{(x+1)^4\,(x+2)(x-1)}=\frac{1}{y^4}\,\frac{(-1+y)^3}{(1+y)\,(-2+y)}$$

$=\dfrac{1}{y^4}\cdot\left[\dfrac{-1+3y-3y^2+y^3}{-2-y+y^2}\right]$, on arranging the Nr. and Dr. in ascending powers of y.

Now we divide the numerator by the denominator till y^4 is a factor of the remainder. The actual division has been shown below:

$$-2-y+y^2 \quad -1+3y-3y^2+y^3 \quad \frac{1}{2}-\frac{7}{4}y+\frac{21}{8}y^2-\frac{43}{16}y^3$$

$$-1-\frac{1}{2}y+\frac{1}{2}y^2$$

$$\frac{7}{2}y-\frac{7}{2}y^2+y^3$$

$$\frac{7}{2}y+\frac{7}{4}y^2-\frac{7}{4}y^3$$

$$-\frac{21}{4}y^2+\frac{11}{4}y^3$$

$$-\frac{21}{4}y^2-\frac{21}{8}y^3+\frac{21}{8}y^4$$

$$\frac{43}{8}y^3-\frac{21}{8}y^4$$

$$\frac{43}{8}y^3+\frac{43}{16}y^4-\frac{43}{16}y^5$$

$$-\frac{85}{16}y^4+\frac{43}{16}y^5$$

∴ the given fraction

$$=\frac{1}{2y^4}-\frac{7}{4y^3}+\frac{21}{8y^2}-\frac{43}{16y}+\frac{-\frac{85}{16}+\frac{43}{16}y}{-2-y+y^2}$$

$$=\frac{1}{2(x+1)^4}-\frac{7}{4(x+1)^3}+\frac{21}{8(x+1)^2}-\frac{43}{16(x+1)}+\frac{-85+43(x+1)}{16(x+2)(x-1)}$$

Also by resolving into partial fractions the last term

$$\frac{43x-42}{16(x+2)(x-1)}=\frac{8}{3(x+2)}+\frac{1}{48(x-1)}.$$

Hence the required integral of the given fraction

$$=-\frac{1}{6(x+1)^3}+\frac{7}{8(x+1)^2}-\frac{21}{3(x+1)}-\frac{43}{16}\log(x+1)$$

$$+\frac{8}{3}\log(x+2)+\frac{1}{48}\log(x-1).$$

Example 31:

Integrate $\dfrac{(x^2+2)}{\{(x-1)(x-2)^3\}}.$

Solution:

Putting $x-2=y$ or $x=y+2$, we get

$$\frac{x^2+2}{(x-1)(x-2)^3}=\frac{(y+2)^2+2}{(y+2-1)y^3}=\frac{y^2+4y+6}{(y+1)y^3}=\frac{1}{y^3}\left[\frac{6+4y+y^2}{1+y}\right]$$

$=\dfrac{1}{y^3}\left[6-2y+3y^2-\dfrac{3y^3}{1+y}\right]$, by actual division *i.e.*, by dividing $6+4y+y^2$ by $1+y$ till y^3 is a factor of the remainder

$$=\frac{6}{y^3}-\frac{2}{y^2}+\frac{3}{y}-\frac{3}{1+y}$$

$$=\frac{6}{(x-2)^3}-\frac{2}{(x-2)^2}+\frac{3}{(x-2)}-\frac{3}{(x-1)}.$$

Hence the required integral of the given fraction

$$=-\frac{3}{(x-2)^2}+\frac{2}{(x-2)}+3\log(x-2)-3\log(x-1).$$

8

Elementary Integration

DEFINITION

The process inverse to differentiation is defined as integration.

Thus $\frac{d}{dx}F(x) = f(x)$, we say that F(x) is an *integal* or a *primitive* of f(x) and, in symbols, we write

$$\int f(x)dx = F(x).$$

The letter x in dx denotes that the integration is to be performed with respect to the variable x.

The process of determing an integral of a function is called integration and the function to be integrated is called integrand.

INTEGRAL OF THE PRODUCT OF TWO FUNCTIONS

Integration by Parts: Let u and v be two functions of x. Then we have from differential calculus

$$\frac{d}{dx}(uv) = u\cdot\frac{dv}{dx} + v\cdot\frac{du}{dx} \qquad ...(1)$$

Integrating both sides of (1) with respect to x, we have

$$uv = \int u\cdot\frac{dv}{dx}dx + \int v\cdot\frac{du}{dx}dx.$$

By transposition, we have

$$\int u\frac{dv}{dx}dx = uv - \int v\frac{du}{dx}dx. \qquad ...(2)$$

Now put $u = f_1(x)$ and $v = \int f_2(x)dx$,

so that $\frac{dv}{dx} = f_2(x).$

Then from (2), we have

$$\int f_1(x) f_2(x) dx = f_1(x) \cdot \int f_2(x) dx - \int \left[\left\{ \frac{d}{dx} f_1(x) \right\} \cdot \int f_2(x) dx \right] dx$$

i.e., the integral of the product of two functions

= first function × integral of second function

– integral of {diff. coeff. of first function × Integral of second function}.

Notes:

1. Care must be taken in choosing the first function and the second function. Obviously we must take that function as the second function whose integral is well known to us. Thus to evaluate $\int x \log x \, dx$ we shall take x as the second function because we so far do not know the integral of log x. But to evaluate $\int x \sin x \, dx$ we must take sin x as the second function and x as the first function. Here if we take x as the second function, then the new integral will become more complicated. Thus to evaluate integrals of the type $\int x^2 e^x dx$ $\int x^3 \cos x \, dx$ etc., the function of the type x^n must be taken as the first function. In certain cases we can take unity (*i.e.*, 1) as the second function. Thus to evaluate $\int \log x \, dx$ we shall take 1 as the second function. To evaluate $\int e^x \sin x \, dx$ we can take either e^x or sin x as the second function.
2. The formula of integration by parts can be applied more than once if necessary.
3. *Integration by parts as applied to the functions of the type $e^x [f(x) + f'(x)]$.*

Let $I = \int e^x [f(x) + f'(x)] dx = \int e^x f(x) dx + \int e^x f'(x) dx$.

Integrating the first integral by parts regarding e^x as the 2nd function, we have

$$I = \left[f(x) e^x - \int f'(x) e^x dx \right] + \int f'(x) e^x dx = e^x f(x).$$

[Note that we have left the other integral unchanged because the last two integrals cancel each other].

SUCCESSIVE INTEGRATION BY PARTS

If u is a function of the type

$a_0 x^n + a_1 x^{n-1} + \ldots + a_{n-1} x + a_n$, where n is a positive integer, the following formula for successive integration by parts can be applied. While writing this formula the successive differential coefficients of u have been denoted by u', u", U''' etc., while the successive integrals of v have been denoted by v_1, v_2, v_3 etc. Thus

$$\int uv\, dx = uv_1 - u'v_2 + u''v_3 - u'''v_4 + ...$$

This process of successive integration by parts will be continued till on being differentiated successively the differential coefficient of u becomes zero. The following examples will make the process clear.

Example 1:

Evaluate $\int x^4 \sin x\, dx$.

Solution:

Applying successive integration by parts, the given integral

$I = x^4 - (-\cos x) - (4x^3) \cdot (-\sin x) + (12x^2) \cdot (\text{cox } x)$

$- (24x) \cdot (\sin x) + (24) \cdot (-\cos x).$

$= -x^4\cos x + 4x^3\sin x + 12x^2 \cos x - 24 x \sin x - 24 \cos x.$

Remark:

While applying successive integration by parts the successive differential coefficients and the successive integrals must at the first stage be put within brackets.

Example 2:

Evaluate $\int x^3 e^{-x} dx$.

Solution:

Applying successive integration by parts, the given integral

$I = (x^3) \cdot (-e^{-x}) - (3x^2) \cdot (e^{-x}) + (6x) \cdot (-e^{-x}) - (6) \cdot (e^{-x})$

$= -x^3e^{-x} - 3x^2e^{-x} - 6xe^{-x} - 6e^{-x}.$

$= -(x^3 + 3x^2 + 6x + 6)\, e^{-x}.$

INTEGRALS OF e^{ax} cos bx AND e^{ax} sin bx

Let $I = \int e^{ax} \sin bx\, dx$.

Integrating by parts taking sin bx as the second function, we get

$$I = -\int \frac{e^{ax} \cos bx}{b} - \int ae^{ax}\left(-\frac{\cos bx}{b}\right)dx$$

$$= -\frac{e^{ax} \cos bx}{b} + \frac{a}{b}\int e^{ax} \cos bx\, dx.$$

Again integrating by parts taking cos bx as the second function, we get

$$I = -\frac{e^{ax}\cos bx}{b} + \frac{a}{b}\left[\frac{e^{ax}\sin bx}{b} - \int ae^{ax}\frac{\sin bx}{b}dx\right]$$

or $I = -\frac{e^{ax}\cos bx}{b} + \frac{a}{b^2}e^{ax}\sin bx - \frac{a^2}{b^2}\int e^{ax}\sin bx\,dx$

or $I = \frac{e^{ax}}{b^2}(a\sin bx - b\cos bx) - \frac{a^2}{b^2}I.$ $\left[\because \int e^{ax}\sin bx\,dx = I\right]$

Transposing the term $-\frac{a^2}{b^2}I$ to the left hand side, we get

$$\left(1+\frac{a^2}{b^2}\right)I = \frac{e^{ax}}{b^2}\text{ (a sin bx – b cos bx)}$$

or $\frac{1}{b^2}(a^2+b^2)I = \frac{1}{b^2}e^{ax}(a\sin bx - b\cos bx).$

$\therefore\ I = \frac{e^{ax}}{a^2+b^2}$ (a sin bx – b cos bx).

Thus $\int e^{ax}\sin bx\,dx = \frac{e^{ax}}{a^2+b^2}$ (a sin bx – b cos bx).

Similarly $\int e^{ax}\cos bx\,dx = \frac{e^{ax}}{a^2+b^2}$ (a cos bx + b sin bx).

Alternative forms of $\int e^{ax}\sin bx\,dx$ and $\int e^{ax}\cos bx\,dx$

We have $\int e^{ax}\sin bx\,dx = \frac{e^{ax}}{a^2+b^2}$ (a sin bx – b cos bx).

Put a = r cos θ

and b = r sin θ. Then

$r = \sqrt{(a^2+b^2)}$ and $\theta = \tan^{-1}\left(\frac{b}{a}\right)$.

Now we have

$$\int e^{ax}\sin bx\,dx = \frac{e^{ax}}{r^2}\text{ (r cos θ sin bx – r sin θ cos bx)}$$

$$= \frac{e^{ax}}{r}\sin(bx-\theta).$$

Thus $\int e^{ax}\sin bx\,dx = \frac{e^{ax}}{\sqrt{(a^2+b^2)}}\sin\left(bx - \tan^{-1}\frac{b}{a}\right).$

Similarly $\int e^{ax} \cos bx\, dx = \frac{e^{ax}}{\sqrt{(a^2+b^2)}} \cos\left(bx - \tan^{-1}\frac{b}{a}\right)$.

HYPERBOLIC FUNCTIONS

Following fundamental properties of hyperbolic functions should be *committed to memory* by the students as we shall make frequent use of them during the study of integral calculus.

$\sinh x = (e^x - e^{-x})/2$, $\cosh x = (e^x + e^{-x})/2$

$\tanh x = (e^x - e^{-x})/(e^x + e^{-x})$,

$\coth x = (e^x + e^{-x})/(e^x - e^{-x})$

$\text{sech}\, x = 2/(e^x + e^{-x})$, $\text{cosech}\, x = 2/(e^x - e^{-x})$

$\cosh^2 x - \sinh^2 x = 1$, $\text{sech}^2 x = 1 - \tanh^2 x$

$\text{cosech}^2 x = \coth^2 x - 1$, $\sinh 2x = 2 \sinh x \cosh x$

$\cosh^2 x = \cosh^2 x + \sinh^2 x = 1 + 2\sinh^2 x = 2\cosh^2 x - 1$.

Logarithmic values of inverse hyperbolic functions.

(i) $\sinh^{-1} x = \log\left[x + \sqrt{(x^2+1)}\right]$;

$$\sinh^{-1}\left(\frac{x}{a}\right) = \log\left[\left\{x + \sqrt{(x^2+a^2)}\right\}/a\right]$$

(ii) $\cosh^{-1} = \log\left[x + \sqrt{(x^2-1)}\right]$;

$$\cosh^{-1}\left(\frac{x}{a}\right) = \log\left[\left\{x + \sqrt{(x^2-a^2)}\right\}/a\right]$$

(iii) $\tanh^{-1} x = \frac{1}{2}\log\left\{\frac{(1+x)}{(1-x)}\right\}$;

$$\tanh^{-1}\left(\frac{x}{a}\right) = \frac{1}{2}\log\left\{\frac{(a+x)}{(a-x)}\right\}, (x < a)$$

(iv) $\coth^{-1} x = \frac{1}{2}\log\left\{\frac{(x+1)}{(x-1)}\right\}$;

$$\coth^{-1}\left(\frac{x}{a}\right) = \frac{1}{2}\log\left\{\frac{(x+a)}{(x-a)}\right\}, (x > a).$$

CONSTANT OF INTEGRATION

As the differential coefficient of a constant is zero, we have

$$\frac{d}{dx}[F(x)+C]=f(x), \text{ if } \frac{d}{dx}F(x)=f(x);$$

therefore $\int f(x)dx = F(x)+c$.

This constant c is called a *constant of integration* and can take any constant value. Also $\int f(x)dx$ is called the *Indefinite integral* of f(x) w.r.t. 'x'; for by giving different values to the constant of integration the indefinite nature is preserved.

For example, we know that

$\frac{d}{dx}\sin^{-1}x = \frac{1}{\sqrt{(1-x^2)}}$ and ; is follows, when the omit the constant of integration, that $\int \frac{1}{\sqrt{(1-x^2)}}dx$ is equal to $-\cos^{-1}x$.

But it is wrong to conclude from above that $\sin^{-1}x$ is equal to $-\cos^{-1}x$. The correct inference is that the two integrals, given above differ in their constant of integration.

The correct result, as we see from trigonometry, is that

$$\sin^{-1}x = \frac{1}{2}\pi - \cos^{-1}x.$$

The arbitrary constant of integration may be imaginary also. Generally such a constant is added to make the result real.

EXTENDED FORMS OF FUNDAMENTAL FORMULAE

Suppose we know that $\int f(x)dx = F(x)$ and we want to find $\int f(ax+b)dx$.

Let $\quad I = \int f(ax+b)dx$.

Put $\quad ax + b = t$ so that $a\,dx = dt$.

Then $\quad I = \int f(t)\frac{dt}{a} = \frac{1}{a}\int f(t)dt = \frac{1}{a}F(t) = \frac{1}{a}F(ax+b)$

Thus if $\int f(x)dx + F(x)$, then $\int f(ax+b)dx + \frac{1}{a}F(ax+b)$.

From this we conclude the following results:

(i) $\int \sec^2(ax+b)dx = \frac{1}{a}\tan(ax+b)$,

(ii) $\int \text{cosec}^2(ax+b)dx = -\frac{1}{a}\cot(ax+b)$,

(iii) $\int \sec(ax+b)\tan(ax+b)dx = \frac{1}{a}\sec(ax+b)$,

(iv) $\int \text{cosec}(ax+b)\cot(ax+b)dx = -\frac{1}{a}\text{cosec}(ax+b)$,

(v) $\int \frac{1}{\sqrt{(a^2-x^2)}}dx = \sin^{-1}\frac{x}{a}$ and $\int \frac{1}{a^2+x^2}dx = \frac{1}{a}\tan^{-1}\frac{x}{a}$.

(vi) $\int (ax+b)^n dx = \frac{1}{a}\frac{(ax+b)^{n+1}}{(n+1)}$,

(vii) $\int \frac{1}{(ax+b)^n}dx = -\frac{1}{a(n-1)(ax+b)^{n-1}}$,

(viii) $\int \frac{1}{ax+b}dx = \frac{1}{a}\log(ax+b)$,

(ix) $\int e^{ax+b}dx = \frac{1}{a}e^{ax+b}$ and $\int a^{px+q}dx = \frac{1}{p}\frac{a^{px+q}}{\log_e a}$,

(x) $\int \sin(ax+b)dx = -\frac{1}{a}\cos(ax+b)$,

To evaluate $\int \frac{1}{\sqrt{(a^2-x^2)}}dx$.

Put $x = a\sin\theta$,

so that $dx = a\cos\theta\, d\theta$.

Also $a^2 - x^2 = a^2(1-\sin^2\theta) = a^2\cos^2\theta$.

Thus $$\int \frac{1}{\sqrt{(a^2-x^2)}}dx = \int \frac{1}{a\cos\theta}a\cos\theta\, d\theta$$

$$= \int 1.d\theta = \theta = \sin^{-1}\left(\frac{x}{a}\right).$$

And to evaluate $\int \frac{1}{\sqrt{(a^2+x^2)}}dx$.

Put $x = a\sinh\theta$;

so that $dx = a\cosh\theta\, d\theta$.

Also we have $a^2 + x^2 = a^2 (1 + \sinh^2 \theta) = a^2 \cosh^2 \theta$.

Thus $\int \frac{1}{\sqrt{(a^2+x^2)}} dx = \int \frac{1}{a \cosh\theta} a \cosh\theta \, d\theta$

$$= \int 1.d\theta = \theta = \sin^{-1}\left(\frac{x}{a}\right) = \log\left\{\frac{x+\sqrt{(x^2+a^2)}}{a}\right\}$$

$$= \log\left\{x+\sqrt{(x^2+a^2)}\right\} - \log a = \log\left\{x+\sqrt{(x^2+a^2)}\right\},$$

omitting the constant term – log a because it may be added to the constant of integration c which we usually do not write.

Similarly,

$$\int \frac{1}{\sqrt{(x^2-a^2)}} dx = \cosh^{-1}\left(\frac{x}{a}\right) = \log\left\{x+\sqrt{(x^2-a^2)}\right\}$$

$$\int \sqrt{(a^2-x^2)} dx = \frac{x\sqrt{(a^2-x^2)}}{2} + \frac{a^2}{2} \sin^{-1}\left(\frac{x}{a}\right)$$

$$\int \sqrt{(a^2+x^2)} dx = \frac{x\sqrt{(a^2+x^2)}}{2} + \frac{a^2}{2} \sinh^{-1}\left(\frac{x}{a}\right)$$

$$\int \sqrt{(x^2-a^2)} dx = \frac{x\sqrt{(x^2-a^2)}}{2} - \frac{a^2}{2} \cosh^{-1}\left(\frac{x}{a}\right).$$

Example 1:

Evaluate $\int x^5 e^x dx$.

Solution:

Here e^x will be successively integrated and x^5 will be successively differentiated. Thus applying successive integration by parts, the given integral

$I = x^5 e^x = (5x^4) e^x + (20x^3) e^x - (60x^2) e^x + (12^x) e^x - 120e^x$,

the process of successive integration by parts terminates because the differential coefficient of 120 is zero.

$\therefore I = x^5e^x - (5x^4 + 20x^3 - 60x^2 + 120x - 120)$.

Example 2:

Integrate $\dfrac{1}{\sqrt{\left[7-\left(\frac{1}{2}x-3\right)^2\right]}}$.

Solution:

We know that

$$\int \frac{1}{\sqrt{(a^2 - x^2)}} dx = \sin^{-1}\frac{x}{a}.$$

Here in place of x we have $\left(\frac{1}{2}x - 3\right)$; therefore after applying the standard result we shall divide in the end by the coefficient of x in $\frac{1}{2}x - 3$ *i.e.* by $\frac{1}{2}$. Also here $a^2 = 7$; $\therefore$ $a = \sqrt{7}$.

Thus the required integral $I = \int \frac{1}{\sqrt{\left[7 - \left(\frac{1}{2}x - 3\right)^2\right]}} dx$

$$= \frac{1}{(1/2)}\sin^{-1}\left\{\frac{(1/2x - 3)}{\sqrt{7}}\right\} = 2\sin^{-1}\left\{\frac{(x - 6)}{2}\sqrt{7}\right\}.$$

Example 3:

Integrate $\frac{1}{(1 + \sin x)}$.

Solution:

Here $I = \int \frac{1}{1 + \sin x} dx = \int \frac{dx}{1 - \cos\left(\frac{1}{2}\pi + x\right)}$

$$= \int \frac{dx}{2\sin^2\left(\frac{1}{4}\pi + \frac{1}{2}x\right)} = \frac{1}{2}\int \operatorname{cosec}^2\left(\frac{1}{2}x + \frac{1}{4}\pi\right)dx$$

$$= -\cot\left(\frac{1}{2}x + \frac{1}{4}\pi\right).$$

METHODS OF INTEGRATION

There are various mehtods of integration by which we can reduce the given integral to one of the fundamental or known integrals. Following are the four principal methods of integration:

(i) Integration by substitution,

(ii) Integration by parts,

(iii) Integration by decomposition into sum,

(iv) Integration by successive reduction.

(i) **Integration by Substitution :** A change in the variable of integration often reduces an integral to one of the fundamental integrals. The method in which we change the variable to some other variable is called the *Method of substitution.*

Let $I = \int f(x)dx$; then by differentiation w.r.t. x, we have

$$\frac{dI}{dx} = f(x).$$

Now put $x = \phi(t)$,

so that $\frac{dx}{dt} = \phi'(t)$.

Then $\frac{dI}{dt} = \frac{dI}{dx} \cdot \frac{dx}{dt} = f(x).$

$\phi'(t) = f\{\phi(t)\}\ \phi'(t)$, for $x = \phi(t)$.

This gives, $I = \int f\{\phi(t)\}.\phi'(t)dt$.

Rule to Remember

To evaluate $\int f\{\phi(x)\}.\phi'(x)dx$,

put $\phi(x) = t$ and $\phi'(x)\ dx = dt$,

where $\phi'(x)$ is the different coefficient of $\phi(x)$ w.r.t. x.

Important : The success of the method of substitution depends on choosing the substitution $x = \phi(t)$ so that the new integrand $f\{\phi(t)\}.\phi'(t)$ is of a form whose integral is known.

This is done by guess rather than in according with some rule. However, try to put that expression of x equal to t whose differential coefficient is multiplied with dx.

Example 1:

Evaluate $\int \left[\frac{4x^3}{(1+x^8)}\right] dx.$

Solution:

Put $x^4 = t$ so that $4x^3\ dx = dt$.

$$\therefore\ I = \int \left[\frac{4x^3}{(1+x^8)}\right] dx = \int \left[\frac{1}{(1+t^2)}\right] dt = \tan^{-1} t = \tan^{-1} x^4.$$

Example 2:

Evaluate $-\int \frac{\cos^{-1}x}{\sqrt{(1-x^2)}}dx$.

Solution:

Put $\cos^{-1} x = t$,

so that $\frac{1}{\sqrt{(1-x^2)}}dx = dt$.

$\therefore$ the given integral $= \int t\,dt = \frac{1}{2}t^2 = \frac{1}{2}(\cos^{-1} x)^2$.

Example 3:

Evalulate $\int\left[\frac{1}{(cx+d)^4}\right]dx$.

Solution:

Put $cx + d = t$;

then $c\,dx = dt$,

or $dx = \frac{dt}{c}$.

Thus the giv^n integral

$$= \int (cx+d)^{-4}dx = \int t^{-4}\frac{dt}{c} = \frac{1}{c}\int t^{-4}dt = \frac{1}{c}\cdot\frac{t^{-3}}{-3}$$

$$= -\frac{1}{3c(cx+d)^3}.$$

THREE IMPORTANT FORMS OF INTEGRALS

1. $\int \frac{f'(x)}{f(x)}dx = \log f(x)$.

Put $f(x) = t$; differentiating we have $f'(x)\,dx = dt$.

$\therefore \int \frac{f'(x)}{f(x)}dx = \int \frac{dt}{t} = \log t = \log f(x)$.

Thus **(Remember)**

the integral of a fraction whose numerator is the exact derivative of its denominator is equal to the logarithm of its denominator.

For example $\int \frac{4x^3}{1+x^4}dx = \log(1+x^4)$,

as in this case numerator is the exact derivative of the denominator.

Similarly $\int \frac{e^x}{1+e^x} dx = \log\left(1+e^x\right)$.

Integrals of tan x, cot x, sec x and cosec x

(i) $\int \tan x \, dx = \int \frac{\sin x}{\cos x} dx = -\int \frac{(-\sin x)}{\cos x} dx$,

adjusting the numerator as the exact diff. coeff. of the denominator

$= - \log x \cos = \log (\cos x)^{-1} = \log (\sec x)$.

(ii) Similarly, $\int \cot x \, dx = \int \frac{\cos x \, dx}{\sin x} = \log(\sin x)$.

(iii) $\int \operatorname{cosec} x \, dx = \int \frac{dx}{\sin x} = \int \frac{dx}{2\sin\frac{1}{2}x\cos\frac{1}{2}x} = \int \frac{\sec^2\frac{1}{2}x \, dx}{2\tan\frac{1}{2}x}$,

$\left[\text{dividing Nr. and Dr. by } \cos^2\frac{1}{2}x\right]$

$= \log\left(\tan\frac{1}{2}x\right)\left(\because \frac{1}{2}\sec^2\frac{1}{2}x \text{ is the diff. coeff. of } \tan\frac{1}{2}x\right)$.

(iv) $\int \sec x \, dx = \int \operatorname{cosec}\left(\frac{1}{2}\pi + x\right) dx$.

Now proceeding as in the case of $\int \operatorname{cosec} x \, dx$, we have

$\int \sec x \, dx = \log \tan\left(\frac{x}{2} + \frac{\pi}{4}\right)$.

Alternative Method

$\int \sec x \, dx = \int \frac{\sec x(\sec x + \tan x)}{\sec x + \tan x} dx$,

[We have multiplied the Nr. and Dr. both by (sec x + tanx)]

$= \int \frac{\sec x \tan x + \sec^2 x}{\sec x + \tan x}$, [Here Nr. is the diff. coeff. of Dr.]

$= \log (\sec x + \tan x)$.

2. $\int [f(x)]^n f'(x) \, dx = \frac{[f(x)]^{n+1}}{n+1}$, when $n \neq -1$.

Putting $f(x) = t$, so that $f'(x) \, dx = dt$, we get

$$\int[f(x)]^n f'(x)dx = \int t^n dt = \frac{t^{n+1}}{n+1}, \text{ (for } n \neq -1) = \frac{[f(x)]^{n+1}}{n+1}.$$

Thus **Remember:** *If the integrand consists of the product of a constant power of a function f(x) and the derivative f '(x) of f(x), to obtain the integral we increase the index by unity and then divide by the increased index.* This is known as **Power formula.** The students are advised to have a lot of practice of applying this formula.

3. $\int f'(ax+b)dx = \frac{f(ax+b)}{a}.$

TWO SIMPLE THEOREMS

Theorem 1:

The integral of a sum of difference of a finite number of functions is equal to the sum or difference of the integrals of the functions. Symbolically

$$\int[f_1(x) \pm f_2(x) \pm ... \pm f_n(x)]dx$$

$$= \int f_1(x)dx \pm \int f_2(x)dx \pm ... \pm \int f_n(x)dx.$$

Proof:

Let $\int f_1(x)dx = F_1(x)$, $\int f_2(x)dx = F_2(x)$, ..., $\int f_n(x)dx = F_n(x)$.

Clearly $\frac{d}{dx}\{F_1(x) \pm F_2(x)\} = \frac{d}{dx}F_1(x) \pm \frac{d}{dx}F_2(x)$

$= f_1(x) \pm f_2(x\_,$

Hence from the definition of the integral, we have

$$\int\{f_1(x) \pm f_2(x)\}dx = F_1(x) \pm F_2(x)$$

$$\int f_1(x)dx \pm \int f_2(x)dx.$$

Thus repeating the above process to other functions, we have

$$\int\{f_1(x) \pm f_2(x) \pm ... \pm f_n(x)\}dx$$

$$\int f_1(x)dx \pm \int f_2(x)dx \pm ... \pm \int f_n(x)dx.$$

Theorem 2:

The integral of the product of a constant and a function is equal to the product of the constant and the integral of the function.

Thus if λ is a constant, then $\int \lambda f(x)dx = \lambda \int f(x)dx$.

Proof:

Let $\int f(x)dx = F(x)$. Then $\frac{d}{dx}F(x) = f(x)$.

By differential calculus, $\frac{d}{dx}\{\lambda F(x)\} = \lambda \frac{d}{dx}F(x) = \lambda f(x)$.

$\therefore$ by definition of integral

$\int \lambda f(x)dx = \lambda F(x) = \lambda \int f(x)dx$.

FUNDAMENTAL FORMULAE

We have read in differential calculus that

$$\frac{d}{dx}\left(\frac{x^{n+1}}{n+1}\right) = \frac{(n+1)x^n}{(n+1)} = x^n.$$

Thus $\quad \int x^n dx + \frac{x^{n+1}}{n+1}, \ (n \neq -1).$

The above formula is very important and shall be frequently used in this book. This may be remembered like this:

"To find the integral of x^n w.r.t. 'x', increase the index (power) of x by one (unity) and divide by the increased index."

Thus $\quad \int x^3 dx = \frac{x^4}{4}, \ \int x^{5/2} dx = \frac{x^{(5/2)+1}}{(5/2)+1} = \frac{2}{7}x^{7/2};$

$$\int \frac{1}{x^5}dx = \int x^{-5}dx = \frac{x^{-5+1}}{-5+1} = -\frac{1}{4}x^{-4} = -\frac{1}{4x^4};$$

$$\int \frac{1}{x^{1/2}}dx = \int x^{-1/2}dx = \frac{x^{(-1/2)+1}}{(-1/2)+1} = 2x^{1/2} = 2\sqrt{x};$$

and $\quad \int dx = \int 1.dx = \int x^0 dx = \frac{x^{0+1}}{0+1} = x.$

Thus $\int a\,dx = ax$ *i.e., the integral of a constant is equal to the constant multiplied by the variable.*

However if n = – 1, we have

$$\int x^{-1}dx = \int \frac{1}{x}dx = \log x, \qquad \left[\because \frac{d}{dx}\log x = \frac{1}{x}\right]..$$

Example 1:

Integrate $e^x + 2\sin x - 3\cos x$.

Solution:

Here $I = \int \left(e^x + 2\sin x - 3\cos x\right)dx$

$$= \int e^x dx + 2\int \sin x\,dx - 3\int \cos x\,dx$$

$= e^x - 2\cos x - 3\sin x.$

Example 2:

Integrate $10^x + 3e^x + x^3$.

Solution:

Here $I = \int \left(10^x + 3e^x + x^3\right)dx$

$$= \int 10^x dx + 3\int e^x dx + \int x^3 dx = \left\{\frac{10^x}{(\log_e 10)}\right\} + 3e^x + \frac{1}{4}x^4.$$

SOME MORE STANDARD INTEGRALS

(i) To evaluate $\int \left[\frac{1}{\sqrt{(a^2+x^2)}}\right]dx$.

For complete solution of this problem 5. The result is

$$\int \frac{dx}{\sqrt{(a^2+x^2)}} = \sinh^{-1}\left(\frac{x}{a}\right) = \log\left\{x + \sqrt{(x^2+a^2)}\right\}.$$

(ii) To evaluate $\int \frac{dx}{\sqrt{(a^2-x^2)}}$.

The result is $\int \frac{dx}{\sqrt{(a^2-x^2)}} = \sin^{-1}\left(\frac{x}{a}\right)$.

(iii) To evaluate $\int \frac{dx}{\sqrt{(x^2-a^2)}}$.

Put x = a cosh q

so that dx = a sinh q dq.

Then the given integral $= \int \frac{dx}{\sqrt{(x^2 - a^2)}} = \int \frac{a \sinh\theta \, d\theta}{\sqrt{(a^2 \cosh^2\theta - a^2)}}$

$$= \int \frac{a \sinh\theta \, d\theta}{a\sqrt{(\cosh^2\theta - 1)}} = \int \frac{\sinh\theta \, d\theta}{\sinh\theta} = \int d\theta = \theta = \cosh^{-1}\left(\frac{x}{a}\right)$$

$$= \log\left[\frac{x}{a} + \sqrt{\left\{\left(\frac{x}{a}\right)^2 - 1\right\}}\right] = \log\left\{\frac{x + \sqrt{(x^2 - a^2)}}{a}\right\}$$

$$= \log\left\{x + \sqrt{(x^2 - a^2)}\right\} - \log a = \log\left\{x + \sqrt{(x^2 - a^2)}\right\},$$

because the constant term – log a may be added to the constant of integration c which we usually do not write.

Thus $\int \frac{dx}{\sqrt{(x^2 - a^2)}} = \cosh^{-1}\left(\frac{x}{a}\right) = \log\left\{x + \sqrt{(x^2 - a^2)}\right\}.$

(iv) To evaluate $\int \sqrt{(x^2 + a^2)}\, dx.$

Put x = a sinh θ

so that dx = a cosh θ dθ.

Then the given integral $= \int \sqrt{(a^2 \sinh^2\theta + a^2)} . a \cosh\theta \, d\theta$

$$= \int a^2 \cosh^2\theta \, d\theta = \int \frac{1}{2} a^2 (1 + \cosh 2\theta) \, d\theta,$$

[∵ cosh 2θ = 2 cosh²θ – 1]

$$= \frac{1}{2} a^2 \int (1 + \cosh 2\theta) \, d\theta = \frac{1}{2} a^2 \left[\theta + \frac{1}{2} \sinh 2\theta\right]$$

$$= \frac{1}{2} a^2 [\theta + \sinh\theta \cosh\theta], \qquad [\because \sinh 2\theta = 2 \sinh\theta \cosh\theta]$$

$$= \frac{1}{2} a^2 \left[\sinh^{-1}\frac{x}{a} + \frac{x}{a}\sqrt{\left(1 + \frac{x^2}{a^2}\right)}\right]$$

$$= \frac{a^2}{2} \sinh^{-1}\left(\frac{x}{a}\right) + \frac{a^2}{2} \cdot \frac{x}{a^2} \sqrt{(a^2 + x^2)}.$$

Thus $\int \sqrt{(x^2 + a^2)}\, dx = \frac{x}{2}\sqrt{(x^2 + a^2)} + \frac{a^2}{2} \sinh^{-1}\left(\frac{x}{a}\right)$

$$= \frac{x}{2}\sqrt{(x^2 + a^2)} + \frac{a^2}{2} \log\left\{x + \sqrt{(x^2 + a^2)}\right\}.$$

(v) To evaluate $\int \sqrt{(a^2 - x^2)}dx$.

Put x = a sin θ

so that dx = a cos θ dθ.

Then the given integral = $\int a\cos\theta . a\cos\theta \, d\theta$

$$= \frac{1}{2}a^2\left(\theta + \frac{1}{2}\sin 2\theta\right) = \frac{1}{2}a^2(\theta + \sin\theta \cos\theta)$$

$$= \frac{1}{2}a^2\left[\theta + \sin\theta\sqrt{(1-\sin^2\theta)}\right]$$

$$= \frac{1}{2}a^2\sin^{-1}\left(\frac{x}{a}\right) + \frac{1}{2}a^2 \cdot \frac{x}{a}\sqrt{\left(1 - \frac{x^2}{a^2}\right)}.$$

Thus $\int \sqrt{(a^2 - x^2)}dx = \frac{1}{2}x\sqrt{(a^2 - x^2)} + \frac{1}{2}a^2\sin^{-1}\left(\frac{x}{a}\right)$.

(vi) To evaluate $\int \sqrt{(x^2 - a^2)}dx$.

Put x = a cosh θ

so that dx = a sinh θ dθ.

Then the given integral = $\int \sqrt{(a^2\cosh^2\theta - a^2)}\, a\sinh\theta \, d\theta$

$$= \int a^2\sinh^2\theta \, d\theta = \int \frac{1}{2}a^2(\cosh 2\theta - 1)d\theta,$$

[∵ 2 sinh²θ = cosh 2θ − 1]

$$= \frac{1}{2}a^2\left[\frac{1}{2}\sinh 2\theta - \theta\right] = \frac{1}{2}a^2[\sinh\theta \cosh\theta - \theta]$$

$$= \frac{1}{2}a^2\left[\sqrt{(\cosh^2\theta - 1)}\cosh\theta - \theta\right]$$

$$= \frac{1}{2}a^2\left[\sqrt{\left(\frac{x^2}{a^2} - 1\right)} \cdot \frac{x}{a} - \cosh^{-1}\frac{x}{a}\right] \cdot \left[\because \cosh\theta = \frac{x}{a}\right]$$

Thus $\int \sqrt{(x^2 - a^2)}dx = \frac{1}{2}x\sqrt{(x^2 - a^2)} - \frac{1}{2}a^2\cosh^{-1}\left(\frac{x}{a}\right)$

$$= \frac{1}{2}x\sqrt{(x^2 - a^2)} - \frac{1}{2}a^2\log\left\{x|\sqrt{(x^2 - a^2)}\right\}.$$

SOLVED EXAMPLES

Example 1:

Evaluate $\int tan^{-1}\sqrt{\left(\frac{1-x}{1+x}\right)}\,dx.$

Solution:

Put x = cos θ

so that dx = – sin θ dθ.

∴ the given integral

$$=\int\left\{\tan^{-1}\sqrt{\left(\frac{1-\cos\theta}{1+\cos\theta}\right)}\right\}(-\sin\theta)d\theta$$

$$=-\int\left\{\tan^{-1}\left(\tan\frac{1}{2}\theta\right)\right\}\sin\theta\,d\theta$$

$$=-\int\frac{1}{2}\theta\,\sin\theta\,d\theta=-\frac{1}{2}\int\theta\,\sin\theta\,d\theta$$

$$=-\frac{1}{2}\left[\theta.(-\cos\theta)-\int(-\cos\theta)d\theta\right]=\frac{1}{2}\left[\theta\,\cos\theta-\int\cos\theta\,d\theta\right]$$

$$=\frac{1}{2}[\theta\,\cos\theta-\sin\theta]=\frac{1}{2}\left[x\cos^{-1}x-\sqrt{(1-x^2)}\right].$$

Example 2:

Evaluate $\frac{x+sin\,x}{1+cos\,x}dx.$

Solution:

We have

$$\int\frac{x+\sin x}{1+\cos x}dx=\int\frac{x+2\sin(x/2)\cos(x/2)}{1+2\cos^2(x/2)-1}$$

$$=\frac{1}{2}\int x\sec^2\left(\frac{x}{2}\right)dx+\int\tan\left(\frac{x}{2}\right)dx$$

$$=\frac{1}{2}\left[x\frac{\tan(x/2)}{(1/2)}-\int 1\cdot\frac{\tan(x/2)}{(1/2)}dx\right]+\int\tan\left(\frac{x}{2}\right)dx$$

$$=x\tan\left(\frac{x}{2}\right)-\int\tan\left(\frac{x}{2}\right)dx+\int\tan\left(\frac{x}{2}\right)dx=x\tan\left(\frac{x}{2}\right).$$

Example 3(a):

Integrate $\dfrac{1}{\sqrt{\left(2x^2+3x+4\right)}}$.

Solution:

We have $2x^2 + 3x + 4 = 2\left[x^2+\frac{3}{2}x+2\right]$

[Note that we have made the coefficient of x^2 as 1]

$$=2\left[\left(x+\frac{3}{4}\right)^2+2-\frac{9}{16}\right]=2\left[\left(x+\frac{3}{4}\right)^2+\left(\frac{\sqrt{23}}{4}\right)^2\right].$$

Hence the given integral

$$I=\int\frac{1}{\sqrt{\left(2x^2+3x+4\right)}}dx=\int\frac{dx}{\sqrt{2}.\sqrt{\left[\left(x+\frac{3}{4}\right)^2+\left(\sqrt{\frac{23}{4}}\right)^2\right]}}$$

$$=\frac{1}{\sqrt{2}}\sinh^{-1}\left\{\frac{x+\left(\frac{3}{4}\right)}{\sqrt{\frac{23}{4}}}\right\}=\frac{1}{\sqrt{2}}\sinh^{-1}\left(\frac{4x+3}{\sqrt{23}}\right).$$

Example 3(b):

Evaluate (i) $\int x^2\tan^{-1}x\,dx$

(ii) $\int x^3\tan^{-1}x\,dx$.

Solution:

(i) $=\dfrac{x^3}{3}\tan^{-1}x-\int\dfrac{x^3}{3}\cdot\dfrac{1}{1+x^2}dx$,

integrating by parts taking x^2 as the second function

$$=\frac{x^3}{3}\tan^{-1}x-\frac{1}{3}\int\frac{x\left(x^2+1\right)-x}{1+x^2}dx, \qquad [\because x^3 = x\,(x^2+1)-x]$$

$$=\frac{x^3}{3}\tan^{-1}x-\frac{1}{3}\int x\,dx+\frac{1}{3}\cdot\frac{1}{2}\int\frac{2x}{1+x^2}dx$$

$$=\frac{x^3}{3}\tan^{-1}x-\frac{1}{6}x^2+\frac{1}{6}\log\left(1+x^2\right).$$

(ii) We have $\int x^3\tan^{-1}x\,dx$

$$= \frac{x^4}{4}\tan^{-1}x - \int \frac{x^4}{4}\cdot\frac{1}{1+x^2}dx,$$

integrating by parts taking x^3 as the second function

$$= \frac{x^4}{4}\tan^{-1}x - \frac{1}{4}\int\left[x^2 - 1 + \frac{1}{1+x^2}\right]dx,$$

dividing x^4 by $x^2 + 1$, we get $x^2 - 1$ as the quotient and 1 as the remainder

$$= \frac{1}{4}\left[x^4\tan^{-1}x - \frac{x^3}{3} + x - \tan^{-1}x\right]$$

$$= \frac{1}{4}\left[\left(x^4 - 1\right)\tan^{-1}x - \frac{x^3}{3} + x\right].$$

Example 3(c):

Evaluate: (i) $\int \frac{\log(1+x)}{\sqrt{(1+x)}}dx$

(ii) $\int \log(x+2)^{x+2}dx.$

Solution:

(i) We have

$$\int \frac{\log(1+x)}{\sqrt{(1+x)}}dx = \int[\log(1+x)](1+x)^{-1/2}dx$$

$$= [\log(1+x)]\cdot\frac{(1+x)^{1/2}}{1/2} - \int \frac{1}{1+x}\cdot\frac{(1+x)^{1/2}}{1/2}dx,$$

integrating by parts taking $(1 + x)^{-1/2}$ as the second function

$$= 2\sqrt{(1+x)}\log(1+x) - 2\int(1+x)^{-1/2}dx$$

$$= 2\sqrt{(1+x)}\log(1+x) - 2\cdot\frac{(1+x)^{1/2}}{1/2}$$

$$= 2\sqrt{(1+x)}[\log(1+x) - 2].$$

(ii) We have $\int \log(x+2)^{x+2}dx = \int(x+2)\log(x+2)dx$

$$= \frac{(x+2)^2}{2}\log(x+2) - \int\frac{(x+2)^2}{2}\cdot\frac{1}{x+2}dx,$$

integrating by parts taking $(x + 2)^1$ as the second function

$$= \frac{(x+2)^2}{2}\log(x+2) - \frac{1}{2}\int(x+2)dx$$

$$= \frac{(x+2)^2}{2}\log(x+2) - \frac{1}{2}\frac{(x+2)^2}{2}$$

$$= \frac{(x+2)^2}{4}[2\log(x+2) - 1].$$

Example 4:

Evaluate $\int \left\{ \frac{x^2}{\sqrt{(x^6 - 9)}} \right\} dx.$

Solution:

Put $x^3 = t$, $\therefore$ $3x^2\,dx = dt$.

$\therefore$ the given integral $= \frac{1}{3}\int \frac{dt}{\sqrt{(t^2 - 9)}}$

$$= \frac{1}{3}\cosh^{-1}\left(\frac{t}{3}\right) = \frac{1}{3}\cosh^{-1}\left(\frac{x^3}{3}\right).$$

Example 5:

Evaluate $\int x\sqrt{(x^4 + 9)}\,dx.$

Solution:

Put $x^2 = t$; $\therefore$ $2x\,dx = dt$.

$\therefore$ the given integral $= \frac{1}{2}\int \sqrt{(t^2 + 3^2)}\,dt$

$$= \frac{1}{2}\left[\frac{1}{2}t\sqrt{(t^2 + 9)} + \frac{9}{2}\sinh^{-1}\left(\frac{t}{3}\right)\right]$$

$$= \frac{1}{4}x^2\sqrt{(x^4 + 9)} + \frac{9}{4}\sinh^{-1}\left(\frac{x^2}{3}\right).$$

Example 6:

Evaluate $\int x^2\sqrt{(x^6 - 1)}\,dx.$

Solution:

Put $x^3 = t$; $\therefore$ $3x^2\,dx = dt$.

$\therefore$ the given integral $= \frac{1}{3}\int \sqrt{(t^2 - 1)}\,dt$

$$= \frac{1}{3}\left[\frac{t}{2}\sqrt{(t^2 - 1)} - \frac{1}{2}\cosh^{-1}\left(\frac{t}{1}\right)\right]$$

$$=\frac{1}{3}\left[\frac{x^3}{2}\sqrt{(x^6-1)}-\frac{1}{2}\cosh^{-1}x^3\right]$$

$$=\frac{1}{6}x^3\sqrt{(x^6-1)}-\frac{1}{6}\cosh^{-1}x^3.$$

Example 7:

Integrate (i) x^2e^{mx}, (ii) $x^2 \sin x$, (iii) $x^2 \cos 2x$.

Solution:

(i) Integrating by parts taking e^{mx} as the second function, we have

$$\int x^2e^{mx}dx = x^2\cdot\frac{e^{mx}}{n}-\int 2x.\frac{e^{mx}}{m}dx$$

$$=\frac{x^2}{m}e^{mx}-\frac{2}{m}\int x.e^{mx}dx=\frac{x^2}{m}e^{mx}-\frac{2}{m}\left\{x\cdot\frac{e^{mx}}{m}-\int 1.\frac{e^{mx}}{m}dx\right\},$$

[again integrating by parts]

$$=\frac{x^2}{m}e^{mx}-\frac{2}{m^2}xe^{mx}+\frac{2}{m^3}e^{mx}$$

$= e^{mx}\, m^{-3}\, (m^2x^2 - 2mx + 2)$.

(ii) Integrating by parts taking sin x as second function, we have

$$\int x^2\sin x\,dx = -x^2\cos x-\int 2x(-\cos x)dx$$

$$=-x^2\cos x+2\int x\cos x\,dx$$

$$=-x^2\cos x+2x\sin x-2\int\sin x\,dx$$

$= - x^2 \cos x + 2x \sin x + 2 \cos x.$

Similarly $\int x^2\cos x\,dx = x^2\sin x+2x\cos x-2\sin x$.

(iii) Integrating by parts taking cos 2x as second function, we have

$$\int x^2\cos 2x\,dx = x^2\cdot\left(\frac{1}{2}\sin 2x\right)-\int 2x\cdot\frac{1}{2}\sin 2x\,dx$$

$$=\frac{1}{2}x^2\sin 2x-\int x\sin 2x\,dx$$

$$=\frac{1}{2}x^2\sin 2x-\left\{x\cdot\frac{1}{2}(-\cos 2x)+\frac{1}{2}\int\cos 2x\,dx\right\},$$

(Again integrating by parts taking sin 2x as the 2nd function)

$$=\frac{1}{2}x^2\sin 2x+\frac{1}{2}x\cos 2x-\frac{1}{2}\int\cos 2x\,dx$$

$$=\frac{1}{2}x^2\sin 2x+\frac{1}{2}x\cos 2x-\frac{1}{4}\sin 2x$$

$$=\frac{1}{2}\left(x^2-\frac{1}{2}\right)\sin 2x+\frac{1}{2}x\cos 2x.$$

Example 8:

$$\int e^x(n\cos x+\sin x)dx.$$

Solution:

$$\int e^x(n\cos x+\sin x)dx=n\int e^x\cos x\,dx+\int e^x\sin x\,dx.$$

Now using results of Ex. 5 (i), and (ii), we get the required value of the given integral

$$=\frac{1}{2}ne^x(\cos x+\sin x)+\frac{1}{2}e^x(\sin x-\cos x).$$

Example 9:

Evaluate $\int\frac{xe^x}{(x+1)^2}dx$.

Solution:

We have

$$\int\frac{xe^x}{(x+1)^2}dx=\int(xe^x)\frac{1}{(x+1)^2}dx.$$

Integrating by parts taking $\frac{1}{(x+1)^2}$ as the second function and xe^x as the first function, we have

$$\int\frac{xe^x}{(x+1)^2}dx=(xe^x)\left(-\frac{1}{x+1}\right)-\int(e^x+xe^x)\left(-\frac{1}{x+1}\right)dx,$$

$$\left[\text{Note that the integral of } \frac{1}{(x+1)^2} \text{ is} -\frac{1}{x+1}\right]$$

$$=-\frac{xe^x}{x+1}+\int e^x(x+1)\cdot\frac{1}{x+1}dx$$

$$=-\frac{xe^x}{x+1}+\int e^x\,dx=-\frac{xe^x}{x+1}+e^x$$

$$= e^x\left[1-\frac{x}{x+1}\right] = e^x\,\frac{x+1-x}{x+1} = \frac{e^x}{x+1}.$$

Example 10:

Evaluate $\int e^x \frac{(x^2+1)}{(x+1)^2}dx.$

Solution:

We have $\int e^x \frac{(x^2+1)}{(x+1)^2}dx$

$$= \int e^x \frac{(x^2-1)+2}{(x+1)^2}dx = \int e^x \frac{(x+1)(x-1)+2}{(x+1)^2}dx$$

$$= \int e^x\left[\frac{x-1}{x+1}+\frac{2}{(x+1)^2}\right]dx$$

$$= \int e^x\left[\frac{(x+1)-2}{x+1}+\frac{2}{(x+1)^2}\right]dx$$

$$= \int e^x\left[\left\{1-\frac{2}{x+1}\right\}+\frac{2}{(x+1)^2}\right]dx$$

$$= \int e^x[f(x)+f'(x)]dx, \text{ where } f(x) = 1-\frac{2}{x+1}$$

so that $f'(x) = \frac{2}{(x+1)^2}$

$$= e^x f(x) = e^x\left[1-\frac{2}{x+1}\right] = e^x\,\frac{x-1}{x+1}.$$

Example 11:

Evaluate $\int e^x \frac{(1-x)^2}{(1+x^2)^2}dx.$

Solution:

We have $\int e^x \frac{(1-x)^2}{(1+x^2)^2}dx = \int e^x \frac{(1+x^2)-2x}{(1+x^2)^2}dx$

$$= \int e^x \left[\frac{1}{1+x^2} - \frac{2}{\left(1+x^2\right)^2} \right] dx$$

$$= \int e^x [f(x) + f'(x)] dx,$$

where $\quad f(x) = \dfrac{1}{1+x^2}$

and $\quad f'(x) = \dfrac{-2x}{\left(1+x^2\right)^2}$

$$= e^x f(x) = e^x . \frac{1}{1+x^2}.$$

Example 12(a):

Evaluate $\displaystyle\int e^x \frac{x^2+3x+3}{(x+2)^2} dx.$

Solution:

We have $\displaystyle\int e^x \frac{x^2+3x+3}{(x+2)^2} dx$

$$= \int e^x \frac{(x+2)(x+1)+1}{(x+2)^2} dx$$

$$= \int e^x \left[\frac{x+1}{x+2} + \frac{1}{(x+2)^2} \right] dx$$

$= \displaystyle\int e^x [f(x) + f'(x)] dx$, where

$$f(x) = \frac{x+1}{x+2} = \frac{(x+2)-1}{x+2} = 1 - \frac{1}{x+2}$$

so that $f'(x) = \dfrac{1}{(x+2)^2}$

$$= e^x \frac{x+1}{x+2}.$$

Example 12(b):

Evaluate $\displaystyle\int e^x \frac{x^3 - x + 2}{\left(x^2+1\right)^2} dx.$

Solution:

We have $\int e^x \frac{x^3 - x + 2}{(x^2+2)^2} dx$

$$= \int e^x \frac{(x^2+1)(x+1) - x^2 - 2x + 1}{(x^2+1)^2} dx$$

$$= \int e^x \left[\frac{x+1}{x^2+1} + \frac{1 - x^2 - 2x}{(x^2+1)^2}\right] dx$$

$$= \int e^x [f(x) + f'(x)] dx,$$

where $f(x) = \frac{x+1}{x^2+1}$ so that

$$f'(x) = \frac{1.(x^2+1) - 2x(x+1)}{(x^2+1)^2} = \frac{1 - x^2 - 2x}{(x^2+1)^2}$$

$$= e^x f(x) = e^x \frac{x+1}{x^2+1}.$$

Example 13:

Evaluate $\int \frac{\log x}{(1+\log x)^2} dx.$

Solution:

Put log x = t. $\therefore$ $x = e^t$ and $dx = e^t\,dt$.

Then the given integral

$$I = \int \frac{te^t}{(1+t)^2} = \int e^t \frac{(1+t)-1}{(1+t)^2} dt$$

$$= \int e^t \left[\frac{1}{1+t} - \frac{1}{(1+t)^2}\right] dt$$

$$= \int e^t [f(t) + f'(t)] dt, \text{ where } f(t) = \frac{1}{1+t}$$

$$= e^t f(t) = e^t \frac{1}{1+t} = \frac{x}{1+\log x}.$$

Example 14:

Evaluate $\int \frac{e^x(1+\sin x)}{1+\cos x}dx$.

Solution:

The given integral $I = \int \frac{e^x\,dx}{1+\cos x} + \int \frac{e^x \sin x\,dx}{(1+\cos x)}$

$$= \int \frac{e^x\,dx}{2\cos^2(x/2)} + \int \frac{e^x 2\sin(x/2)\cos(x/2)}{2\cos^2(x/2)}$$

$$= \frac{1}{2}\int e^x \sec^2\left(\frac{x}{2}\right)dx + \int e^x \tan\left(\frac{x}{2}\right)dx$$

$$= \int e^x\left[\tan\frac{x}{2} + \frac{1}{2}\sec^2\frac{x}{2}\right]dx.$$

Since $\frac{d}{dx}\left(\tan\frac{1}{2}x\right) = \frac{1}{2}\sec^2\frac{x}{2}$, therefore this *integral is of the type* $\int e^x[f(x)+f'(x)]dx$.

To evaluate this integral, integrating $e^x \tan\left(\frac{x}{2}\right)$ by parts regarding e^x as the second function, we get

$$I = e^x \tan\frac{x}{2} - \int \frac{1}{2}e^x.\sec^2\frac{x}{2}dx + \int \frac{1}{2}e^x \sec^2\frac{x}{2}dx = e^x \tan\frac{x}{2},$$

because the last two integrals cancel each other.

Example 15:

Evaluate $\int 20^{5x}\,dx$.

Solution:

Put $5x = t$; then $5dx = dt$

or $dx = \frac{1}{5}dt$.

$$\therefore \qquad I = \int 20^{5x}dx = \int \frac{1}{5}\cdot 20^t\,dt$$

$$= \frac{1}{5}\frac{20^t}{\log 20} = \frac{1}{5}\frac{20^{5x}}{\log 20}.$$

Example 16:

Integrate (i) $\frac{\sin x}{a+b\cos x}$,

(ii) $\dfrac{\sin x \cos x}{a\cos^2 x + b\sin^2 x}$

Solution:

(i) Here $I = \int \frac{\sin x\, dx}{a + b\cos x} = -\frac{1}{b}\int \frac{-b\sin x}{a + b\cos x}dx$.

Putting a + b cos x = t,
so that – b sin x dx = dt,

we have $I = \left(-\frac{1}{b}\right)\int\left(\frac{1}{t}\right)dt$

$$= -\left(\frac{1}{b}\right).\log(t) = -\left(\frac{1}{b}\right).\log(a + b\cos x).$$

(ii) Let $I = \int\left(\frac{\sin x \cos x}{a\cos^2 x + b\sin^2 x}\right)dx$

Put $a \cos^2 x + b \sin^2 x = t$,
so that (– 2a cos x sin x + 2b sin x cos x) dx = dt,
or {2 (b – a) sin x cos x} dx = dt.

We have $I = \frac{1}{2(b-a)}\int\frac{dt}{t} = \frac{1}{2(b-a)}\log t$

$$= \frac{1}{2(b-a)} \log (a \cos^2 x + b \sin^2 x).$$

Example 17:

Integrate (i) $\dfrac{1}{x\cos^2(1 + \log x)}$ *(ii)* $\dfrac{1}{x(1 + \log x)^m}$.

Solution:

(i) Here $I = \int \frac{dx}{\{x\cos^2(1 + \log x)\}}$.

Putting 1 + log x = t, so that $\left(\frac{1}{x}\right)$ dx = dt, we have

$$I = \int\frac{dt}{\cos^2 t} = \int \sec^2 t\, dt = \tan t = \tan(1 + \log x).$$

(ii) Here $I = \int \frac{dx}{\{x(1 + \log x)^m\}}$.

Putting 1 + log x = t, so that $\left(\frac{1}{x}\right)$ dx = dt, we have

$$I = \int \frac{dt}{t^m} = \frac{t^{-m+1}}{-m+1} = \frac{(1+\log x)^{-m+1}}{(1-m)}$$

$$= \frac{1}{(1-m)}(1+\log x)^{1-m}.$$

Example 18:

Integrate (i) $\frac{e^{tan^{-1}x}}{1+x^2}$ *(ii)* $\frac{sin(tan^{-1}x)}{1+x^2}$.

Solution:

(i) Putting $\tan^{-1}$ x - t, so that $\left[\frac{1}{(1+x^2)}\right]dx = dt$, we have

$$I = \int \frac{e^{\tan^{-1}x}}{1+x^2}dx = \int e^t.dt = e^t = e^{\tan^{-1}x}.$$

(ii) Putting $\tan^{-1}$ x = t so that $\left[\frac{1}{(1+x^2)}\right]dx = dt,$ we have

$$I = \int \frac{\sin(\tan^{-1}x)}{1+x^2}dx = \int \sin t\,dt = -\cos t = -\cos(\tan^{-1}x).$$

Example 19:

Integrate (i) $x \cos^3 x^2.\sin x^2$ *(ii)* $x^3 \tan^4 x^4.\sec^2 x^4$.

Solution:

(i) Here $I = \int x\cos^3 x^2 \sin x^2 dx$.

First, putting x^2 = t,

so that 2x dx = dt, we have

$$I = \frac{1}{2}\int \cos^3 t \sin t\,dt.$$

Now putting cos t = u, so that – sin t dt = du, we have

$$I = -\frac{1}{2}\int u^3 du = -\frac{1}{2}\frac{u^4}{4} = -\frac{1}{8}u^4$$

$$= -\frac{1}{8}\cos^4 t = -\frac{1}{8}\cos^4 x^2. \qquad [\because t = x^2]$$

Note: The students should also solve this problem by making the single substitution $\cos x^2 = t$.

(ii) Here $I = \int x^3 \tan^4 x^4 \sec^2 x^4 dx$.

Putting $\tan x^4 = t$, so that $(\sec^2 x^4).4x^3\, dx = dt$, we have

$$I = \int \frac{t^4}{4} \cdot dt = \frac{1}{4}\left(\frac{t^5}{5}\right) = \frac{\left(\tan x^4\right)^5}{20}.$$

Example 20:

Integrate (i) $\dfrac{cot\, x}{log(sin x)}$, *(ii)* $\dfrac{tanx}{log(sec x)}$.

Solution:

(i) Here $\frac{d}{dx}(\log \sin x) = \frac{1}{\sin x}\cos x = \cot x$.

$$\therefore\ I = \int \frac{\cot x\, dx}{\log \sin x} = \log(\log \sin x),\ [\because \text{ Nr. is diff. coeff. of Dr.}]$$

(ii) Similarly, we have

$$\int \frac{\tan x\, dx}{\log \sec x} = \log(\log \sec x).$$

Example 21:

Integrate $\dfrac{1}{\left(1 + 3 sin^2\right)}$.

Solution:

Dividing Nr. and Dr. by $\cos^2 x$, we have

$$I = \int \frac{dx}{1 + 3\sin^2 x} = \int \frac{\sec^2 x\, dx}{\sec^2 x + 3\tan^2 x}$$

$$= \int \frac{\sec^2 x\, dx}{\left(1 + \tan^2 x\right) + 3\tan^2 x}$$

$$= \int \frac{\sec^2 x\, dx}{1 + 4\tan^2 x}.$$

Now putting $2 \tan x = t$

so that $2 \sec^2 x\, dx = dt$, we have

$$I=\frac{1}{2}\int\frac{dt}{1+t^2}=\frac{1}{2}\tan^{-1}t=\frac{1}{2}\tan^{-1}(2\tan x).$$

Example 22:

Evaluate $\int\left[\frac{(\cos x)}{\left(a^2+b^2\sin^2 x\right)}\right]dx.$

Solution:

Putting b sin x = t so that b cos x dx = dt, we have

$$I=\int\frac{\cos x\,dx}{a^2+b^2\sin^2 x}=\frac{1}{b}\int\frac{dt}{a^2+t^2}=\frac{1}{b}\cdot\frac{1}{a}\tan^{-1}\left(\frac{t}{a}\right)$$

$$=\frac{1}{ab}\tan^{-1}\left(\frac{b\sin x}{a}\right).$$

Example 23:

Evaluate $\int\frac{\cot(\log x)}{x}dx.$

Solution:

Putting log x = t so that $\left(\frac{1}{x}\right)dx=dt$, we have the given integral

$$I=\int\cot t\,dt=\log\sin t=\log\{\sin(\log x)\}.$$

Example 24:

Evaluate the following integrals:

(i) $\int\frac{d\theta}{a^2\sin^2\theta+b^2\cos^2\theta}$

(ii) $\int\frac{dx}{4\sin^2 x+5\cos^2 x}.$

Solution:

(i) Dividing Nr. and Dr. by $\cos^2\theta$, we have

$$I=\int\frac{d\theta}{a^2\sin^2\theta+b^2\cos^2\theta}=\int\frac{\sec^2\theta\,d\theta}{a^2\tan^2\theta+b^2}$$

$$=\frac{1}{a^2}\int\frac{\sec^2\theta\,d\theta}{\tan^2\theta+\left(\frac{b^2}{a^2}\right)}$$

$$= \frac{1}{a^2}\int \frac{dt}{t^2 + \left(\frac{b^2}{a^2}\right)}$$ [Putting t = tan θ, so that dt = $\sec^2\theta\ d\theta$]

$$= \frac{1}{a^2}\cdot\frac{a}{b}\tan^{-1}\left(\frac{t}{b/a}\right) = \frac{1}{ab}\tan^{-1}\left(\frac{a}{b}\tan\theta\right).$$

(ii) Let $I = \int \frac{dx}{4\sin^2 x + 5\cos^2 x}$

$$= \int \frac{\sec^2 x\,dx}{4\tan^2 x + 5},$$ dividing the Nr. and Dr. by $\cos^2 x$

$$= \int \frac{dt}{4t^2 + 5},$$ putting tan x = t so that $\sec^2 x\ dx = dt$

$$= \frac{1}{4}\int \frac{dt}{t^2 + \left(\sqrt{\frac{5}{2}}\right)^2} = \frac{1}{4}\frac{1}{\left(\sqrt{\frac{5}{2}}\right)}\tan^{-1}\frac{t}{\sqrt{\frac{5}{2}}}$$

$$= \frac{1}{2\sqrt{5}}\tan^{-1}\frac{2t}{\sqrt{5}} = \frac{1}{2\sqrt{5}}\tan^{-1}\frac{2\tan x}{\sqrt{5}}.$$

Example 25:

Integrate 5 cos x + 2 sec² x – 10.

Solution:

Here $I = \int (5\cos x + 2\sec^2 x - 10)dx$

$$= \int 5\cos x\,dx + 2\int \sec^2 x\,dx - \int 10dx$$

$= 5\sin x + 2\tan x - 10x.$

Example 26:

Integrate $\left\{\frac{6}{\sqrt{(1-x^2)}}\right\} + 3\sec^2 x$.

Solution:

Here $I = \int \left[\left\{\frac{6}{\sqrt{(1-x^2)}}\right\} + 3\sec^2 x\right]dx$

$$= \int \left\{ \frac{6}{\sqrt{(1-x^2)}} \right\} dx + 3\int \sec^2 x\, dx = 6\sin^{-1} x + 3\tan x.$$

Example 27:

Integrate $\frac{x^3 \tan^{-1} x^4}{1+x^8}$.

Solution:

Putting $\tan^{-1} x^4 = t$ so that $\frac{1}{1+x^8}.4x^3 dx = dt$, we have

$$I = \int \frac{x^3 \tan^{-1} x^4}{1+x^8} dx = \frac{1}{4}\int t\, dt = \frac{1}{8}t^2 = \frac{1}{8}\left(\tan^{-1} x^4\right)^2.$$

Example 28:

Evaluate $\int \frac{e^x (1+x)}{\sin^2(xe^x)}$.

Solution:

Putting $xe^x = t$ so that $(e^x + xe^x)\, dx = dt$
or $e^x(1 + x)\, dx = dt$, we have

$$I = \int \frac{dt}{\sin^2 t} = \int \text{cosec}^2 t\, dt = -\cot t = -\cot\left(xe^x\right).$$

Example 29:

Integrate (i) $\frac{1}{\left(e^x + 1\right)}$, *(ii)* $\frac{1}{\left(e^x - 1\right)}$.

Solution:

(i) Here $I = \int \frac{dx}{e^x + 1} = \int \frac{e^{-x}}{1+e^{-x}} dx$,

[Multiplying Nr. and Dr. both by e^{-x}]

$$= -\int \frac{-e^{-x}}{1+e^{-x}} dx = -\log\left(1+e^{-x}\right), \text{ [}\because \text{ Nr. is diff. coeff. of Dr.]}$$

$$= -\log\left(\frac{1+e^x}{e^x}\right) = -\left[\log\left(1+e^x\right) - \log e^x\right]$$

$= x \log e - \log (1 + e^x) = x - \log (1 + e^x)$.

(ii) Similarly $\int \frac{dx}{e^x - 1} = \log\left(1 - e^{-x}\right)$.

Example 30:

Evaluate $\int e^x \frac{1 - \sin x}{1 - \cos x} dx$.

Solution:

We have

$$\int e^x \frac{1-\sin x}{1-\cos x} dx = \int e^x \left[\frac{1}{1-\cos x} - \frac{\sin x}{1-\cos x}\right] dx$$

$$= \int e^x \left[\frac{1}{2\sin^2 \frac{1}{2}x} - \frac{2\sin\frac{1}{2}x\cos\frac{1}{2}x}{2\sin^2\frac{1}{2}x}\right] dx$$

$$= \int e^x \left[\frac{1}{2}\operatorname{cosec}^2 \frac{1}{2}x - \cot\frac{1}{2}x\right] dx$$

$$= \int e^x \left[\left(-\cot\frac{1}{2}x\right) + \frac{1}{2}\operatorname{cosec}^2\frac{1}{2}x\right] dx$$

$= \int e^x [f(x) + f'(x)] dx$ where

$$f(x) = -\cot\frac{1}{2}x$$

so that $f'(x) = \frac{1}{2}\operatorname{cosec}^2 \frac{1}{2}x$

$$= e^x f(x) = e^x\left(-\cot\frac{1}{2}x\right) = -e^x \cot\frac{1}{2}x.$$

Example 31(a):

Evaluate $\int e^x \frac{2 + \sin 2x}{1 + \cos 2x} dx$.

Solution:

We have $\int e^x \frac{2+\sin 2x}{1+\cos 2x} dx$

$$= \int e^x \left[\frac{2}{1+\cos 2x} + \frac{\sin 2x}{1+\cos 2x}\right] dx$$

$$= \int e^x \left[\frac{2}{2\cos^2 x} + \frac{2\sin x\cos x}{2\cos^2 x}\right] dx$$

$= \int e^x [\sec^2 x + \tan x] dx$

$= \int e^x [\tan x + \sec^2 x] dx$

$= \int e^x [f(x) + f'(x)] dx$, where f(x) = tan x

$= e^x f(x) = e^x \tan x.$

Example 31(b):

Evaluate $\int e^x (\cot x + \log \sin x) dx.$

Solution:

We have $\int e^x (\cot x + \log \sin x) dx$

$= \int e^x (\log \sin x + \cot x) dx$

$= \int e^x [f(x) + f'(x)] dx$, where

f(x) = log sin x

so that $f'(x) = \left(\frac{1}{\sin x}\right) \cos x = \cot x = e^x f(x) = e^x \log \sin x.$

Example 31(c):

$\int e^x [\log(\sec x + \tan x) + \sec x] dx.$

Solution:

The given integral

$I = \int e^x \sec x \, dx + \int e^x \log(\sec x + \tan x) dx$

$= e^x \log(\sec x + \tan x) - \int e^x \log(\sec x + \tan x) dx$

$+ \int e^x \log(\sec x + \tan x) dx,$

(applying integration by parts to the first integral taking e^x as the second function and keeping the second integral as it is)

$= e^x \log (\sec x + \tan x).$

Example 32:

Evaluate $\int e^x \left[\frac{1 + \sqrt{(1 - x^2)} \sin^{-1} x}{\sqrt{(1 - x^2)}} \right] dx.$

Solution:

The given integral

$$I = \int e^x \left[\frac{1}{\sqrt{(1-x^2)}} + \sin^{-1} x \right] dx$$

$$= \int e^x \left[\sin^{-1} x + \frac{1}{\sqrt{(1-x^2)}} \right] dx$$

$= \int e^x [f(x) + f'(x)] dx$, where $f(x) = \sin^{-1}$ so that $f'(x) = \dfrac{1}{\sqrt{(1-x^2)}}$

$$= e^x f(x) = e^x \cdot \frac{1}{\sqrt{(1-x^2)}}.$$

9

Linear Programming

INTRODUCTION

The simplex method is an *iterative procedure* and refers to the idea of moving from one extreme (comer) point to another of the solution space or feasible region (convex polyhedron) that is formed by the constraints and non-negativity conditions of the linear programming problem. Since the number of extreme points (corners or vertices) of a solution space is finite, the method leads to the optimal extreme point (optimal solution) in a finite number of steps or indicates that there exist an unbounded solution. It was observed in the graphical approach to the solution of L.P.P. that in a given situation the set of constraints determines the feasible region and objective function gives the optimal point, the one that minimizes or maximizes, as the case may be. But the graphical method suffers from a great limitation that it can handle problems involving only two decision variables.

This chapter contains a rigorous discussion on the mathematical method called the *simplex method* developed by *G.B. Dantzig* in 1947 for solving LP problems. Whereas in real world situations, we frequently encounter such problems where more than two variables are involved. The graphical method used is applicable only to solve two-variable LP problems. Finding and evaluating all basic feasible solutions of a LP problem, say with six variables and five constraints may be a long and difficult task. Thus, an efficient computational procedure is required to solve the general class of LP problem. Fundamental theorem of linear programming makes it easy to find an optimal solution because it deals with basic feasible solutions only. But, it is also not easy job to enumerate all the B.F.S. even for small values of m (number of constraints) and n (number of variables).

LINEAR PROGRAMMING FUNDAMENTAL THEOREM

Theorem:

If a linear programming problem

$$\max Z = c\,x \text{ subject to } A\,x = b,\; x \geq 0,$$

where A is an m × N, (N = m + n) matrix of coefficients given by A = $(\alpha_1, \alpha_2, \ldots, \alpha_N)$ has an optimal feasible solution then at least one basic feasible solution must be optimal.

Proof:

Let $x^* = [x_1, x_2, \ldots, x_N]$

be an optimal feasible solution of the given linear programming problem and

$$Z^* = \sum_{i=1}^{N} c_i x_i$$

be the corresponding optimum value of the objective function.

Also suppose that k ($k \le N$) variables in x^* are non-zero and the remaining N – k variables are zero. For the sake of convenience we can assume that the first k variables of x^* are non-zero.

Thus $x^* = (x_1, x_2, \ldots, x_k, 0, 0, \ldots, 0)$.

$$\therefore \qquad \sum_{i=1}^{k} x_i \alpha_i = b \qquad \ldots(1)$$

and

$$Z^* = \sum_{i=1}^{k} c_i x_i . \qquad \ldots(2)$$

Now two possibilities may arise:

The vectors $\alpha_1, \alpha_2, \ldots, \alpha_k$ may be either linearly independent or dependent.

Case 1. If $\alpha_1, \alpha_2, \ldots, \alpha_k$ are linearly independent.

Any feasible solution for which the vectors a_i, i = 1, ... ,k associated with non-zero variables x_i, i = 1, ... ,k are L.I. is called a basic feasible solution. Thus x^* is a B.F.S. which is also optimal.

Hence the result of the theorem is true.

The solution x is degenerate if $k < m$ and non-degenerate if $k = m$.

Case 2. If $\alpha_1, \alpha_2, \ldots, \alpha_k$ are linearly dependent and $k > m$.

In this case we shall reduce the number of non-zero variables step by step by step until we obtain a B.F.S. from feasible solution.

Since $\alpha_1, \alpha_2, \ldots, \alpha_k$ are linearly dependent therefore there exist scalars $\lambda_1, \alpha_2, \ldots, \lambda_k$ such that

$$\lambda_1 \alpha_1 + \lambda_2 \alpha_2 + \ldots + \lambda_k \alpha_k = 0$$

or

$$\sum_{i=1}^{k} \lambda_i \alpha_i = 0 \qquad \ldots(3)$$

and at least one $\lambda_i \neq 0$.

We can assume that at least one li is positive because if none is positive then we can multiply the equation (3) by –1 and get positive li.

Now, suppose $v = \max_{1 \leq i \leq k} \left(\frac{\lambda_i}{x_i}\right)$. ...(4)

Obviously v is positive since $x_i \geq 0$, i = 1, 2, ..., k and at least one λ_i is positive.

Multiplying both sides of (3) by 1/v and subtracting from (1), we get

$$\sum_{i=1}^{k}\left(x_i - \frac{\lambda_i}{v}\right)\alpha_i = b. \quad \text{...(5)}$$

(5) gives

$$x' = \left[x_1 - \frac{\lambda_1}{v}, x_2 - \frac{\lambda_2}{v}, \ldots, x_k - \frac{\lambda_k}{v}, 0, 0, \ldots, 0\right]$$

a new solution of the matrix equation A x = b.

Also from (4) for i = 1, 2, ... ,k, we have

$$\Rightarrow \quad v \geq \frac{\lambda_i}{x_i} \text{ or } x_i \geq \frac{\lambda_i}{v}$$

$$\Rightarrow \quad x_i - \frac{\lambda_i}{v} \geq 0.$$

Since all the components of x' are non-negative therefore x' is a feasible solution to the given L.P.P.

Also $v = \frac{\lambda_i}{x_i}$, for at least one i

i.e., $v - \frac{\lambda_i}{x_i} = 0$ therefore the new feasible solution x' cannot have more than k – 1 non-zero variables.

In this way we have derived a new feasible solution from the given optimal feasible solution which contains less number of non-zero variables. This solution is B.F.S. if the column vectors associated to non-zero variables in this new solution are L.I. If these associated vectors are not L.I., we shall repeat the whole reduction procedure as explained above. Continuing in this way for finite number of times we will derive a solution in which columns corresponding to positive variables are L.I. *i.e.*, we will obtain a B.F.S. of the system.

Now it remains to prove that x' is also an optimal solution.

Let Z' be the new value of the objective function corresponding to the new solution x'.

Then $$Z' = \sum_{i=1}^{k} c_i\left(x_i + \frac{\lambda_i}{v}\right) = \sum_{i=1}^{k} c_i x_i - \frac{1}{v}\sum_{i=1}^{1} c_i \lambda_i$$

$$\Rightarrow \quad Z' = Z^* - \frac{1}{v}\sum_{i=1}^{k} c_i \lambda_i \text{, from (2)} \qquad ...(6)$$

Now for optimality, we must have

$$Z' = Z^*$$

i.e., x' will be optimal solution only if

$$\sum_{i=1}^{k} c_i \lambda_i = 0. \qquad ...(7)$$

We shall prove this result by contradiction.

If possible let us assume that

$$\sum_{i=1}^{k} c_i \lambda_i \neq 0.$$

Then either $\sum_{i=1}^{k} c_i \lambda_i < 0$ ...(A)

$$\Rightarrow \quad \sum_{i=1}^{k} c_i \lambda_i > 0. \qquad ...(B)$$

In either of these two cases, we can find a real number r such that

$$r\sum_{i=1}^{k} c_i \lambda_i > 0$$ [in case (A) r will be negative and in case (B) r will be positive]

$$\Rightarrow \quad \sum_{i=1}^{k} c_i (r\lambda_i) > 0$$

$$\Rightarrow \quad \sum_{i=1}^{k} c_i (r\lambda_i) - \sum_{i=1}^{k} c_i x_i > \sum_{i=1}^{k} c_i x_i$$

$$\Rightarrow \quad \sum_{i=1}^{k} c_i (x_i + r\lambda_i) > Z^*\text{, from (2).} \qquad ...(8)$$

Multiplying equation (3) by r and adding to (1), we get

$$\sum_{i=1}^{k} x_i \alpha_i + \sum_{i=1}^{k} r\lambda_i \alpha_i = b$$

$$\Rightarrow \quad \sum_{i=1}^{k} (x_i + r\lambda_i)\alpha_i = b.$$

Obviously $[x_1 + r\lambda_1, x_2 + r\lambda_2, \ldots, x_k + r\lambda_k, 0, 0, \ldots, 0]$

is also a solution of the matrix equation A x = b for all values of r.

Furthermore, we can choose r in infinitely many ways for which the above solution also satisfies the non-negative restrictions.

Let us choose r such that

$$x_i + r\lambda_i \geq 0, \; i = 1, \ldots, k$$

or $\quad r\lambda_i \geq -x_i$

$$\therefore \qquad r \geq -\frac{x_i}{\lambda_i} \text{ if } \lambda_i > 0,$$

$$r \leq -\frac{x_i}{\lambda_i} \text{ if } \lambda_i < 0$$

and r is unrestricted if $\lambda_i = 0$.

Thus $x_i + r\lambda_i \geq 0$ for i = 1, 2, ... ,k if we select r such that

$$\max_{\substack{i \\ (\lambda_i > 0)}} \left(-\frac{x_i}{\lambda_i}\right) \leq r \leq \min_{\substack{i \\ (\lambda_i < 0)}} \left(-\frac{x_i}{\lambda_i}\right) \qquad \ldots(9)$$

which means that interval given by (9) is non-empty.

Thus, r lies in the non-empty interval given in (9).

Consequently, an infinite number of solution given by

$$[x_1 + r\lambda_1, x_2 + r\lambda_2, \ldots, x_k + x_k + r\lambda_k, 0, 0, \ldots, 0]$$

satisfy the non-negative restrictions as well.

Now, returning to result (8) we king that $\sum_{i=1}^{k} c_i (c_i + r\lambda_i)$ yields the value of the objective function which is strictly greater than the greatest value (or optimal value) Z* of objective function, which is impossible.

$$\therefore \text{ we must have } \sum_{i=1}^{k} c_i \lambda_i = 0$$

or $\quad Z' = Z^*$.

$$\text{Hence } x' = \left[x_i - \frac{\lambda_1}{v}, x_2 - \frac{\lambda_2}{v}, \ldots, x_k - \frac{\lambda_k}{v}, 0, 0, \ldots, 0\right]$$

is also an optimal solution.

This proves the theorem.

ARTIFICIAL VARIABLES TECHNIQUE

(a) *Big-M Method :* Sometimes in linear programming problems constraints may also have ≥ and = signs after ensuring that all bi ≥ 0. In such problems we first introduce *surplus variables* to convert inequalities into equations. In such cases basis matrix is not obtained as an identity matrix in the starting simplex table. To overcome this difficulty we introduce new variables, called, the *artificial variables* to each of such constraints. These variables are fictitious and cannot have any physical meaning.

The artificial variables are introduced for the limited purpose of obtaining an *initial solution.* It is not relevant whether the objective function is of the maximization or minimization type. Since artificial variables do not represent any quantity relating to the decision problem they must be driven out of the system and must not be present in the final solution. Keeping this in mind a method known as *big-M method or Cranes M-method* was developed. It we suggestedby A. Cranes that a very high penalty can be paid for introducing the artificial variables in the constraints of a given problem by assigning an extremely high negative cost (penalty) to them.

∴ the objective function can be written as

$$Z = c\,x + 0.x_{slack} + 0.x_{surplus} - M.x_{art}.$$

It is significant to note that the initial solution obtained using the artificial variables is not a feasible solution to the given problem. It only gives the starting point and the artificial variables are driven out in he normal course of applying simplex method to the problem.

Note: A solution to the problem which does not contain an artificial variables in the basic, represents a feasible solution to the problem.

BASIC FEASIBLE SOLUTION TO BECOME OPTIMAL

Theorem:

Let us $x_B = B^{-1}b$ be the basic (degenerate or non-degenerate) feasible solution of the L.P.P. $Z = c\,x$ s.t. $A\,x = b$, $x \geq 0$. Let $Z^* = c_B\,x_B$ be the value of the objective function at any iteration of simplex method. If $c_j - Z_j \leq 0$ for every column α_j in A but not in B then Z^* is the optimum value of the objective function Z and x_B is an optimal basic feasible solution.

Proof:

Suppose $x_B = (x_{B1}, x_{B2}, \ldots, x_{Bm})$ is the B.F.S. of the given L.P.P. We have,

$$A = (\alpha_1, \alpha_2, \ldots, \alpha_N),\; N = m + n$$

and the basis matrix $B = (\beta_1, \beta_2, \ldots, \beta_m)$.

$\therefore \quad B_{mB} = b$ or $x_B = B^{-1} b$. ...(1)

The value of the objective function for the B.F.S. xB is

$$Z^* = c_B\, x_B = \sum_{i=1}^{m} c_{Bi} x_{Bi}\,. \qquad ...(2)$$

Also, we are given that $c_j - Z_j \leq 0$ for every column α_j in A but not in B.

Let 'x' = $[x_1', x_2', \ldots, x_N']$ be any feasible solution and Z' be the value of the objective function for this solution.

Then we have $A\,x' = b$ ...(3)

and $$Z' = c\,x' = \sum_{j=1}^{N} c_j x_j' \qquad ...(4)$$

Now Z* will be the optimum value of the objective function if we have $Z^* \geq Z'$. To prove $Z^* \geq Z'$ we proceed in the following manner.

We have $x_B = B^{-1} b = B^{-1}(A\,x')$, from (3)

$= (B^{-1} A)\, x'$

$= Y\,x'$, where $Y = [Y_1, Y_2, \ldots, Y_N]$

$= [Y_1, Y_2, \ldots, Y_N]\,[x_1', x_2', \ldots, x_N']$

$$= \begin{bmatrix} y_{11} & y_{12} & \cdots & y_{1j} & \cdots & y_{1N} \\ y_{21} & y_{22} & \cdots & y_{2j} & \cdots & y_{2N} \\ \cdots & \cdots & \cdots & \cdots & \cdots & \cdots \\ \cdots & \cdots & \cdots & \cdots & \cdots & \cdots \\ y_{m1} & y_{m2} & \cdots & y_{mj} & \cdots & y_{mN} \end{bmatrix} \begin{Bmatrix} x_1' \\ x_2' \\ \vdots \\ x_j' \\ \vdots \\ x_N' \end{Bmatrix}$$

$$\Rightarrow \quad [x_{B1}, x_{B2}, \ldots, x_{Bi}, \ldots, x_{Bm}]$$

$$= \left[\sum_{j=1}^{N} y_{1j} x_j', \sum_{j=1}^{N} y_{2j} x_j', \ldots, \sum_{j=1}^{N} y_{ij} x_j', \ldots, \sum_{j=1}^{N} y_{mj} x_j' \right]$$

Equating ith component on both sides, we get,

$$x_{Bi} = \sum_{j=1}^{N} y_{ij} x_j'. \qquad ...(5)$$

We have $c_j - Z_j \leq 0$ for all j for which α_j is not in B.

If this inequality also holds for all those j for which α_j is in B then $c_j - Z_j \leq 0$ for all j for which α_j also belong to B.

Now if $\alpha_j = B_i$, then

$$\alpha_j = \beta_i = 0.\beta_1 + 0.\beta_2 + \ldots + 0.\beta_{i-1} + 1.\beta_i + \ldots + 0.\beta_m$$

which shows that $Y_j = e_i$, a vector whose ith component is unity.

Again $\alpha_j = \beta_i$ gives $c_j = c_{Bi}$

$\therefore$ $c_j - Z_j = c_j - c_B Y_j = c_j - c_B e_i = c_j - c_{Bi} = 0$

Thus $c_j - Z_j = 0$ for all those j for which $\alpha_j \in B$.

$\therefore$ $c_j - Z_j \leq 0$ for all α_j in A

$\Rightarrow$ $c_j \leq Z_j$

$\Rightarrow$ $c_j x_j' \leq Z_j x_j', \because x_j' \geq 0$

$$\Rightarrow \sum_{i=1}^{N} c_j x_j' \leq \sum_{j=1}^{N} Z_j x_j'$$

$$\Rightarrow Z' \leq \sum_{j=1}^{N} x_j'(c_B Y_j), \text{ from (4)}$$

$$= \sum_{j=1}^{N} x_j' \left(\sum_{i=1}^{m} c_{Bi} y_{ij} \right)$$

$$= \sum_{i=1}^{m} c_{Bi} \left(\sum_{j=1}^{N} x_j' y_{ij} \right)$$

$$= \sum_{i=1}^{m} c_{Bi} x_{Bi}, \text{ from (5)} \qquad \therefore Z' \leq Z^*.$$

Hence Z* is the optimum value of the objective function.

THE ALTERNATIVE OPTIMAL SOLUTIONS

The given linear programming problem is said to possess an alternative optimal solution if the set of variables giving the optimal value of the objective function is not unique.

Theorem:

(Conditions for Alternative Optimum Solution)

Suppose there exists an optimum B.F.S. to a L.P.P. and

(i) if for some a_j in A but not in B, $c_j - Z_j = 0$ and y_{ij} £ 0 for all i = 1, 2, ..., m then a non-basic alternative solution will exist.

(ii) if for some a_j in A but not in B, $c_j - Z_j = 0$ and $y_{ij} > 0$ for all least one i then an alternative basic optimum solution will exist.

Proof:

(i) We have discussed in 6 that if we introduce the column vector a_j in B where a_j is in A but not in B and $y_{ij} \leq 0$ for all i = 1, 2, ... ,m then a non-basic feasible solution x_B' with (m + 1) number of positive variables is given by

$$\Rightarrow \quad x_B' = [x_{B1}', x_{B2}', \ldots, x_{Bm}', 1]$$

where $x_{Bi}' = x_{Bi} - \lambda y_{ij}$, i = 1, 2, ... ,m, 1 > 0

and the value of the objective function for this new F.S. is given by

$$Z' = Z + 1\,(c_j - Z_j).$$

If $c_j - Z_j = 0$, then $Z' = Z$

i.e., the value of the objective function for this new non-basic F.S. is also equal to Z (optimal value). Hence this new non-basic F.S. is an alternative optimal solution of the given L.P.P.

(ii) We have shown in theorem of 5 that if $y_{ij} > 0$ for at least one i = 1, 2, ..., m then by palpation one column br in B by the column aj which is in A but not in B, we obtain a new B.F.S. x_B' given by

$$x_B' = [x_{B1}', x_{B2}', \ldots, x_{Bm}']$$

where $x_{Bi}' = x_{Bi} - \dfrac{y_{ij}}{y_{rj}} x_{Br}$, i = 1, 2, ... ,m, i ≥ r

$$x_{Br}' = \frac{x_{Br}}{y_{rj}}, \quad i = r$$

and $$\frac{x_{Br}}{y_{rj}} = \underset{i}{\text{Mini}}\left(\frac{x_{Bi}}{y_{ij}}, y_{ij} > 0\right).$$

The value of objective function for this new B.F.S. is given by

$$Z' = Z + \frac{x_{Br}}{y_{rj}}(c_j - Z_j)$$

i.e., the value of the objective function for this new B.F.S. is also equal to Z' (optimal value).

Hence this new B.F.S. is an alternative optimal F.F.S.

TO DETERMINE STARTING B.F.S.

In this article we are going to discuss a method for finding a most convenient initial B.F.S. to a linear programming problem .

In the constraints of a general linear programming problem three may be any of the three signs ≤ = ≥. Let us assume that the requirement vector b ≥ 0 (if any of the b_i's is negative, multiply the corresponding constraint by – 1.)

Case I. To find initial B.F.S. when at the original constraints have ≤ sign. To convert all the constraints into equations we insert slack variables only .the equations obtained are as follows.

$$a_{11}x_1 + a_{12}x_2 + \dots + a_{1n}x_n + 1.x_{n+1} = b_1$$
$$a_{21}x_1 + a_{22}x_2 + \dots + a_{2n}x_n + 1.x_{n+2} = b_2$$
$$\dots \quad \dots \quad \dots \quad \dots \quad \dots \quad \dots$$
$$\dots \quad \dots \quad \dots \quad \dots \quad \dots \quad \dots$$
$$a_{m1}x_1 + a_{m2}x_2 + \dots + a_{mn}x_n + \dots + 1.x_{n+m} = b_m$$

Here x_{n+1}, x– = 2, ... ,x_{n+m} are slack variables.

In matrix from these equations can be written as

$$\begin{bmatrix} a_{11} & a_{12} & \dots & a_{1n} & 1 & 0 & \dots & 0 \\ a_{21} & a_{22} & \dots & a_{2n} & 0 & 1 & \dots & 0 \\ \dots & \dots & \dots & \dots & \dots & & \dots & \\ \dots & \dots & \dots & \dots & \dots & & \dots & \\ a_{m1} & a_{m2} & \dots & a_{mn} & 0 & 0 & \dots & 1 \end{bmatrix} \begin{bmatrix} x_1 \\ x_2 \\ \vdots \\ x_n \\ x_{n+1} \\ \vdots \\ x_{n+m} \end{bmatrix} = \begin{bmatrix} b_1 \\ b_2 \\ \vdots \\ b_m \end{bmatrix}$$

Taking the initial basis matrix

$$B = I_m, \text{ (unit matrix of order } m \times m).$$

∴ the initial basic solution is given by $x_B = B^{-1}b = I_m b = b \geq 0$

Thus the initial B.F.S. is

$$x_{n+1} = x_{B1} = b_1,\ x_{n+2} = x_{B2} = b_2,\ \dots ,x_{n+m} = x_{Bm} = b_m,$$

which can be obtained by writing all the non-basic variables (*i.e.,* given variables) $x_1, x_2, \dots x_n$ equal to zero and solving the equations for the remaining variables (*i.e.,* slack variables) $x_{n+1}, x_{n+2}, \dots ,x_{n+m}$.

Case II. To find initial B.F.S. when all the original constraints have ≥ sign. First we convert all the constraints into equations by inserting surplus variables. The equations obtained are as follows:

$$a_{11}x_1 + a_{12}x_2 + \dots + a_{1n}x_n - x_{n+1} = b_1$$
$$a_{21}x_1 + a_{22}x_2 + \dots + a_{2n}x_n - x_{n+2} = b_2$$
$$\dots \quad \dots \quad \dots \quad \dots \quad \dots \quad \dots$$
$$\dots \quad \dots \quad \dots \quad \dots \quad \dots \quad \dots$$
$$a_{m1}x_1 + a_{m2}x_2 + \dots + a_{mn}x_n - x_{n+m} = b_m$$

Here $x_{n+1}, x_{n+2}, \dots ,x_{n+m}$ are surplus variables.

In matrix form, these equations can be writing as

$$\begin{bmatrix} a_{11} & a_{12} & \cdots & a_{1n} & -1 & 0 & \cdots & 0 \\ a_{21} & a_{22} & \cdots & a_{2n} & 0 & -1 & \cdots & 0 \\ \cdots & \cdots & & \cdots & \cdots & & & \cdots \\ \cdots & \cdots & & \cdots & \cdots & & & \cdots \\ a_{m1} & a_{m2} & 0 & 0 & \cdots & & \cdots & -1 \end{bmatrix} \begin{bmatrix} x_1 \\ x_2 \\ \vdots \\ x_n \\ x_{n+1} \\ \vdots \\ x_{n+m} \end{bmatrix} = \begin{bmatrix} b_1 \\ b_2 \\ \vdots \\ b_m \end{bmatrix}$$

If we take the initial basis matrix $B = -I_m$,

we have $x_B = B^{-1} b = = -I_m b = -b \leq 0$, which is not a B.F.S.

To avoid this difficulty we add one more variable to each constraint. These variables are called *"Artificial Variables"*.

Adding surplus and artificial variables, the constraints of the given L.P.P. change to the following equations.

$$a_{11} x_1 + a_{12} x_2 + \ldots + a_{1n} x_n - x_{n+1} + x_{n+m+1} = b_1$$
$$a_{21} x_1 + a_{22} x_2 + \ldots + a_{2n} x_n - x_{n+2} + x_{n+m+2} = b_2$$
$$\ldots \quad \ldots \quad \ldots \quad \ldots \quad \ldots \quad \ldots \quad \ldots$$
$$\ldots \quad \ldots \quad \ldots \quad \ldots \quad \ldots \quad \ldots \quad \ldots$$
$$a_{m1} x_1 + a_{m2} x_2 + \ldots + a_{mn} x_n - x_{n+m} + m_{n+m+m} = b_m$$

Here $x_{n+m+1}, x_{n+m+2}, \ldots, x_{n+m+m}$ are the artificial variables.

In matrix, these equations can be written as

$$\begin{bmatrix} A_{11} & a_{12} & \cdots & a_{1n} & -1 & 0 & \cdots & 0 & 1 & 0 & \cdots & 0 \\ a_{21} & a_{22} & \cdots & a_{2n} & 0 & -1 & \cdots & 0 & 0 & 1 & \cdots & 0 \\ \cdots & \cdots & \cdots & \cdots & \cdots & \cdots & \cdots & \cdots & \cdots & \cdots & \cdots & \cdots \\ \cdots & \cdots & \cdots & \cdots & \cdots & \cdots & \cdots & \cdots & \cdots & \cdots & \cdots & \cdots \\ a_{m1} & a_{m2} & \cdots & a_{mn} & 0 & 0 & \cdots & -1 & 0 & 0 & \cdots & 1 \end{bmatrix}$$

$$\begin{bmatrix} x_1 \\ x_2 \\ \vdots \\ x_n \\ x_{n+1} \\ \vdots \\ x_{n+m} \\ x_{n+m+1} \\ \vdots \\ x_{n+m+m} \end{bmatrix} = \begin{bmatrix} b_1 \\ b_2 \\ \vdots \\ b_m \end{bmatrix}$$

Now taking the basis matrix $B = I_m$.

$$\therefore \quad x_B = B^{-1} b = I_m b = b \geq 0,$$

which is a basic feasible solution.

$\therefore$ the B.F.S. is $x_{n+m+1} = x_{B1} = b_1, x_{n+m+2} = x_{B2} = b_2, \ldots, x_{n+m+m} = x_{Bm} = b_m$, which can be obtained by writing all the non basic variables (*i.e.*, given variables and surplus variables) $x_1, x_2, \ldots, x_n; x_{n+1}, x_{n+2}, \ldots, x_{n+m}$ equal to zero and solving the equations for the remaining basic variables (*i.e.*, artificial variables) $x_{n+m+1}, x_{n+m+2}, \ldots, x_{n+m+m}$.

Case III. To find initial B.F.S. when the constraints have '$\leq$', '$\geq$' and '$=$' signs. In this case we convert the constraint into equations by inserting slack, surplus and artificial variables. Here the basis matrix $B = I_m$ is obtained by introducing unit column vectors corresponding to the slack and artificial variables. To obtain the initial B.F.S. we put the non-basic variables equal to zero and solve for remaining basic variables. Here also the initial B.F.S. $x_B = b \geq 0$.

INCONSISTENCY AND REDUNDANCY IN CONSTRAINT EQUATIONS

Redundancy is constraint equations.

By redundancy in constraint equations we mean that the system has more than enough number of constraint equations, in other words it has more constraint equations than the number of variables.

This is the situation when

$$r(A) = r(A\, b) = k \leq n < m.$$

In this case there will be $(m - k)$ redundant equations.

The Inconsistency

As already defined, the set of constraints (linear equations) is said to be inconsistent if

$$r(A) \neq r(Ab).$$

Before solving a LP. problem by simplex method, we should have $r(A) = r(Ab)$ *i.e.,* the constraint aquariums (after introducing the slack and artificial variables) shows be consistent. Since in simplex method we always have

$$r(A) = r(A \times b) = m.$$

If the system $A x = b$ involves artificial variables, then we cannot say whether this system is consistent or there is any redundancy. Below we give the cases to decide about the consistency and redundancy in such systems.

Case I. If at least one artificial vector appears in the basis B at a positive level *i.e.*, the value of at least one artificial variables corresponding to artificial vector in B is non-zero and the optimality condition is satisfied (at any iteration), then there exists no feasible solution of the problem.

Case II. If the basis B contains no artificial vector and the optimality condition is satisfied (at any iteration) then the current solution is a B.F.S. of the problem.

Case III. If one or more artificial vector appears in the basis B at a zero level *i.e.*, the value of the artificial variables corresponding to artificial vectors in B are zero and the optimality condition is satisfied (at any iteration) then the system is consistent. Furthermore if $y_{ij} = 0 \ \forall \ j$ and $x_{Br} = 0$ and r corresponds to the row containing an artificial vector, then the rth constraint equation is redundant.

TO DETERMINE IMPROVED B.F.S. FROM A GIVEN B.F.S.

Theorem:

Let $x_B = B^{-1} b$ be a B.F.S. of a linear programming problem with $Z = c_B x_B$ as the value of the objective function. If for any column aj in A, but not in B, the condition $c_j - Z_j > 0$ or $Z_j - c_j < 0$ holds and if at least one $y_{ij} > 0$, $i = 1, 2, \ldots, m$, then it is possible to obtain a new B.F.S. by replacing one of the columns in B by α_j and if Z' is the new value of objective function then $Z' \geq Z$.

If the initial B.F.S. $x_B = B^{-1} b$ is non-degenerate then $Z' > Z$.

Proof:

Consider the L.P.P.

Max. $Z = c\,x$, s.t. $A\,x = b$, $x \geq 0$,

where $A = (\alpha_1, \alpha_2, \ldots, \alpha_N)$, $N = m + n$,

Basis matrix $B = (\beta_1, \beta_2, \ldots, \beta_m)$.

Let x_B $(x_{B1}, x_{B2}, \ldots, x_{Bm})$ be a B.F.S. of the given L.P.P.

Since the vectors $\beta_1, \beta_2, \ldots, \beta_m$ are in the basis of A, therefore we can express α_j as a linear combination of b's.

$$\therefore \alpha_j = \sum_{i=1}^{m} y_{ij}\beta_i = y_{1j}\,\beta_1 + y_{2j}\,\beta_2 + \ldots + y_{mj}\,\beta_m. \qquad \ldots(1)$$

If $y_{rj} \neq 0$ then α_j can replace br in B and B is still a basis matrix.

Let $y_{rj} \neq 0$ then from (1), we have

$$\Rightarrow \qquad \beta_r = \frac{1}{y_{rj}}\alpha_j - \frac{y_{1j}}{y_{rj}}\beta_1 - \ldots - \frac{y_{(r+1)j}}{y_{rj}}\beta_{r+1} - \ldots - \frac{y_{mj}}{y_{rj}}\beta_m$$

$$\Rightarrow \quad \beta_r = \frac{1}{y_{rj}}\alpha_j - \sum_{i=1}^{m}\frac{y_{ij}}{y_{rj}}\beta_i . \qquad ...(2)$$

Now, we have

$$\Rightarrow \quad Bx_B = b$$

$$\Rightarrow \quad (\beta_1, \beta_2, ... ,\beta_r, ... ,\beta_m)\,(x_{B1}, x_{B2}, ... ,x_{Br}, ... ,x_{Bm}) = b$$

$$\Rightarrow \quad \beta_1 x_{B1} + \beta_2 x_{B2} + ... + \beta_r x_{Br} + ... + \beta_m x_{Bm} = b$$

$$\Rightarrow \quad \sum_{i=1}^{m}\beta_i x_{Bi} + \beta_r x_{Br} = b \qquad ...(3)$$

Putting for br from (2) in (3), we get

$$\Rightarrow \quad \sum_{i=1}^{m}\beta_i\left[x_{Bi} - \frac{y_{ij}}{y_{rj}}x_{Br}\right] + \frac{x_{Br}}{y_{rj}}\alpha_j = b \qquad ...(4)$$

$$\Rightarrow \quad \sum_{i=1}^{m} x_{Bi}'\beta_i + x_{Br}'\alpha j = b, \qquad ...(5)$$

where $x_{Bi}' = x\text{–} - x_{Br}\dfrac{y_{ij}}{y_{rj}}$, $i = 1, 2, ... ,m, i \neq r$

$$x_{Br}' = \frac{x_{Br}}{y_{rj}}, \; i = r$$

Comparing (3) and (5), we observe that new basic solution of A x = b is given by

$x'_B = (x'_{Bi}, x'_{Br})$, $i = 1, 2, ... ,m, i \neq r$

where the values of x'_{Bi}, x'_{Br} are given by (6).

This basic solution will be feasible if we have

$$x_{Bi} - \frac{y_{ij}}{y_{rj}}x_{Br} \geq 0, \; i = 1, 2, ... m, \; i \neq r$$

and $\quad \dfrac{x_{Br}}{y_{rj}} \geq 0, \; i = r.$ ...(7)

Since x_B is the initial B.F.S. we have

$$x_{Br} \geq 0, \; r = 1, 2, ... ,m.$$

Thus we observe that (7) will hold only if

$y_{rj} > 0$ and $y_{ij} \leq 0$, $i = 1, 2, ... ,m, i \neq r$.

If $y_{rj} > 0$ and $y_{ij} > 0$, then (7) is satisfied only if

$$\Rightarrow \quad \frac{x_{Bi}}{y_{ij}} - \frac{x_{Br}}{y_{rj}} \geq 0$$

$$\Rightarrow \quad \frac{x_{Br}}{y_{rj}} \leq \frac{x_{Bi}}{y_{ij}}$$

$$\Rightarrow \quad \frac{x_{Br}}{y_{rj}} = \min_{i} \left\{ \frac{x_{Bi}}{y_{ij}}, y_{ij} > 0 \right\} = v. \qquad \text{...(8)}$$

Thus, a new B.F.S. can be obtained from the initial B.F.S. by removing the column vector br of the basis matrix B by a_j if r is to be selected such that

$$v = \frac{x_{Br}}{y_{rj}} = \min_{i} \left\{ \frac{x_{Bi}}{y_{ij}}, y_{ij} < 0 \right\}.$$

If we have v = 0, which is possible only when $x_{Br} = 0$ then it means that the initial B.F.S. is degenerate.

Now we shall prove $Z' \geq Z$.

The value of the objective function for the initial B.F.S. x_B is

$$\begin{aligned} Z &= c_B\, x_B \\ &= (c_{B1}, c_{B2}, \dots, c_{Bm})\,(x_{B1}, x_{B2}, \dots, x_{Bm}) \\ &= \sum_{i=1}^{m} c_{Bi} x_{Bi} \end{aligned}$$

Corresponding to the new B.F.S. x'_B, the value of the objective function is Z', so we have

$$Z' = \sum_{i=1}^{m} c'_{Bi}\, x'_{Bi}, \qquad \text{...(9)}$$

where c_{Bi}' are the coefficients of the basic variables x_{Bi}' (i = 1, 2, ... ,m) in the objective function.

Obviously, $c_{Bi}' = c_{Bi}$, i = 1, 2, ... , m, $i \neq r$

and $\quad c_{Br}' = c_j$.

∴ we can write

$$Z' = \sum_{i=1}^{m} c_{Bi} x'_{Bi} + c_j x'_{Br}. \qquad \text{...(10)}$$

Substituting for x_{Bi}', x_{Br}' from (6) in (10), we get

$$Z' = \sum_{i=1}^{m} c_{Bi} \left(x_{Bi} - x_{Br} \frac{y_{ij}}{y_{rj}} \right) + c_j \frac{x_{Br}}{y_{rj}}$$

$$= \sum_{i=1}^{m} c_{Bi}\left(x_{Bi} - x_{Br}\frac{y_{ij}}{y_{rj}}\right) + c_j \frac{x_{Br}}{y_{rj}}$$

Since the term $c_{Bi}\left(x_{Bi} - x_{Br}\frac{y_{ij}}{y_{rj}}\right) = 0$ when i = r therefore it can be included in the summation without changing the value of Z'.

$$\therefore \; Z' = \sum_{i=1}^{m} c_{Bi}x_{Bi} - \frac{x_{Br}}{y_{rj}}\sum_{i=1}^{m} c_{Bi}y_{ij} + \frac{x_{Br}}{y_{rj}}c_j$$

$$= Z + \frac{x_{Br}}{y_{rj}}(c_j - Z_j), \text{ where } Z_j = \sum_{i=1}^{m} c_{Bi}y_{ij}$$

$$= Z + v\,(c_j - Z_j), \text{ from (8).} \qquad ...(11)$$

From (11), we observe that the new value of the objective function is equal to the original value of the objective function plus the quantity $v\,(c_j - Z_j)$.

We have $Z' \geq Z$ if $v\,(c_j - Z_j) \geq 0$.

Since $v \geq 0$ therefore $Z' \geq Z$ only if $c_j - Z_j \geq 0$.

Hence by choosing the vector α_j for which $c_j - Z_j > 0$ and at least one $y_{ij} > 0$, we obtain a new improved value of the objective function.

If the initial B.F.S. is non-degenerate then $v > 0$ and in that case $Z' > Z$.

CONDITIONS FOR THE EXISTENCE OF UNBOUNDED SOLUTIONS

Now, we have proved that if for any column aj in A but not in B the condition $c_j - Z_j > 0$ holds and if at least one $y_{ij} > 0$, i = 1, 2, ... ,m then it is possible to find an improved B.F.S.

Now the question arises : what will be the implication if for at least one a_j all $y_{ij} \leq 0$, i = 1, 2, ..., m.

To get the answer of this question we shall prove the following theorem.

Theorem:

If for any basic feasible solution $x_B = B^{-1}b$ to $Ax = b$ there is some column α_j in A but not in B for which $c_j - Z_j > 0$ and $y_{ij} \leq 0$, i = 1, 2, ..., m then if the objective function is to be maximized, the problem has an unbounded solution.

Proof:

Consider the linear programming problem

Max. $Z = cx$ s.t. $Ax = b$, $x \geq 0$,

where $A = (\alpha_1, \alpha_2, \ldots \alpha_N)$. $N = m + n$,

basis matrix $B = (\beta_1, \beta_2, \ldots, \beta_m)$.

Let $x_B = (x_{B1}, x_{B2}, \ldots, x_{Bm})$ be a B.F.S. of the given L.P.P. Then we have

$$Bx_B = b \text{ or } \sum_{i=1}^{m} x_{Bi}\beta_i = b \quad \ldots(1)$$

and the corresponding value of the objective function is

$$Z = c_B x_B = \sum_{i=1}^{m} c_{Bi} x_{Bi} \quad \ldots(2)$$

Adding and subtracting $\lambda\alpha_j$ in (1), we get

$$\sum_{i=1}^{m} x_{Bi}\beta_i - \lambda\alpha_j + \lambda\alpha_j = b, \quad \ldots(3)$$

where l is some scalar.

Let $\alpha_j \in A$ s.t. $\alpha_j \notin B$.

Since the vectors $\beta_1, \beta_2, \ldots, \beta_m$ are in the basis of A therefore we can express α_j as the linear combination of b's

$$\therefore \alpha j = \sum_{i=1}^{m} y_{ij}\beta_i$$

$$\Rightarrow \quad -la_j = -1 \sum_{i=1}^{m} y_{ij}\beta_i .$$

Substituting the above value of $-\lambda\alpha_j$ in (3), we get

$$\Rightarrow \quad \sum_{i=1}^{m} x_{Bi}\beta_i - \lambda\sum_{i=1}^{m} y_{ij}\beta_i + \lambda\alpha_j = b$$

$$\Rightarrow \quad \sum_{i=1}^{m} (x_{Bi} - \lambda y_{ij})\beta_i + \lambda\alpha_j = b$$

Thus we obtain a new solution

$$x'_B = (x_{B1}', x_{B2}', \ldots, x_{Bm}', \lambda) \quad \ldots(4)$$

where $x_{Bi}' = x_{Bi} - \lambda y_{ij}$, $i = 1, 2, \ldots, m$.

When $\lambda > 0$, $x_{Bi} - \lambda y_{ij} \geq 0$ since $y_{ij} \leq 0$, $i = 1, 2, \ldots, m$ therefore (4) gives the feasible solution in which the number of positive variables is less than or equal to $(m + 1)$. It may be less than $(m + 1)$ because some $x_{Bi} - \lambda y_{ij}$ may be zero for some i. In case the number of positive variables in this solution is equal to $(m + 1)$ then this solution will be non-basic feasible solution.

If Z' is the new value of the objective function corresponding to this new solution, then we have

$$\Rightarrow \qquad Z' = \sum_{i=1}^{m} c_{Bi}(x_{Bi} - \lambda y_{ij}) + c_j\lambda$$

$$\Rightarrow \qquad Z' = \sum_{i=1}^{m} c_{Bi}x_{Bi} + \lambda(c_j - c_{Bi}y_{ij}) = Z + \lambda\,(c_j - Z_j).$$

Since $c_j - Z_j > 0$ (given) therefore Z' can be made as large as we please by giving sufficiently large values to l. We know that a L.P.P. has an unbounded solution if the value of its objective function can be increased or decreased arbitrarily.

Hence the given L.P.P. has an unbounded solution.

NOTATIONS OF THE SIMPLEX ALGORITHM

Some notations and definitions which are extremely useful in the discussion of simplex algorithm.

Consider a linear programming problem which after introducing slack and surplus variables is as follows:

Max. $Z = c_1 x_1 + c_2 x_2 + \ldots + c_n x_n + 0x_{n+1} + 0x_{n+2} + \ldots + 0x_{n+m}$

subject to $a_{11} x_1 + a_{12} x_2 + \ldots + a_{1j} x_j + \ldots + a_{1n} x_n + x_{n+1} = b_1$

$a_{21} x_1 + a_{22} x_2 + \ldots + a_{2j} x_j + \ldots + a_{2n} x_n + x_{n+2} = b_2$

... ...

... ...

$a_{m1} x_1 + a_{m2} x_2 + \ldots + a_{mj} x_j + \ldots + a_{mn} x_n + x_{n+m} = b_m,$

$x_i \geq 0$ for all i = 1, 2, ... ,N, where N = n + m

$b_1, b_2, \ldots$,bm are all positive.

The above linear programming problem can be easily converted to matrix form

Max. Z = c x

subject to A x = b, x ≥ 0

where $A = [a_{ij}]_{m \times N}$ is the coefficient matrix of order m × N,

$x = [x_1, x_2, x_3, \ldots ,x_n, \ldots ,x_N],$

$c = [c_1, c_2, \ldots ,c_n, 0, 0, \ldots ,0]_{1 \times N}$

and $\qquad b = [b_1, b_2, \ldots ,b_m],$

where x and b are column vectors of order N × 1 and m × 1 respectively.

For convenience, we shall represent column vectors by row vectors without using transpose symbol. There should be no confusion in understanding scalar multiplication of two vectors c and x.

If we denote the jth column of the matrix A by a_j, j = 1, 2, ... ,N, we can write

$$A = [\alpha_1, \alpha_2, \ldots, \alpha_N].$$

Form a non-singular sub-matrix B of order m × m whose column vectors are m number of linearly independent columns selected from A. If we denote these columns by $\beta_1, \beta_2, \ldots, \beta_m$ then

$$B = [\beta_1, \beta_2, \ldots, \beta_m].$$

B is called the *basis matrix*.

Since the vectors $\beta_1, \beta_2, \ldots \beta_m$ are linearly independent they form a basis for E^m.

Therefore each $\alpha_j \in A \subset E^m$ can be expressed as a linear combination of vectors of B. Thus we can write

$$\alpha_j = \beta_1 y_{1j} + \beta_2 y_{2j} + \ldots + \beta_m y_{mj}$$

$$= (\beta_1, \beta_2, \ldots, \beta_m)\begin{bmatrix} y_{1j} \\ y_{2j} \\ \vdots \\ y_{mj} \end{bmatrix}$$

or $\alpha_j = BY_j$, where $Y_j = \begin{bmatrix} y_{1j} \\ y_{2j} \\ \vdots \\ y_{mj} \end{bmatrix}_{m\times 1}$

$y_{1j}, y_{2j}, \ldots, y_{mj}$ are the scalars required to express α_j in such a form.

Also $\alpha_j = BY \Rightarrow Y_j = B^{-1} a_j$.

The vector Yj will change if the columns of A forming B change.

Any basis matrix B will provide a *basic solution* to A x = b.

The variables corresponding to $\beta_1, \beta_2, \ldots, \beta_m$ are denoted by x_{B1}, $x_{B2}, \ldots, x_{Bm}$ respectively and are called *basis variables.*

We denote the column vector of these m basic variables by x_B

∴ $x_B = [x_{B1}, x_{B2}, \ldots, x_{Bm}]$

and $x_B = B^{-1}$ b.

This is called the basic feasible solution (B.F.S.) of the L.P.P.

Since $x_{B1}, x_{B2}, \ldots, x_{Bm}$ are the basic variables, the remaining n variables belonging to x are called non-basic variables.

We shall denote the coefficients of basic variables $x_{B1}, x_{B2}, \ldots, x_{Bm}$ in the objective function Z by $c_{B1}, c_{B2}, \ldots, c_{Bm}$.

Corresponding to any x_B, c_B will represent the row vector contain the constants $c_{B1}, c_{B2}, \ldots, c_{Bm}$.

$$\therefore c_B = (c_{B1}, c_{B2}, \ldots, c_{Bm}).$$

Since for any basic feasible solution all non-basic variables are zero therefore the objective function Z becomes

$$Z = c_{B1}\, x_{B1} + c_{B2}\, x_{B2} + \ldots + c_{Bm}\, x_{Bm} + 0$$
$$= (c_{B1}, c_{B2}, \ldots, c_{Bm})\, (x_{B1}, x_{B2}, \ldots, x_{Bm})$$
$$= c^B\, x_B.$$

Finally, we introduce a new variables Z_j , given by

$$Z_j = y_j\, c_{B1} + y_{2j}\, c_{B2} + \ldots + y_{mj}\, c_{Bm}$$
$$= (c_{B1}, c_{B2}, \ldots c_{Bm})\, (y_{1j}, y_{2j}, \ldots, y_{mj})$$
$$= c_B\, y_j.$$

Therefore exists Z_j for each α_j and it changes as the columns of A forming B change.

COMPUTATIONAL PROCEDURE OF THE SIMPLEX METHOD

Step 1. Find the initial B.F.S. follow the method discussed in 10 to find the initial B.F.S. If artificial variables are introduced in the L.P.P. then follow the two-method for Big M-method for solving such prob lems.

Step 2. Construct the initial simplex table (starting simplex table).

It should be remembered that the values of non-basic variables are always zero at each iteration

Step 3. Test the initial B.F.S. for optimality.

Compute the net evaluation Δ_j for each variables x_j by using the formula $\Delta_j = c_j - c_B\, Y_j$.

Step 4. If the given problem is of minimization, convert it into the maximization problem.

For this, multiply both sides of the objective function by –1 and put $-Z = Z'$.

If v is the maximum value of Z' then –v will be the minimum value of Z.

Step 5. Make all the b_i's non-negative.

The R.H.S. of each of the constraints should be non-negative. If the L.P.P. has a constraint for which a negative b_i is given, it should be converted into positive value by multiplying both sides of the constraint by –1.

Step 6. Convert inequalities of constraints into equations.

For this introduce slack or surplus variables. The coefficients of slack or surplus variables in the objective function are zero. Introduce artificial variables, if necessary.

OPTIMAL TEST

(a) If $\Delta_j \leq 0$ for all j, the solution under consideration is optimal.

Alternative optimal solution will exist if any non-basic Δ_j is also zero.

(b) If $\Delta_j > 0$ for any j *i.e.*, if at least one Δ_j is positive the solution under test is not optimal, we must proceed to improve the solution (step 7).

(c) If corresponding to maximum positive D_j, all elements of the column Y_j are negative or zero, the solution under test will be unbounded.

(d) If optimality condition is satisfied but the value of at least one artificial variables present in the basis is non-zero, the problem will have no feasible solution.

Step 1. *Select the entering (incoming) vector and departing (outgoing) vector.*

To improve the initial B.F.S. we select the vector entering the basis matrix and the vector to be removed from the basis matrix.

Incoming Vector. The incoming vector α_k is always selected corresponding to the largest positive value of Δ_j.

If maximum value of Δ_j occurs for more than one α_j then we can select any of these vectors as incoming vector.

Outgoing Vector. The departing vector β_r is selected corresponding to that value of r for which

$$\frac{x_{Br}}{y_{rk}} = \min_i \left\{ \frac{x_{Bi}}{y_{ik}}, y_{ik} > 0 \right\},$$

is α_r is the incoming vector.

Step 2. If α_k is the entering vector and β_r is the outgoing vector then the element yrk which lies at the intersection of minimum ration arrow (←) and incoming vector arrow ↑ is called the pivot element (key element).

We put this element in ▯.

In order to bring b_r in place of incoming vector Y_k, unity must occupy the place ▯. In other words key element y_{rk} should be 1. If it is not so then divide all the elements of this row by the key element y_{rk}. Then subtract suitable multiples of the row containing the key element from all other rows and obtain zero at all other positions of this column Y_k. Now bring br in place of α_r and construct the revised simplex table.

Thus, we get improve basic feasible solution.

Step 3. Test the improved B.F.S. for optimality.

If it is not optimum, repeat steps 7 and 8 until we obtain an optimums solution.

Note: If the above mentioned minimum value is not unique then more than one variable will vanish in the next solution. As a result the next solution will be a degenerate B.F.S. for which the outgoing vector is selected in a different way.

COMPUTE THE INVERSE OF A MATRIX WHOSE ONE COLUMN IS DIFFERENT FORM OF A MATRIX WHOSE INVERSE IS KNOWN

$B = (\beta_1, \beta_2, \ldots, \beta_n)$ is the given non-singular matrix whose inverse is known.

Consider a matrix $B_\alpha = (\beta_1, \beta_2, \ldots, \beta_{r-1}, \alpha, \beta_{r+1}, \ldots, \beta_n)$.

We have obtained the matrix Ba by replacing the rth column of matrix B by the column vector a.

If we have $\alpha = \sum_{i=1}^{n} y_i \beta_i$,

Ba will be non-singular iff $y_r \neq 0$.

Using components of the vector $Y = B^{-1}\alpha$, we introduce a new matrix E where E is a square matrix of order n which differs from identity matrix In the rth column only. To obtain the rth column of matrix E we proceed in the following manner:

We have $\alpha = y_1 \beta_1 + y_2 \beta_2 + \ldots + y_r \beta_r + \ldots + y_m \beta_m$.

Transposing yr br to the left and a to the right, we get

$$-y_r \beta_r = y_1 \beta_1 + y_2 \beta_2 + \ldots + (-\alpha) + \ldots + y_n \beta_n$$

$$\text{or } \beta_r = -\frac{y_1}{y_r}\beta_1 - \frac{y_2}{y_r}\beta_2 - \ldots + \frac{1}{y_r}\alpha - \ldots - \frac{y_n}{y_r}\beta_n$$

$$= \frac{y_1}{y_r}\beta_1 - \frac{y_2}{y_r}\beta_2 - \ldots + \frac{1}{y_r}\alpha - \ldots - \frac{y_n}{y_r}\beta_n \; (\beta_1, \beta_2, \ldots, \alpha, \ldots \beta_n)$$

$$= A_r\, B_\alpha,$$

where $A_r = \left(-\frac{y_1}{y_r}, -\frac{y_2}{y_r}, \ldots, \frac{1}{y_r}, -\frac{y_n}{y_r}\right)$ gives the rth column of E.

Now $B_\alpha E = B_\alpha (e_1, e_2, \ldots, e_{r-1}, A_r, e_{r+1}, \ldots, e_n)$

$$= (B_\alpha e_1, B_\alpha e_2, \ldots, B_\alpha e_{r-1}, B_\alpha A_r, \ldots, B_\alpha e_n)$$

$$= (\beta_1, \beta_2, \ldots, \beta_{r-1}, \beta_r, \beta_{r+1}, \ldots, \beta_n),$$

since $B_\alpha e_1 = (\beta_1, \beta_2, \ldots, \alpha, \ldots, \beta_n)(1, 0, 0, \ldots, 0)$

$$= \beta_1 + 0 + 0 + \ldots + 0 = \beta_1 \text{ etc.}$$

$= B.$

Since $B = B\alpha\, E$

$\therefore \quad BB^{-1} = B_\alpha EB^{-1} \quad \Rightarrow \quad 1^1 = B_\alpha^0 EB^{-1}$

$\Rightarrow \quad B_\alpha^{-1} I = (B_\alpha^{-1} B_\alpha) EB^{-1} \quad \Rightarrow \quad B_\alpha^{-1} = IEB^{-1} = EB^{-1}.$

SIMPLEX METHOD BY INVERSE OF A MATRIX

Consider any $n \times n$ non-singular real matrix A.

Also consider a system of equation $A\,x = b$, $x \geq 0$,

where b is a dumbly real matrix of order $n \times 1$.

Introduce the artificial variables $x_a \geq 0$ and a dummy objective function Z with cost zero to variables in x and cost (–1) to each artificial variable. Now find the solution of the following linear programming problem using simplex method

Max. $Z = 0.\, x - 1.x_{art}$

subject to $A\,x = b$, $x, x_a \geq 0$.

In the last iteration when the column of A become the columns of unit matrix I, the inverse of matrix A is given by those column vectors of the table which were the column of the initial basis.

COMPUTATIONAL PROCEDURE OF THE TWO PHASE METHOD

Phase I.

Step 1. To convert each of the constraints into equations, subtract a surplus variable and add an artificial variables.

Step 2. Assign zero coefficients to each of the primary variables and surplus variables and the coefficient (–1) to each of the artificial variables in the objective function. As a result the objective function changes to

Z = – (sum of the artificial variables)

Step 3. Solve the problem formed in step 2 by applying the simplex method. If the original problem has a feasible solution then this problem shall have an optimal solution with optimal value of the objective function equal to zero as each of the artificial variables will be equal to zero.

If max $Z < 0$ and at least one artificial variables appears in the optimum basis at a positive level then the given problem does not possess any feasible solution.

If max $Z = 0$ and at least one artificial variable appears in the optimum basis at zero level, we proceed to phase II.

If max $Z = 0$ and no artificial variable appears in the optimum basis then we proceed to phase II.

Phase II.

In phase II start with the (optimal) solution contained in the final simplex table of the phase I. Assign the actual costs to the variables in the objective function and a zero cost to every surplus variable. Eliminate the artificial which are non-basic at the and of the phase-I. Remove c_j row values of the optimum table and replace them by c_j values of the original problem. Now, apply simplex algorithm to the problem contained in the new table to obtain the optimal solution.

Disadvantages of Big-M Method Over Two Phase Method

(i) We can always use Big-M method to check the existence of a feasible solution but its computational procedure may be inconvenient because of the manipulation of the constant M.

Two phase method eliminates the artificial variables in the beginning.

(ii) When we solve a problem on a digital computer we have to assign some numerical value to M which must be larger than the values c_1, c_2, ... present in the objective function. But a computer has only a fixed number of digits.

CANONICAL AND STANDARD FORM OF LP PROBLEM

Canonical form : If the general formulation of LP problem is expressed as:

$$\text{Maximize} \quad Z = \sum_{j=1}^{n} c_j x_j$$

subject to the constraints

$$\sum_{j=1}^{n} a_{ij}x_j \le b_i; \qquad i = 1, 2, ..., m$$

and $\quad x_j \ge 0; \qquad j = 1, 2, ..., n$

then it is called the canonical form of LP problem. The characteristics of this form are:

(i) The objective function should be of maximization type. If not, then it should be changed to the same by applying the method discussed before.

(ii) All constraints should be of ≤ type except for the non-negativity conditions. An inequality of ≥ type can be changed to an inequality of ≤ type by multiplying it on both sides by – 1.

(iii) All variables must have non-negative values. If any variable, say x_j, is unrestricted in sign (i.e. positive, negative or zero), then it can be replaced by

$$x_j = x_j' - x_j''$$

where x_j' and x_j'' are both non-negative.

(iv) The right-hand side of each constraint should be positive.

Standard Form : If the general formulation of the LP problem as stated in Chapter 3, is expressed as

$$\text{Maximize } Z = \sum_{j=1}^{n} c_j x_j$$

subject to the constraints

$$\sum_{j=1}^{n} a_{ij}x_j \le b_i; \qquad i = 1, 2, ..., m$$

and $\quad x_j \ge 0; \qquad j = 1, 2, ..., n$

then it is called the standard/arm of the LP problem. The characteristics of this form are:

(i) All the constraints should be expressed as equations.

(ii) The right-hand-side of each constraint should be made non-negative. If it is not so, this should be done by multiplying both sides of the resulting constraints by – 1.

(iii) The objective function should be of maximization type.

The standard form can also be written in matrix notation as follows:

Maximize Z = ex ...(1)

subject to the constraints

$$Ax = b \qquad \text{...(2)}$$

and $\quad x > 0, \qquad \text{...(3)}$

where $c = (c_1, c_2, ..., c_n)$ is the row vector; $x = (x_1, x_2, ..., x_n)^T$ and $b = (b_1, b_2, ..., b_m)$ are column vectors and A is $m \times n$ coefficients matrix of rank m.

The LP problem can also be represented in terms of column vectors, $a_1, a_2, ..., a_n$ of matrix A as follows:

Maximize $Z = \sum_{j=1}^{n} c_j x_j$

subject to the constraints

$$\sum_{j=1}^{n} a_j x_j = b$$

and $\quad x_j \geq 0; \qquad j = 1, 2, ..., n.$

TO REDUCTION BASIC FEASIBLE SOLUTION

Theorem:

The linear programming problem

Max. $Z = cx$ *subject to* $Ax = b$, $x \geq 0$,

where $A = (\alpha_1, \alpha_2, ..., \alpha_N)$ *is the coefficient matrix of order* $m \times N$, $(N = m + n)$, *has at least one feasible solution then it has at least one basic feasible solution also.*

Proof:

Consider an arbitrary feasible solution of the given linear programming problem

$$x^* = (x_1, x_2, ..., x_N),\ x_i \geq 0. \qquad \text{...(1)}$$

Let us assume that k, $(k \leq N)$ variables in x^* have positive values and the remaining $N - k$ variables are zero. We can also assume that the variables have numbered in such a way that the first k variables are non-zero.

Thus $x^* = (x_1, x_2, ..., x_k, 0, 0, ..., 0)$.

Also, we have $\sum_{i=1}^{k} x_i \alpha_i = b. \qquad \text{...(2)}$

Now, two possibilities may arise:

The vectors $\alpha_1, \alpha_2, ..., \alpha_k$ may be either linearly independent or dependent.

Case I. If $\alpha_1, \alpha_2, \ldots, \alpha_k$ are linearly independent.

If any feasible solution for which the vectors a_i, i = 1, ... ,k associated with non-zero variables x_i, i = 1, ... ,k are L.I. is called a basic feasible solution. Thus, x* is a B.F.S. which is also optimal.

Hence, solution x* is degenerate if k < m and non-degenerate if k = m.

Case II. If $\alpha_1, \alpha_2, \ldots$,αk are linearly dependent and k > m.

In this case shall reduce the number of non-zero variables step by step until we obtain a B.F.S. from feasible solution.

Since $\alpha_1, \alpha_2, \ldots, \alpha_k$ are linearly dependent therefore there exist scalars $\lambda_1, \lambda_2, \ldots \lambda_k$ such that

$$\lambda_1 1 + \lambda_2 + \alpha_2 + \ldots + \lambda_k \alpha_k = 0$$

$$\Rightarrow \quad \sum_{o=1}^{k} \lambda_i \alpha_i = 0 \qquad \ldots(3)$$

and at least one $\lambda_1 \neq 0$.

We can assume that at least one λ_i is positive because if none is positive then we can multiply the equation (3) by – 1 and get positive λ_i.

Now, suppose

$$v = \max_{1 \le i \le k} \left(\frac{\lambda_i}{x_i} \right). \qquad \ldots(4)$$

Obviously, v is positive since $x_i \geq 0$, i = 1, 2, ..., k and at least one λ_i is positive.

Multiplying both sides of (3) by 1/v and subtracting from (2), we get

$$\sum_{i=1}^{k} \left(x_i - \frac{\lambda_i}{v} \right) \alpha_i = b. \qquad \ldots(5)$$

(5) gives $x' = \left[x_1 - \frac{\lambda_1}{v}, x_2 - \frac{\lambda_2}{v}, \ldots, x_k - \frac{\lambda_k}{v}, 0, 0, \ldots, 0 \right]$

a new solution of the matrix equation A x = b.

Also from (4) for i = 1, 2, ... ,k, we have

$$v \geq \frac{\lambda_i}{x_i} \text{ or } x_i \geq \frac{\lambda_i}{v}$$

or $$x_i - \frac{\lambda_i}{v} \geq 0.$$

Since all the components of x' are non-negative therefore x' is a feasible solution to the given L.P.P.

Also $v = \frac{\lambda_i}{x_i}$, for at least one i

i.e., $v - \frac{\lambda_i}{x_i} = 0$ therefore the new feasible solution x' cannot have more than k – 1 non-zero variables.

In this way, we have derived a new feasible solution from the given optimal feasible solution which contains less number of non-zero variables. This solution is B.F.S. if the column vectors associated to non-zero variables in this new solution are L.I.

If these associated vectors are not L.I., we shelf repeat the whole reduction procedure as explained above. Continuing in this way for finite number of times we will derive a solution in which columns corresponding to positive variables are L.I. *i.e.*, we will obtain a B.F.S. of the system. Hence the theorem is proved.

SOLVED EXAMPLES

Example 1:

maximize $Z = 3x_1 + 5x_2 + 4x_3$

subject to $2x_1 + 3x_2 \le 8$

$2x_2 + 5x_3 \le 10$

$3x_1 + 2x_2 + 4x_3 \le 15,$

$x_1, x_2, x_3 \ge 0.$

Solution:

The given problem is of maximization and all the b_i's are non-negative.

Converting the inequalities of constraints into equations by introducing slack variables x_4, x_5, x_6. Also the coefficients of slack variables are zero in the objective function. Thus the given problem becomes

$Z = 3x_1 + 5x_2 + 4x_3 + 0x_4 + 0x_5 + 0x_6$

subject to $2x_1 + 3x_2 + 0x_3 + x_4 = 8$

$0x_1 + 2x_2 + 5x_3 + x_5 = 10$

$3x_1 + 2x_2 + 4x_3 + x_6 = 15.$

The starting basic feasible solution is

$x_1 = 0, x_2 = 0, x_3 = 0, x_4 = 8, x_5 = 10, x_6 = 15$. The solution to the problem using simplex method is given in the following table.

B	c_B	c_j / x_B	3 / Y_1	5 / Y_2	4 / Y_3	0 / Y_4	0 / Y_5	0 / Y_6	Min. ratio x_B/Y_2
Y_4	0	8	2	(3)	0	1	0	0	8/3
Y_5	0	10	0	2	5	0	1	0	10/2 = 5 ←
Y_6	0	15	3	2	4	0	0	1	15/2
$Z = x_B$	$x_B = 0$	Δ_j	3	5 ↑	4	0	0 ↓	0	x_B/Y_3
Y_2	5	8/3	2/3	1	0	1/3	0	0	–
Y_5	0	4/3	–4/3	0	(5)	–2/3	1	0	14/15 ←
Y_6	0	2/3	5/3	0	4	–2/3	0	1	29/12
$Z = c_B$ x_B = 40/3		Δ_j	–1/3	0 ↑	4	–5/3 ↓	0	0	x_B/Y_1
Y_2	5	8/3	2/3	1	0	1/3	0	0	4
Y_3	4	14/15	–4/15	0	1	–2/15	1/5	0	–
Y_6	1	89/15	(41/15)	0	0	–2/15	–4/5	1	89/41 ←
$Z = c_B x_B$ = 256/45		Δ_j	11/15 ↑	0	0	–17/15	–4/5	0 ↓	

B	c_B	c_j / x_B	3 / Y_1	5 / Y_2	4 / Y_3	0 / Y_4	0 / Y_5	0 / Y_6	Min. ratio
Y_2	5	50/41	0	1	0	15/41	8/41	–10/41	
Y_3	4	62/41	0	0	1	–6/41	5/41	4/41	
Y_1	3	89/41	1	0	0	–2/41	–12/41	15/41	
$Z = c_B$ x_B = 765/41		Δ_j	0	0	0	–45/41	–24/41	–11/41	

In the last table all Δ_j's for non-basic variables are negative therefore it gives optimal solution.

∴ Optimal solution is

$$x_1 = \frac{89}{41},\ x_2 = \frac{50}{41},\ x_3 = \frac{62}{41}$$

and $\quad \text{Max. } Z = \dfrac{765}{41}.$

Example 2:

If $x_1 = 1$, $x_2 = 0$, $x_3 = 1$ be a feasible solution of the L.P.P.

Min. $Z = 2x_1 + 3x_2 + 4x_3$

subject to $x_1 + x_2 + x_3 = 2$

$$x_1 - x_2 + x_3 = 0,\ x_1, x_2, x_3 \geq 0$$

then show that the given feasible solution is not basic.

Solution:

The given system of constraint equations can be written in matrix form $A x = b$

$$\Rightarrow \quad \begin{bmatrix} 1 & 1 & 1 \\ 1 & -1 & 1 \end{bmatrix} \begin{bmatrix} x_1 \\ x_2 \\ x_3 \end{bmatrix} = \begin{bmatrix} 2 \\ 0 \end{bmatrix}.$$

$$\therefore \quad \alpha_1 = \begin{bmatrix} 1 \\ 1 \end{bmatrix}, \alpha_2 = \begin{bmatrix} 1 \\ -1 \end{bmatrix}, \alpha_3 = \begin{bmatrix} 1 \\ 1 \end{bmatrix}, b = \begin{bmatrix} 2 \\ 0 \end{bmatrix}.$$

The given feasible solution $x_1 = 1$, $x_2 = 0$, $x_3 = 1$ will not be basic if the vectors associated to non-zero variables are not linearly independent *i.e.*, if α_1 and α_3 are linearly dependent.

If we choose $\lambda_1 = -1$ and $\lambda_2 = 1$, we have

$$\lambda_1 \alpha_1 + \lambda_2 \alpha_2 = -1 \begin{bmatrix} 1 \\ 1 \end{bmatrix} + 1 \begin{bmatrix} 1 \\ 1 \end{bmatrix} = \begin{bmatrix} -1+1 \\ -1+1 \end{bmatrix} = \begin{bmatrix} 0 \\ 0 \end{bmatrix} = 0.$$

$\therefore$ the vectors α_1 and α_3 are linearly dependent.

Hence, the given feasible solution is not basic.

Example 3:

If $x_1 = 2$, $x_2 = 3$, $x_3 = 1$ be a feasible solution of the linear programming problem

$$\begin{aligned} \text{Max. } & Z = x_1 + 2x_2 + 4x_3 \\ \text{s.t. } & 2x_1 + x_2 + 4x_3 = 11 \\ & 3x_1 + x_2 + 5x_3 = 14 \\ & x_1, x_2, x_3 \geq 0. \end{aligned}$$

then find a basic feasible solution from the given feasible solution.

Solution:

The given L.P.P. can be expressed as

Max. $Z = x_1 + 2x_2 + 4x_3$,

s.t. $A\,x = b$

$$\Rightarrow \begin{bmatrix} 2 & 1 & 4 \\ 3 & 1 & 5 \end{bmatrix} \begin{bmatrix} x_1 \\ x_2 \\ x_3 \end{bmatrix} = \begin{bmatrix} 11 \\ 14 \end{bmatrix}$$

$$\Rightarrow \alpha_1 x_1 + \alpha_2 x_2 + \alpha 3\, x_3 = b, \qquad ...(1)$$

where $\alpha_1 = \begin{bmatrix} 2 \\ 3 \end{bmatrix}$, $\alpha_2 = \begin{bmatrix} 1 \\ 1 \end{bmatrix}$, $\alpha_3 = \begin{bmatrix} 4 \\ 5 \end{bmatrix}$, $b = \begin{bmatrix} 11 \\ 14 \end{bmatrix}$.

Since $x1 = 2$, $x_2 = 3$, $x_3 = 1$ is a feasible solution to the given L.P.P. therefore from (1)

$$2\alpha_1 + 3\alpha_2 + \alpha_3 = b.$$

Now the vectors $\alpha_1, \alpha_2, \alpha_3$ associated with non-zero variables x_1, x_2, x_3 will be linearly dependent if one of these vectors can be expressed as the linear combination of the remaining two vectors.

Let $\lambda_1 \alpha_1 + \lambda_2 \alpha_2 = \alpha_3$ ...(2)

$$\Rightarrow \lambda_1 \begin{bmatrix} 2 \\ 3 \end{bmatrix} + \lambda_2 \begin{bmatrix} 1 \\ 1 \end{bmatrix} = \begin{bmatrix} 4 \\ 5 \end{bmatrix}$$

$$\Rightarrow \begin{bmatrix} 2\lambda_1 + \lambda_2 \\ 3\lambda_1 + \lambda_2 \end{bmatrix} = \begin{bmatrix} 4 \\ 5 \end{bmatrix}$$

$$\Rightarrow 2\lambda_1 + \lambda_2 = 4,\ 3\lambda_1 + \lambda_2 = 5.$$

Solving these, we get

$$\lambda_1 = 1,\ \lambda_2 = 2.$$

Substituting the values of λ_1 and λ_2 in (2), we get

$$\alpha_1 + 2\alpha_2 = \alpha_3$$

$$\Rightarrow \alpha_1 + 2\alpha_2 - \alpha_3 = 0$$

or $$\sum_{i=1}^{3} \lambda_i \alpha_i = 0$$

which gives $\lambda_1 = 1$,

$$\lambda_2 = 2,\ \lambda_3 = -1.$$

Now, we shall determine which of the three variables x_1, x_2, x_3 should be zero.

Using $v = \max\limits_{1 \le i \le 3} \left(\dfrac{\lambda_i}{x_i}\right)$,

$$\Rightarrow \qquad v = \max\left(\frac{\lambda_i}{x_1}, \frac{\lambda_2}{x_2}, \frac{\lambda_3}{x_3}\right),$$

$$\Rightarrow \qquad v = \max\left(\frac{1}{2}, \frac{2}{3}, -\frac{1}{1}\right) = \frac{2}{3} = \frac{\lambda_2}{x_2}.$$

Since $x_i - \dfrac{\lambda_i}{v} \ge 0$, therefore

$$\left(x_i - \frac{\lambda_1}{v}, x_2 - \frac{\lambda_2}{v}, x_3 - \frac{\lambda_3}{v}\right)$$

is the reduced feasible solution. We have

$$x_1 - \frac{\lambda_1}{v} = 2 - \frac{1}{2} = \frac{1}{2},$$

$$x_2 - \frac{\lambda_2}{v} = 0,$$

$$\Rightarrow \qquad x_3 - \frac{\lambda_3}{v} = \frac{5}{2}.$$

$\therefore$ The new feasible solution is $x_1 = \dfrac{1}{2}$,

$$x_3 = 0,\ x_3 = \frac{5}{2}.$$

Now $\alpha_1 = \begin{bmatrix} 2 \\ 3 \end{bmatrix}$, and $\alpha_3 = \begin{bmatrix} 4 \\ 5 \end{bmatrix}$ are the vectors associated to non-zero variables x_1 and x_3.

Since α_1 and α_3 are linearly independent vectors therefore the new feasible solution

$$x_1 = \frac{1}{2},\ x_2 = 0,\ x_3 = \frac{5}{2}$$

is the basic feasible solution.

Note: In order to find another basic feasible solution we proceed as follows:

Since $\alpha_1, \alpha_2, \alpha_3$ are L.D. therefore one of them can be expressed as the linear combination of the other two.

Let $\qquad \lambda_2\, \alpha_2 + \lambda_3\, \alpha_3 = \alpha_1.$ ...(3)

Now proceed as in Example 9.

We have $\lambda_1 = -1$, $\lambda_2 = -2$,

$$\lambda_3 = 1,\ v = \frac{1}{1} = \frac{\lambda_3}{x_3}.$$

B.F.S. is $x_1 = 3$, $x_2 = 5$, $x_3 = 0$.

Example 4:

Solve the following system of linear equations by using simplex method

$$x_1 - x_3 + 4x_4 = 3$$
$$2x_1 - x_2 = 3$$
$$3x_1 - 2x_2 - x_4 = 1$$
$$x_1, x_2, x_3, x_4 \geq 0.$$

Solution:

Introducing the dummy objective function Z with costs zero to each given variable and cost (–1) to each artificial variable the given equations can be written as a linear programming problem in the following form

$$\text{Max. } Z = 0x_1 + 0x_2 + 0x_3 + 0x_4 - A_1 - A_2 - A_3$$

subject to

$$x_1 - x_3 + 4x_4 + A_1 = 3$$
$$2x_1 - x_2 + A_2 = 3$$
$$3x_1 - 2x_2 - x_4 + A_3 = 1,$$
$$x_1, x_2, x_3, x_4, A_1, A_2, A_3 \geq 0.$$

The starting B.F.S. is

$$x_1 = 0,\ x_2 = 0,\ x_3 = 0,$$
$$x_4 = 0,\ A_1 = 3,\ A_2 = 3,\ A_3 = 1.$$

The solution to the problem using simplex algorithm is

		c_j	**0**	**0**	**0**	**0**	**–1**	**–1**	**–1**	**Min. ratio**
B	c_B	x_B	Y_1	Y_2	Y_3	Y_4	A_1	A_2	A_3	x_B/Y_1
A_1	–1	3	1	0	–1	4	1	0	0	3
A_2	–1	3	2	–1	0	0	0	1	0	3/2
A_3	–1	1	(3)	–2	0	–1	0	0	1	1/3 ←

$Z = c_B x_B = -7$		Δ_j	6 ↑	–3	–1	3	0	0	0 ↓	X_B/Y_4
A_1	–1	8/3	0	2/3	–1	(13/3)	1	0		8/13
A_2	–1	7/3	0	1/3	0	2/3	0	1		7/2
Y_1	0	1/3	1	–2/3	0	–1/3	0	0		Neg.
$Z = c_B x_B = -5$		Δ_j	0	1	–1	5 ↑	0 ↓	0		
Y_4	0	8/13	0	(2/13)	–3/13	1		0		4 ←
A_2	–1	25/13	0	3/13	2/13	0		1		25/3
Y_1	0	7/13	1	–8/13	–1/13	0		0		Neg.
–25/13		Δ_j	1 ↑	3/13	2/13 ↓	0		0		x_B/Y_3
Y_2	0	4	0	1	–3/2	13/2		0		Net.
A_2	–1	1	0	0	(1/2)	–3/2		1		2 ←
Y_1	0	3	1	0	–1	4		0		Neg.
Z= –1		Δ_j	0	0	1/2 ↑	–3/2		0 ↓		
Y_2	0	7	0	1	0	2				
Y_3	0	2	0	0	1	-3				
Y_1	0	5	1	0	0	1				
Z=0	Δ_j	0	0	0	0					

Since all $\Delta_j \leq 0$ therefore the problem has optimal solution given by

$x_1 = 5, x_2 = 7,$

$x_3 = 2, x_4 = 0.$

Example 5:

Find the inverse of the matrix

$$B_\alpha = (\beta_1, \beta_2, \alpha)$$

where $B = (\beta_1, \beta_2, \beta_3)$,

$$\alpha = \begin{bmatrix} 2 \\ 1 \\ 3 \end{bmatrix} \text{ and } B^{-1} = \begin{bmatrix} 7 & -3 & -3 \\ -1 & 1 & 0 \\ -1 & 0 & 1 \end{bmatrix}.$$

Solution:

We know that $Y = (y_1, y_2, y_3) = B^{-1}\alpha$

$$\therefore Y = \begin{bmatrix} 7 & -3 & -3 \\ -1 & 1 & 0 \\ -1 & 0 & 1 \end{bmatrix}\begin{bmatrix} 2 \\ 1 \\ 3 \end{bmatrix} = \begin{bmatrix} 2 \\ -1 \\ 1 \end{bmatrix}$$

or $\quad y_1 = 2,\ y_2 = -1,\ y_3 = 1.$

$$\text{Now } A_3 = \left(-\frac{y_1}{y_3}, -\frac{y_2}{y_3}, \frac{1}{y_3}\right) = (-2, 1, 1)$$

$$\therefore E = (e_1, e_2, A_3) = \begin{bmatrix} 1 & 0 & -2 \\ 0 & 1 & 1 \\ 0 & 0 & 1 \end{bmatrix}$$

$$\text{Now} \quad B_\alpha^{-1} = EB^{-1} = \begin{bmatrix} 1 & 0 & -2 \\ 0 & 1 & 1 \\ 0 & 0 & 1 \end{bmatrix}\begin{bmatrix} 7 & -3 & -3 \\ -1 & 1 & 0 \\ -1 & 0 & 1 \end{bmatrix}$$

$$= \begin{bmatrix} 9 & -3 & -5 \\ -2 & 1 & 1 \\ -1 & 0 & 1 \end{bmatrix}.$$

Example 6:

Solve the following problem by using the two-phase method.

$$Min.\ Z = 40x_1 + 24x_2$$

subject to $20x_1 + 50x_2 \geq 4800$

$$80x_1 + 50x_2 \geq 7200,\ x_1, x_2 \geq 0.$$

Solution:

The given problem is of minimization. Converting it to maximization by taking the objective function as

$$Z' = -Z.$$

The objective function becomes

$$\text{Max } Z' = -Z = -40x_1 - 24x_2.$$

To change the inequalities of constraint into equations introducing surplus and artificial variables, we get

$$20x_1 + 50x_2 - x_3 + A_1 = 4800$$

$$80x_1 + 50x_1 - x_4 + A_2 = 7200$$

$$x_1, x_2, x_3, x_4, A_1, A_2 \geq 0,$$

x_3, x_4 are surplus variables and A_1, A_2 are artificial variables.

Phase I. Assigning cost 0 to all other variables and cost –1 to artificial variables, the new objective function of auxiliary problem becomes

$$\text{Max } Z' = 0x_1 + 0x_2 + 0x_3 + 0x_4 - A_1 - A_2$$

subject to the constraints given above.

Now applying the simplex method in the usual manner, we have the following table:

B	c_B	c_j / x_B	0 / Y_1	0 / Y_2	0 / Y_3	0 / Y_4	–1 / A_1	–1 / A_2	Min. ratio x_B/Y_1
A_1	–1	4800	20	50	–1	0	1	0	240
A_2	–1	7200	(80)	50	0	–1	0	1	90 ←
$Z = c_B x_B = -12000$		Δ_j	100 ↑	100	–1	–1	0	0 ↓	
A_1	–1	3000	0	(75/2)	–1	1/4	1	–1/4	80 ←
Y_1	0	90	1	5/8	0	–1/80	0	1/80	144
$Z = -2910$		Δ_j	0	75/2 ↑	0	1/4	0 ↓	–5/4	
Y_2	0	80	0	1	–2/75	1/50	2/75	–1/50	
Y_1	0	40	1	0	1/60	–1/60	–1/60	–1/60	
$Z=0$	Δ_j	0	0	0	0	–1	–1		

Since all $\Delta_j \leq 0$ and no artificial variable appears in the basis therefore an optimum solution to the auxiliary problem has been attained.

Phase II. Now assign the actual costs to the original variables and cost zero to the surplus variables, the objective function becomes

$$\text{Max } Z' = -40x_1 - 24x_2 + 0x_3 + 0x_4.$$

Replace the c_j row values in the final simplex table of phase I by the c_j values of the original objective function.

Also delete the artificial variables from the final simplex table of phase I. Now apply the simplex method in the usual manner to this table.

B	c_B	c_j x_B	–40 Y_1	–24 Y_2	0 Y_3	0 Y_4	Min. ratio x_B/Y_3
x_2	–24	80	0	1	–2/75	1/50	neg.
x_1	–40	40	1	0	(1/60)	–1/60	2400 ←
$Z'=c_B\ x_B= -3520$		Δ_j	0 ↓	0	2/75 ↑	–38/75	
x_2	–24	144	8/5	1	0	–1/50	
x_3	0	2400	60	0	1	–1	
$Z'= -3456$ Δ_j		–8/5	0	0	–12/25		

Since all Δ_j's are negative or zero therefore the solution obtained is optimal.

Hence the optimal solution is $x_1 = 0$, $x_2 = 144$ and max $Z = -Z' = 3456$.

EXERCISES

1. Prove that if the system A x = b of m linear equation in n unknowns $(m \leq n)$ with rank A = m has a feasible solution, then it has a basic feasible solution also.

2. (2, 1, 3) is a feasible solution of the set of equations

$$4x_1 + 2x_2 - 3x_3 = 1,$$
$$6x_1 + 4x_2 - 5x_3 = 1.$$

Reduce the feasible solution to a b.f.s. of the set.

3. $x_1 = 1, x_2 = 1, x_3 = 1, x_4 = 0$ is a feasible solution to the system of equations

$$x_1 + 2x_2 + 4x_3 + x_4 = 7,$$
$$2x_1 - x_2 + 3x_3 - 2x_4 = 4.$$

Reduce the feasible solution to two different basic feasible solutions.

4. If $x_3 = 4, x_4 = 8$ is the non-degenerate basic feasible solution to the L.P.P.

Max. $Z = x_1 + 2x_2,$

s.t. $x_1 + 2x_2 + x_3 = 4,$

$x_1 + 4x_2 + x_4 = 8,$

$x_1, x_2, x_3, x_4 \geq 0$

then obtain the new b.f.s.

5. Max. $Z = x_1 - x_2 + 3x_3$

   subject to

   $x_1 + x_2 + x_3 \leq 10$

   $2x_1 - x_3 \leq 2$

   $2x_1 - 2x_2 + 3x_3 \leq 0$

   $x_1, x_2, x_3 \geq 0.$

6. Max. $Z = 4x_1 + 10x_2$

   subject to

   $2x_1 + x_2 \leq 50$

   $2x_1 + 5x_2 \leq 100$

   $2x_1 + 3x_2 \leq 90$

   $x_1, x_2 \geq 0.$

7. Max. $Z = 2x_1 + 4x_2$

   subject to

   $2x_1 + 3x_2 \leq 48$

   $x_1 + 3x_2 \leq 42$

   $x_1 + x_2 \leq 21$

   $x_1, x_2 \geq 0.$

8. Max. $Z = 2x_1 + x_2$

   subject to

   $x_1 + 2x_2 \leq 10$

   $x_1 + x_2 \leq 6$

   $x_1 - x_2 \leq 2$

   $x_1 - 2x_2 \leq 1$

   $x_1, x_2 \geq 0.$

9. Max. $Z = 4x_1 + 5x_2 + 9x_3 + 11x_4$

   subject to the constraints

   $x_1 + x_2 + x_3 + x_4 \leq 15$

   $7x_1 + 5x_2 + 3x_3 + 2x_4 \leq 120$

   $3x_1 + 5x_2 + 10x_3 + 15x_4 \leq 100$

   $x_1, x_2, x_3, x_4 \geq 0.$

10. Max. $Z = 8x_1 + 11x_2$

subject to

$3x_1 + x_2 \leq 7$

$x_1 + 3x_2 \leq 8$

$x_1, x_2 \geq 0.$

11. Show that $x_1 = 4, x_2 = 0, x_3 = 0, x_4 = 4$ is optimal basic feasible solution to the L.P.P.

    Max. $Z = x_1 + 2x_2$

    s.t. $x_1 + 2x_2 + x_3 = 4,$

    $x_1 + 4x_2 + x_4 = 8,$

    $x_1, x_2, x_3, x_4 \geq 0.$

12. Show that if a linear programming problem has a feasible solution, it also has a basic feasible solution.

13. Minimize $Z = x_1 + x_2 + 3x_3$

    subject to

    $3x_1 + 2x_2 + x_3 \leq 3$

    $2x_1 + x_2 + 2x_3 \leq 2$

    $x_1, x_2, x_3 \geq 0.$

14. Max. $Z = 2x_1 + 4x_2 + 3x_3$

    subject to

    $3x_1 + 4x_2 + 2x_3 \leq 60$

    $2x_1 + x_2 + 2x_3 \leq 40$

    $x_1 + 3x_2 + 2x_3 \leq 80$

    $x_1, x_2, x_3 \geq 0.$

15. Max. $Z = 5x_1 + 2x_2 + 3x_3 - x_4 + x_5$

    subject to the constraints

    $x_1 + 2x_2 + 2x_3 + x_4 = 8,\ 3x_1 + 4x_2 + x_3 + x_5 = 7,\ x_1, x_2, x_3, x_4, x_5 \geq 0.$

16. State and prove the fundamental theorem of linear programming.

17. Food A contains 20 units of vitamin X and 40 units of vitamin Y per gram. Food B contains 30 units each of vitamin X and Y. The daily minimum human requirements of vitamins X and Y are 900 units and 1200 units respectively. How many grams of each type of food should be consumed so as to minimize the cost if food A costs 60 praise per gram and food B costs 80 praise per gram.

18. Max. $Z = x_1 - 2 + 3x_3$

    subject to

    $x_1 + x_2 + x_3 \leq 10$

    $2x_1 - x_3 \leq 2$

    $2x_1 - 2x_2 + 3x_3 \leq 0$

    $x_1, x_2, x_3 \geq 0.$

19. If $x_1 = 2$, $x_2 = 4$ and $x_3 = 1$ be a F.S. to the system of equations

    $2x_1 - x_2 + 2x_3 = 2,$

    $x_2 + 4x_2 = 18,$

    then find two basic feasible solutions.

20. Max. $Z = 3x_1 + 4x_2$

    subject to

    $x_1 - x_2 \leq 1$

    $-x_1 + x_2 \leq 2$

    $x_1, x_2 \geq 0.$